Gould and Lincoln's Publications.

MILLER'S CRUISE OF THE BETSEY; or, a Summer Ramble among the Fossiliferous Deposits of the Hebrides. With Rambles of a Geologist; or, Ten Thousand Miles over the Fossiliferous Deposits of Scotland. 12mo, pp. 524, cloth, 1.75.

MILLER'S ESSAYS, Historical and Biographical, Political and Social, Literary and Scientific. By HUGH MILLER. With Preface by Peter Bayne. 12mo, cloth, 1.75.

MILLER'S FOOT-PRINTS OF THE CREATOR; or, the Asterolepis of Stromness, with numerous Illustrations. With a Memoir of the Author, by LOUIS AGASSIZ. 12mo, cloth, 1.75.

MILLER'S FIRST IMPRESSIONS OF ENGLAND AND ITS PEOPLE. With a fine Engraving of the Author. 12mo, cloth, 1.50.

MILLER'S HEADSHIP OF CHRIST, and the Rights of the Christian People, a Collection of Personal Portraitures, Historical and Descriptive Sketches and Essays, with the Author's celebrated Letter to Lord Brougham. By HUGH MILLER. Edited, with a Preface, by PETER BAYNE, A. M. 12mo, cloth, 1.75.

MILLER'S OLD RED SANDSTONE; or, New Walks in an Old Field. Illustrated with Plates and Geological Sections. NEW EDITION, REVISED AND MUCH ENLARGED, by the addition of new matter and new Illustrations, &c. 12mo, cloth, 1.75.

MILLER'S POPULAR GEOLOGY; With Descriptive Sketches from a Geologist's Portfolio. By HUGH MILLER. With a Resume of the Progress of Geological Science during the last two years. By MRS. MILLER. 12mo, cloth, 1.75.

MILLER'S SCHOOLS AND SCHOOLMASTERS; or, the Story of my Education. AN AUTOBIOGRAPHY. With a full-length Portrait of the Author. 12mo, 1.75.

MILLER'S TALES AND SKETCHES. Edited, with a Preface, &c., by MRS. MILLER. 12mo, 1.50.

Among the subjects are: Recollections of Ferguson — Burns — The Salmon Fisher of Udoll — The Widow of Dunskaith — The Lykewake — Bill Whyte — The Young Surgeon — George Ross, the Scotch Agent — M'Culloch, the Mechanician — A True Story of the Life of a Scotch Merchant of the Eighteenth Century.

MILLER'S TESTIMONY OF THE ROCKS; or, Geology in its Bearings on the two Theologies, Natural and Revealed. "Thou shalt be in league with the stones of the field." — *Job.* With numerous elegant Illustrations. One volume, royal 12mo, cloth, 1.75.

HUGH MILLER'S WORKS. Ten volumes, uniform style, in an elegant box, embossed cloth, 17; library sheep, 20; half calf, 34; antique, 34.

MACAULAY ON SCOTLAND. A Critique from HUGH MILLER'S "Witness." 16mo, flexible cloth. 37 cts.

ESSAYS,

HISTORICAL AND BIOGRAPHICAL, POLITICAL, SOCIAL, LITERARY, AND SCIENTIFIC.

BY

HUGH MILLER,

AUTHOR OF "THE OLD RED SANDSTONE," "MY SCHOOLS AND SCHOOLMASTERS," "THE TESTIMONY OF THE ROCKS," ETC. ETC.

Edited, with a Preface,

BY PETER BAYNE,

AUTHOR OF "THE CHRISTIAN LIFE," ETC.

BOSTON:
GOULD AND LINCOLN,
59 WASHINGTON STREET.
NEW YORK: SHELDON AND COMPANY.
CINCINNATI: GEORGE S. BLANCHARD.
1865.

PREFACE.

UNUSUAL as it is to republish newspaper articles, no apology is deemed necessary in presenting this volume to the public. At the time of Mr. Hugh Miller's death, it was felt that a large proportion of his contributions to the "Witness" deserved a permanent place in the literature of his country. They were recognized as distinguished, both by their literary merit and their sterling value, from the fugitive and ephemeral productions of every-day journalism.

Assuming the conduct of a newspaper in the maturity of his powers, and in the plenitude of his literary and scientific information, Mr. Miller's habit of composition was entirely different from that of ordinary ready writers of the press. As was correctly remarked at the time of his death, "he did not work *easy*, but with laborious special preparation." He meditated his articles as an author meditates his books or a poet his verses, — conceiving them as wholes, working fully out their trains of thought, enriching them with far-brought treasures of fact, and adorning them with finished and apposite illustration. In the quality of *completeness*, those articles stand, so far as I know, alone in the records of journalism. For rough and hurrying vigor they might be matched, or more, from the columns of the "Times;" in lightness of wit and smart lucidity of statement they might be surpassed by the happiest performances of French journalists, — a Prevost Paradol or a St. Marc Girardin; and for occasional brilliancies of imagination, and sudden gleams of piercing thought, neither they nor any other newspaper articles have, I think, been comparable with those of S. T. Coleridge. But as complete journalistic essays, symmetrical in plan, finished in execution, and of sustained and splendid ability, the articles of Hugh Miller are unrivalled. For the most part, the topic suggesting them was but the occasion for a display of the writer's thought and imagination, — the fly round which the precious and imperishable amber of Mr. Miller's genius was accumulated.

I am not prepared to say that these are the most striking or powerful articles published in the "Witness" by Mr. Miller. He conducted that paper for sixteen years; and, on a moderate computation, he wrote for it a thousand articles. Having surveyed this vast field, I retain the impression of a magnificent expenditure of intellectual energy, — an expenditure of which the world will never estimate the sum. By far the larger portion of what Mr. Miller wrote for the "Witness" is gone forever. Admirable disquisitions on social and ethical questions, felicities of humor and sportive though trenchant satire, delicate illustration and racy anecdote from an inexhaustible literary erudition, and crystal jets of the purest poetry, — such things will repay the careful student of the "Witness" file, but can never be known to the general public.

Having done my utmost in the way of compression, there still remained about three volumes of articles, between the claims of which to republication I could not decide. This most difficult and delicate task was performed by Mrs. Hugh Miller, in a way which commanded my entire approval, and which will, I have no doubt, give satisfaction to the public.

Should the present volume meet the reception which, in my humble opinion, it deserves, its issue can be followed up by that of others of closely corresponding character and value.

PETER BAYNE.

CONTENTS.

HISTORICAL AND BIOGRAPHICAL.

POLITICAL AND SOCIAL.

LITERARY AND SCIENTIFIC.

ESSAYS.

HISTORICAL AND BIOGRAPHICAL.

I.

THE NEW YEAR.

Ere our sheet shall have passed from the press into the hands of our readers, we shall have entered on a new year. It is barely ninety degrees distant from us at the present moment. It landed on the eastern extremity of Asia as the 1st of January, 1845, just as we were rising from our breakfasts in Edinburgh, on the 31st of December, 1844; and it has been gliding westwards towards us, in the character of *one o'clock in the morning*, ever since. In a few hours more it will be striding across the backwoods of America, in its seven-league boots, and careering over the Pacific in its canoe. And then, at some undefinable point, not yet fixed by the philosopher, it will find itself transformed from the first into the second day of the year; and thus it will continue to roll on, round and round like an Archimedes screw, picking up at every gyration an additional unit, until the three hundred and sixty-five shall be complete.

The past year has witnessed many curious changes, as a dweller in time; the coming year has already looked down on many a curious scene, as a journeyer over space. It has seen Cochin-China, with all its unmapped islands, and the ancient empire of Japan, with its cities and provinces unknown to Europe. It has heard the roar of a busy population amid the thousand streets of Pekin, and the wild dash of the midnight tides as they fret the rocks of the Indian Archipelago. It has been already with our friends in Hindostan; it has been greeted, we doubt not with the voice of prayer, as the slow iron hand of the city-clock indicated its arrival to the missionaries at Madras; it has swept over the fever jungles of the Ganges, where the scaled crocodile startles the thirsty tiger as he stoops to drink, and the exposed corpse of the benighted Hindoo floats drearily past. It has travelled over the land of pagodas, and is now entering on the land of mosques. Anon it will see the moon in her wane, casting the dark shadows of columned Palmyra over the sands of the desert, and the dim walls of Jerusalem, looking out on a silent and solitary land that has cast forth its interim tenants, and waits unappropriated for the old predestined race, its proper inhabitants. In two short hours it will be voyaging along the cheerful Mediterranean, greeting the rower in his galley among the isles of Greece, and the seaman in his barque embayed in the Adriatic. And then, after marking the red glare of Ætna reflected in the waves that slumber around the moles of Syracuse, — after glancing on the towers of the Seven-hilled City, and the hoary snows of the Alps, — after speeding over France, over Flanders, over the waves of the German Sea, it will be with ourselves; and the tall, ghostly tenements of Dun-Edin will reëcho the shouts of the High Street. Away and away it will cross the broad Atlantic, and visit watchers in their beacon-towers on the deep, and the emigrant in his log-hut among the brown woods of the West; it will see the fire of the red man umbering with its gleam tall trunks and giant branches in some deep glade of the forest; and then mark, on the far shores of the Pacific, the rugged bear stalking sullenly over the snow. Away and away, and the vast globe shall be girdled by the zone of the new-born year.

Many a broad plain shall it have traversed that is still unbroken from the waste, — many a moral wilderness on which the Sun of Righteousness has not yet arisen. Nearly eighteen and a half centuries shall have elapsed since the shepherds first heard the midnight song in Bethlehem, — "Glory to God in the highest, peace on earth, good-will to the children of men." And yet the coming year shall pass, in its first visit, over prisons, and gibbets, and penal settlements, and battle-fields on which the festering dead moulder unburied; it will see the shotted gun and the spear and the crease and the murdering tomahawk, slaves in their huts, and captives in their dungeons. It will look down on uncouth idols in their temples; worshippers of the false prophet in their mosques; the Papist in his confessional; the Puseyite in his stone allegory; and on much idle and bitter controversy among those holders of the true faith whose proper work is the conversion of the world. But the years shall pass, and a change shall come: the sacrifice on Calvary was not offered up in vain, nor in vain hath the adorable Saviour conquered, and ascended to reign as King and Lord over the nations. The kingdoms shall become his kingdoms, the people his people. The morning rises slowly and in clouds, but the dawn has broken; and it shall shine forth more and more, until the twilight shadows shall have dispersed and the sulphurous fogs shall have dissipated, and all shall be peace and gladness amid the blaze of the perfect day.

II.

ROYAL PROGRESSES.

A BANQUET-HALL gay with lights and crowded with revellers, and the same banquet-hall lying silent in the dim gray of morning, the lights all extinguished, and the revellers all gone, — such is the contrast which the Edinburgh of the present week presents to the Edinburgh of the last. The living tide is receding even more suddenly than it arose, — ebbing by its hundred outlets, — roads, canals, railways, and the sea; and already do our streets, in both the ancient and modern portions of the city, present the characteristic aspect of the season. In the older thoroughfares, long appropriated to trade and labor, the current flows languidly, save at the hours when warehouses and workshops pour out their numerous inmates. In the more fashionable streets and squares it has altogether ceased to flow; and as solitude ever seems deeper amid sunshiny lines of deserted buildings than among even rocks and woods, however lonely, in no parts of the city or its neighborhood have the late scenes of noisy bustle and excitement been followed by scenes of more striking contrast than amid the more splendid streets of the New Town, with their few unemployed chairmen here and there sauntering about corners, or their single domestics here and there tripping leisurely along the pavement. Parade and pageantry seem over for the time; and the royal visit to Edinburgh has taken its place among other royal progresses of the past, as a thing of history, — as an event to which future chroniclers will refer, agreeably to their character as writers, either as a trivial fact, deserving of but its single brief sentence, or as interesting incident, suited, from its picturesque accompaniments, to relieve the dry narrative of contemporary occurrences.

Viewed in connection with the character of the respective ages to which they belong, these progresses form no uninteresting passages in our annals. We find them peculiarly impressed by the stamp of their time, and linked in most instances with the main events and more striking traits of the national history. We see a series of them rising in succession before us even now, like a series of pictures in a show-box. Shall we not just once or twice pull the string, and exhibit some of at least their more prominent features to our readers?

A youthful monarch wends his way northwards through a wild, trackless country, surrounded by a band of cowled and shaven monks. His lay attendants have doffed the gay attire of the court for dingy black or sober gray,—for the stole of coarse serge and the shirt of hair. The monarch himself is meanly wrapped in robes of the order of St. Francis, bound with a girdle of rope, and with a huge belt of hammered iron pressing uneasily on his loins. In that lugubrious assemblage all is assumed heaviness and well-simulated sorrow: not a trace of the splendor of royalty is visible. For the gratulatory shout, or the joyous burst of music, we hear only the sound of the whip plied in self-inflicted flagellation, or the chant of the penitential psalm. To what very distant age can this royal progress belong? Surely to the dark obscure of history, — to some uncertain era, at least a thousand years back. Not at all; not further back than one third of that period. That becowled and begirdled bigot is the *grandfather* of the royal lady whose progress we witnessed on Saturday last,—her grandfather *just ten times removed.* We see James IV. passing on his pilgrimage to the shrine of St. Dothus, to do idle penance, in the far wilds of Ross, for the unnatural part taken by him, in well-nigh his childhood, against his unfortunate father at Bannockburn. Nor are the effects of the deplorable superstition which has stamped its impress on that mean pageant less palpably evident in the uncultivated wildness of the surrounding country, or in the servile condition and savage ignorance of the inhabitants, than in the royal progress itself. Wherever superstition wakes, intellect and industry slumber. Popery, wherever it obtains, overlays the national mind like a nightmare; not only inducing sleep, but also rendering hideous the sleep which it induces. And what is the nature of the morality which grows up under its fostering influences? Look on that pageant. Could the repentance which bemoans itself in the confessional, and then expends itself in sore penances and long journeys, be in any instance more sincere? The haircloth, the whip, the iron belt, the shoeless foot, the weary pilgrimage, — these are all realities. In a few brief days, however, the season of penance will be over, and that devoted prince, laying down his repentance with his cowl, shall have engaged, undisturbed by a single compunctious qualm, in all the grosser debaucheries of an immoral and semi-barbarous court. And such is invariably the sort of connection which exists between the religion of penances, pilgrimages, and masses, and purity of life and conduct.

The scene changes, and a lady, as now, has become the centre of the pageant. The rank dew lies heavy on grass and stone; a deep gloom hangs over the landscape, — a thick, unwholesome fog, unstirred by the wind; but we can see the huge back of Arthur Seat faint and gray amid the haze, with the unaltered outline of the crags below; and yonder are the two western towers of Holyrood, and yonder the Abbey, with its stone roof entire, and the hoar damps settling on its painted glass. The scene is that of the pageant of Saturday last, in all its more prominent features: nought has changed, save man and his puny labors. Nature seems to have no sympathy with the general joy. The sun has not shone for five days, nor the moon for five nights; the boom of the cannon from the distant harbor, where the French galleys lie, falls dead and heavy on the ear, like the echoes of a sepulchral vault; the mingled shouts and music from the half-seen crowds sound drearily amid the chill and dripping damps, like tones of the winter wind in a ruin at midnight; and yonder comes the pageant of the day, enwrapped in fog, like a drifting vessel half-enveloped in the spray of a lee-shore. Mark these gay and volatile strangers, the *élite* of the French Court. Yonder are the three Maries, and yonder the two Guises; and here comes the Queen herself, encircled by her iron barons. And who is that Queen? Mary, — the gay, the fascinating, the exquisitely beautiful, — a true sovereign of the imagination, — a choice heroine of poetry and romance, — a woman whose loveliness still exerts its influence over hearts, — a monarch whose misfortunes and sorrows still command tears; Mary, — the loose, the voluptuous, the unprincipled, — alike fitted to enchant a lover or to destroy a husband, — the victim of her own unregulated passions, — the canonized martyr of Popery, — in no degree less surely the martyr of adultery and murder. But none of the darker traits yet appear; and with all the enthusiasm of the national character, the Scotch welcome their Queen. And yet motto and device speak to her in a strange language as she passes on: the very signs that indicate the general joy at her arrival are fraught with unpalatable truth. Nor will she be left to guess merely at their meaning when, after matins shall be sung and the Host elevated in yonder chapel, the echoes of that ancient High Church — a building so peculiarly associated with all that is truly great in Scottish history — shall be awakened by the stormy indignation of Knox; nay, in the very presence-chamber shall the sovereign be told that her reformed people have determined to brook no revival of the blood-stained idolatry of Rome. Mary's grandfather rode unquestioned on his pilgrimage, to mumble unprofitable prayers over the bones of dead men, to prostrate himself before stone saints, and to worship flour wafers. And yet, though thus blind and ignorant himself, he possessed a power of controlling and prescribing the beliefs of his subjects. But a principle of tremendous energy has arisen among the masses, — a principle destined to convulse empires and overthrow dynasties, — to curb the tyranny of rulers, and to spread wide among the people the blessings of freedom and the light of civilization. Kings are no longer to prescribe the beliefs of their subjects; subjects, on the contrary, are virtually to prescribe the beliefs of their kings. Monarchs are to profess the religion of their people, or to resign their thrones. Is the doctrine challenged? Mary might well challenge it; nor was she left long without the opportunity. It darkened her brief reign, and rendered the gloom of that dreary procession exactly what a few melancholy spirits had deemed it, — a gloom too significantly ominous of the long troubles which followed. It convulsed the country for more than a century, reddening many a battle-field, and staining many a scaffold, from the scaffold of the infatuated monarch who died at Whitehall, to that of our noble covenanting peasants and mechanics who suffered scarce two hundred yards from where we write, and whose honored bones moulder in the neighboring churchyard. But whatever it might be in Mary's days, it is surely no disputable doctrine now. It is the doctrine of the "Protestant Succession," of the "Coronation Oath," of the "Revolution Settlement." Except for this doctrine, the royal personage whose progress through the city on Saturday drew together so vast an assemblage, would not now be the Queen of Great Britain. She could have come among us merely as a highborn, but not the less obscure, Continental lady, who, were she to be pointed out to some curious spectator, could only be pointed out as the niece of a German prince.

The progress of James to the borders, to hold justice courts at the head of an army, sufficiently indicated the wild and unsettled character of the age. It was an age in which all power, judicial or monarchial, existed in its first elements; the authority of the judge, though a king, was nothing apart from the terrors of the military. Nor were the scenes of sudden execution which followed — scenes the recollection of which still survives in song and ballad — in any degree less characteristic. Even justice itself, infected by the savageism of the period, seems to have existed as but a stern principle of violence and revenge. The progress of Mary to the north bears a similar impress. It seems pregnant with the character of the age. We see the royal escort dogged in its course by the retainers of a turbulent and ambitious noble; scarce a dell without its ambuscade, — scarce a hill-top without its hostile horde of observation and annoyance; royal fortresses shut against royalty, until reduced by siege; chiefs and their septs hastily arming either to assail or to defend the sovereign; and the whole terminating in a hard-contested and bloody conflict, execution, confiscation, and exile. There is scarce a prominent trait in the old character and condition of the country, or scarce an influential event in its history, which some one of the royal progresses does not serve to illustrate.

There were none of them more characteristic, however, than the progress of Charles I., when he visited Scotland in 1633, to "reduce the kirk to conformity." James IV. brought his shavelings with him to the far north, to patter masses and chant matins. Charles brought with him a much more dangerous man than all the shavelings of James united. He brought with him, — the Pusey, the Newman, the Archdeacon Wilberforce of those days, — he brought with him Archbishop Laud. Rarely in Edinburgh has there been a more profuse or tasteful display of all the various symbols by which the public indicate cordial joy and welcome, than on the evening of Friday. There was the rich firework, the brilliant device visible by its own-tinted light, the motto, the bonfire, the blaze of torches, lamps, and tapers. The age of Charles, however, was, much more than the present, an age of mysteries and emblems; it was the age of the masque and the allegory, — an age in which even a Bacon could write of such things, and a Quarles of scarce anything else; and we question whether Edinburgh was not as interesting a sight when Charles visited it rather more than two hundred years ago, as when Victoria visited it last week. "The streets on both sides," says Stevenson, "were lined by the citizens in their best apparel and arms, from the West Port to Holyrood." At one "theatre, exquisitely adorned," where the Lord Provost presented the keys to his Majesty, there was a "painted description of the city." At another, near the Luckenbooths, were arranged the portraits of all the kings of Scotland, from Fergus downward. A fountain at the cross ran with wine for the benefit of the lieges; and Bacchus, large as life, superintended the distribution of the liquor. The Muses made themselves visible in Hunter Square; the heavenly bodies danced harmoniously at the Netherbow. Bells chimed, cannons rattled, and "all sorts of music that could be invented" mingled their tones with the booming of the guns, the pealing of the bells, the melody of the planets, the speeches of Fergus, Bacchus, and the Provost, and the songs of Apollo, the Burghers, and the Muses. We are further told that the streets were actually "sanded," and that the "chief places were set out with stately triumphal arches, obelisks, pictures, artificial mountains, and other costly shows." It must have been altogether a bizarre scene. Parnassus, with all its rocks, trees, and fountains, leaned against the old weigh-house. When the Muses sung, the nymphs of the Cowgate joined in the chorus; the genius of Scotland discoursed of war and conquest in the middle of the West Bow; classic arches of lath strided over the odoriferous Cranes; festoons of flowers hung romantically above the unsullied waters of the Nor'-Loch; obelisks of pasteboard shot up their taper pinnacles among the gray chimneys of the Grassmarket; — the entire city must have not a little resembled its defunct patron saint of blackened wood, "old St. Gyle," when bedizzened on a holiday with colored glass, tinsel, and cut paper. And then, the handsome, imperious, melancholy Charles, with violent death impressed, according to the belief of the age, in the very lines of his countenance, and the withered, diminutive Laud, perplexed by some half-restored recollection of his last night's dream, or bent to the full stretch of his faculties in originating some new religious form suggested by the surrounding mummeries, or in determining whether his cope might not possibly be improved by the addition of a few spangles, must have looked tolerably picturesque as they passed along the lines of the grave, whiskered burghers stretching on either hand, surmounted by all the beauty of the place, as it hung gaping and curious from the windows above. On Sunday, Charles, unlike our present monarch, attended the High Church. We fain trust the presence of the one and the absence of the other did not indicate the same thing. "The ordinary reader began to sing, as usual," says the historian; "whereat his Majesty, displeased, despatched the Bishop of Ross to turn him out. And the bishop straightway did so, with no few menaces, and introduced into his place two English choristers in their vestments, who, with the help of the dignitaries, performed their service after the English manner." "That being ended," adds the historian, "Bishop Guthrie of Moray went up to the pulpit to preach; but, instead of making divine truth his theme, he had little else than some flattering panegyrics, which made the king to blush, mingled with bitter scoffs at those who scrupled the use of the vestments." Poor, hapless king! With so many flatterers and so few friends, with the Bishop of Ross for his very humble servant, the Bishop of Moray for his chaplain, and the Archbishop of Canterbury for his adviser, was it a wonder he should have lost his head? The storm broke out only four years after, — broke out in that very High Church, — which overturned both throne and altar.

Surely, a curious subject of reflection! The reigning

monarch derives her lineage, not from Charles, but from the sister of Charles. The legitimate branch was lopped off, and left to wither and die, and the collateral branch grafted in. Why? What could have led to an event so contrary to the first principles of the law of succession, as embodied in statute by our legislative assemblies, and expounded in our civil courts? A question easily answered. The germ of the whole transaction might be seen in the royal progress, in which Laud and the infatuated Charles passed down the High Street together, and in the scene enacted on the following Sunday in the High Church. The abdication of James VII. was not less intimately connected with the infection communicated by the Archbishop, than the troubles and death of James's father. The Laudism of the one terminated in the Popery of the other. No one thinks it at all strange that the Puseyism of a Sibthorp or a Miss Gladstone should land them full in the Romish Church. A hundred other such conversions of the present day from Puseyism to Popery show us that such is the natural tendency of the rivived doctrines; they constitute no resting-place; they form merely a passage from one state of mind to another, — a sort of inclined plane, by which reluctant Protestantism scales inch by inch the transcendental heights of Popery. It was exactly a similar process that produced one of the most remarkable revolutions recorded in history. Two princes, educated in the *transition* beliefs of the king, their father, followed up these to their legitimate consequences, and so died members of the infallible Church. They did exactly what Sibthorp and Miss Gladstone have done. The one, a careless debauchee, declined sacrificing anything for the sake of a creed loosely held by him at best, and which, in his gayer moods, he occasionally abandoned for the indifferency of infidelity; the other, an honest bigot, acted up to his adopted beliefs, and so forfeited the crown. And hence the claim and standing of the high-born lady who now occupies the supreme place in the government of the country.

Can there be a more legitimate object of solicitude to all in a time like the present, — to all, at least, who are at once loyal subjects and true Protestants, — than her preservation from the dangerous contagion of the *transition* beliefs and doctrines, and from that perilous process of change which produced of old so great a national revolution, and which is so palpably operative in the apostasies of the present day? If, as indicated by the course of events, popery be fast rising by the deceitful slope which Puseyism supplies, and rising, as prophecy so clearly intimates, only to fall for ever, it were well, surely, that the daughter of our ancient kings should be on her guard against its insidious approaches. It involved princes of her blood in its former fall; nor is it a thing impossible that, misled by the counsel of other Lauds, other princes may share in its final ruin. But we digress.

There is little of an intrinsically pleasing character in the visit of George IV., and not much in it particularly characteristic, except perhaps of the monarch himself. It was much a matter of show, — a masque on a large scale. There was little that was real in it, save the enthusiasm of the people. It was, however, a masque enacted under the superintendence of a great genius, — the first scene-painter in the world, — not very worthily employed, perhaps, in designing mere *tableaux vivans*, as on this occasion, but not without an apology of Bacon when he wrote of "Masques and Triumphs." "These things are but toys," said the philosopher; "but yet, since princes will have such things, it is better they should be graced with elegance than daubed with cost." What was chiefly remarkable in the visit of George was the tact with which the monarch avoided every occasion of offence, and how, trading on so very slender a stock of real worth as that which he possessed, and in the face of so large an amount of adverse feeling as that which he had previously excited, he should have contrived to render himself popular by the exercise of the "petty moralities" alone. Never did the "mere gentleman," abstracted from the good, great, generous qualities of our nature, accomplish more. He came to Edinburgh only two years after the trial of Queen Caroline, and, without exhibiting anything higher than the urbanity of a thoroughly well-bred man, he taught all Scotland to forget for the time the result of that trial. We regret that Sir Robert Peel should not have availed himself of the advantages of having served under so accomplished a master in the art of pleasing. George IV. came to Edinburgh under every disadvantage, and regained there much of the popularity which he had previously lost; Sir Robert came to Edinburgh in the train of his royal mistress, — a monarch in whose favor the partialities of the nation had been largely awakened; and after losing well-nigh all that remained of his own popularity, he would have lost for her, had the thing been possible, her popularity too.

The recent royal progress through Edinburgh has had its many striking scenes; but the chronicler who may have to concentrate himself on one description as a specimen of the whole, would do well to select the scene of Saturday last, as exhibited in the upper part of the High Street, when her Majesty, after just receiving the city keys, passed on to the Castle. As a pageant the thing was nothing; it had the disadvantage, too, which the Queen's passage through the city on Thursday morning had not, of being *artificial*, — a projected piece of parade, with but the *parade* itself for its ostensible object. The Queen rode along the streets just that people might see the Queen. There is sublimity, however, in the appearance of vast multitudes animated by some overpowering feeling; and we know not that crowds could be better disposed for effect, or in a locality richer in historic recollection, than along the High Street of Edinburgh, with its old Parliament Hall, its venerable High Church, and its double line of tall, antique houses, some of which must have cast their shadows over the pageant of Mary, and not a few of them over the pageant of Charles. The morning, though not bright, was pleasant; the rack flew high over head, showing that a smart breeze blew in the upper regions, but all was comparatively calm beneath. Now and then an occasional gleam of the sun lighted up the tall gray fronts on the western side, or played among the fantastic tracery and lofty pinnacles of St. Giles; but it passed as suddenly as it flashed out, and the general tone was a subdued, smoky gray. A dense and ever-increasing crowd occupied the space below; direct through the middle there ran a narrow passage, that reminded one of a river, with steep, erect banks, winding its way through a flat alluvial meadow. At one point it expanded into what seemed a small lake: 'twas where the city magistracy awaited her Majesty, clad in long scarlet cloaks of office; and here a few dragoons flitted across the open space, or paced along the winding passage, — the shallops of this lake and river. Every window was crowded, story on story, from the windows immediately over the street to the casements of the attics eighty feet above head. Even the roofs had their clustering groups. We marked a few ragged boys perilously grouped round a chimney full ninety feet from the pavement; and to this dizzy eminence the urchins had contrived to bring with them the tattered fragment of a flag, which ever and anon they waved with huge glee. The group was one in which a Hogarth would have delighted. The roof of St. Giles's seemed scarce less densely occupied than the street below; and the effect of the whole was striking in the extreme. Blair, in his "Grave," speaks of "overbellying crowds." The spectators of the scene of Saturday must have been able to appreciate the picturesqueness of the phrase. The living masses, hanging from every corner and coigne of vantage, seemed, if we may so express ourselves, to project the antique architecture of the High Street against the sky. Almost every snugger corner, too, had its temporary scaffold or balcony. There was in particular one scaffold that greatly gratified us; the object of its erection showed both good taste and good feeling. It had been raised for the accommodation of the boys and girls of Heriot's; and never was there a group of happier faces than that which it exhibited. Such was the scene, when, shortly after eleven, a solitary horseman came spurring up the street, and, pausing for a moment in the centre of the open space where the magistracy of the city were assembled, he intimated that the Queen had reached Holyrood. The whisper passed along the crowd, and was caught from balcony to window, and from window to roof. The bells of the city had been rung at intervals from morning; they now broke out into a merry peal; and the near boom of cannon from the neighboring Castle suddenly awoke the echoes of the High Street. There was a movement in the close-wedged crowds beneath, — a murmur expressive of the general excitement, — a swaying to and fro; and then for a space all was still as before. From our point of observation we could catch a view of the roofs and upper stories of the tenements in the lower part of the street, with their dimly-seen groups of spectators. We could mark a sudden waving of handkerchiefs, — a deep, though distant cheer; a cry of "The Queen, the Queen," passed along the crowd. The masses opened heavily and slowly, as if compressed by the lateral weight; a train of coaches was seen advancing: there was the gleam of helmets, the flash of swords; the shout rose high; and as the vehicle in front moved on, there was a fluttering of scarfs and kerchiefs at every casement and in every gallery, as if a stiff breeze had swept by and shaken them as it passed. The city magistrates, in their scarlet robes, had formed a group in front of the Exchange; and here the royal vehicle paused, and the Lord Provost went through the ceremony of delivering the city keys into the hands of the sovereign. We sat within less than twenty yards of her Majesty at the time, and employed ourselves in marking how thoroughly the countenance is a German one, — how very much of Brunswick there is in it, and how little of the Stuarts. It bears trace of the Guelphs in every feature and lineament. As a family face, it has its historic associations speaking of Revolution principles and Protestant succession. The pageant moved on, and disappeared as, passing from where the street terminates in front of the Castle, it entered on the esplanade.

Such is a faint and imperfect outline of the one prominently-striking scene connected with the recent progress. We have said that the progresses of James, Mary, and Charles were characteristically impressed by the stamp of their time, and linked to the main events and more striking traits of the national history. May the recent progress be regarded as also characteristic? Time alone can show. It may be found to speak all too audibly of the revived superstition to which the troubles of Charles were mainly owing, — the superstition which conducted him ultimately to the front of Whitehall, and his younger son to a French palace in St. Germains. But we shall meanwhile hope for the best, without, however, attempting to conceal from ourselves that one cloud more seems to have arisen on the already darkened horizon of the Church of Scotland.

III.

THE INFANT PRINCE.

A PRINCE born to the throne of Great Britain! The firing of cannon, the ringing of bells, the crackling of fire-works, the blazing of bonfires, holiday dresses and holiday faces everywhere, all testify to the general joy.

We are reminded of a day which must have mingled with the first recollections of even the most aged of our readers, and which men in the prime of early manhood are quite old enough to remember too, — that happy fourth of June, the birthday of the good George III., on which, for two whole generations, and a little longer, there used to be such waving of flags and flashing of gunpowder, and, notwithstanding all our wars abroad and all our difficulties and troubles at home, so large an amount of hearty national enjoyment. Is the ninth of November to be just such another day to the generations of the future? Shall flags be flaunting gaily in the sun to welcome the birthday of the reigning monarch, — the child of our Victoria, — at a time when our tombstones shall be casting their shadows across the withered November sward of silent churchyards; and shall bonfires be blazing on the hills, as the stars twinkle out one by one from amid the deepening blue to look down upon our graves?

The future belongs to One only, — to that adorable Being who has made His great goodness so manifest to our country for ages and centuries, and rarely more vividly manifest than in the present happy event. He alone sees the end from the beginning, and He more than sees it; for in His unchanging righteousness, and infinite goodness and wisdom, has He ordered and determined it all. *Our* histories relate to but the past; in *His* the chronicles of all the future are also recorded. We write in *ours* as their latest event, that there has been born an heir-apparent to the British crown, and our remoter hills still reverberate the echoes which our gratulations have awakened; in *His* the circumstances of the birth are not more minutely laid down than the details of the funeral. There is a coffin in the distance that lies in the gloomy solitude of a royal vault; and the golden tablet that rests on the lid bears a date and an age well known to Him, for His own finger hath inscribed it. To us all is dark; but what so natural for creatures whose birthright is hope, whose privilege and whose nature it is to look both before and behind, to dwell upon the past, and anticipate a hereafter! What so natural for them as to let their thoughts out upon the future, and to imagine where they cannot see!

Our children are around us, — the bright eyes, and silken locks, and rosy cheeks of infancy. Is there no pleasure in saying to them, Listen to these sounds, — to that distant peal of the city bells, and that measured, sullen boom of the cannon: there has been a king born who is to be your king, though, we can trust, not ours, for we are old enough to remember the birth of the Queen, his mother. But he is to be *your* king, and in happier days, we would fain hope, than those of either the present or the past. The world will not be always what it has been; misery will not be for ever the prevailing state, nor unhappiness the o'ermastering feeling, nor evil the dominant power. There is a time coming, foretold by the Spirit of God, when wars, and violence, and crime, and misery, shall cease, — when men shall live together as brethren, as the children of one family, — and the knowledge of the Lord shall be everywhere. That time cannot now be far distant; and if good and wise men have calculated aright, — studious and venerable fathers of the church, who, in poring over the sacred oracles, have arrived, each apart, at conclusions singularly alike, — the dawn may break with no doubtful flush of promise during the reign of the monarch at whose birth three kingdoms are now gladdened; the eastern sky may be reddened by the first glories of that millennial light which shall continue to shine more and more till the perfect day shall have arisen; and even he himself, made wise through the teaching of the Spirit, may be one of those nursing fathers of the church whose happy reigns prophets have foretold. Are these but the wild dreams of the enthusiast? We may, indeed, err widely in attempting to fix the *time*, but be it remembered that God himself has fixed the *events*.

It were little wonder though men should weary of the present. There are, we doubt not, some of our readers who can look back on the events of sixty years. How has the space been filled? A sullen and doubtful peace had just succeeded the disastrous — we must add unjust — war with our American brethren. It was broken by the fierce and bloody tumults of the French Revolution. Atheism and murder stalked abroad; nation rose up against nation; Europe bristled over with arms; and for eighteen years together, during which millions perished by famine, fire, and the sword, manslaying was the trade of the *civilized* and *Christian* world! Men, as little wise as their rude ancestors, were playing at the old vulgar trick of hero-making, and the progress of the species stood still till the disastrous game was finished. In our own country, times of hardship and discontent succeeded, and poor, hunger-bitten men, maddened and blinded by their misery, snatched hold of uncouth weapons, in the vain hope of bettering their condition by violence. The madness passed, and a period of political heats and animosities ensued. Civil right was regarded as but another name for national happiness. The delirium of this second fever is over for the time. The rights have been gained; but the poor, overtoiled man who wrought sixteen hours every day ere the struggle began, works sixteen hours still, and hunger and the sense of hapless degradation presses upon him as sorely as ever. The present, in the main, is assuredly no happy time. Never were there such frightful accumulations of misery in our cities, and rarely have the sullen murmurs of the masses evinced deeper discontent. In our own country we have witnessed the revival of the evils of an earlier period, — superstition stalking abroad unquestioned; persecution assailing the truth; the spiritual nature, the eternal concerns of man, made the game of quibbling lawyers impressed by no true sense of a hereafter; consciences outraged; and the care of souls transferred by an abuse of law to the charge of wretched hirelings. It is well to believe there are better times in store; that the right shall eventually prevail, whatever may be the fate of those who contend for it in the present; that Christ reigns; and that the day is assuredly coming, though it must rise on the tombs of the present generation, when his sovereignty shall be universally acknowledged, and the influences of his Spirit everywhere felt.

IV.

REMAINS OF NAPOLEON.

There are no people in Europe who bear a better marked character than the French, and no people whose peculiar tastes and dispositions seem to have been so closely studied by their more sagacious statesman. They are employed at present, heart and soul, in adding, at the suggestion of Thiers, a supplementary paragraph to the posthumous history of Napoleon. The wily politician has applied to our English Government for permission to remove the remains of the great hero of France from St. Helena to Paris; the English have acceded to the request with the best possible grace; and the French people, brimful of sentiment and enthusiasm, and on tiptoe expectation of the coming pageant, are lauding Thiers to the skies as the best possible of all good ministers, and the English as the most generous of all old enemies, made friends for evermore. When a Roman general wished to conciliate the people of Rome, he turned loose a score or two of wild beasts in the amphitheatre, or hired a few hundred gladiators to fight together till the one-half of them were dead. One general, however, was content just to imitate another general; and though they squandered their bronze and silver in immense sums, there was no expense of invention. Thiers is immensely more original: he has got a dead Napoleon for the French to bury, and will probably command majorities, on the strength of their gratitude and respect, for a twelvemonth or two to come. Even the classes with discernment enough to see through his policy will admire him for the great tact and ability which it displays, — and there is perhaps no civilized people in the world whom the mere admiration of talent or of greatness influences more. The French as a people are followers rather of great men than of great principles. Nature does not seem to have intended them for republicans; they were content of old to be little individually that their kings might be great; and in after days they were equally content to lose their individuality in the glory of Napoleon. But is it not well, for the sake of peace, that the policy of Thiers tells, on the present occasion, as powerfully in favor of the English Government as in that of the sagacious politician himself?

It matters little whether the remains of Napoleon lie in a gorgeous sepulchre amid the multitudes of Paris, or raised high over the sea on a lonely rock of the Atlantic, like an eagle dead in his eyry. The scourge which vexed the nations has been laid by; the purpose of mingled wrath and mercy which it was called into existence to accomplish has been fully performed. The last lesson taught regarding it was to show how utterly passive and powerless a thing it was in itself, when flung aside by the Omnipotent Hand which had wielded it. The melancholy prisoner of the rock, — the fretful invalid, so unhappy in society, and yet so unfitted for solitude, — the petty squabbler with officials and underlings about forms of etiquette and modes of address, — was the terrible Napoleon, the hero of a hundred fields, the dispenser of crowns and sceptres, the warrior who had borne down the congregated soldiery of civilized Europe, the conqueror of powerful kingdoms, whom the united might of a Cæsar and an Alexander might have assailed in vain. Never was there greatness so great, or littleness on a smaller scale; and it will be long ere the people of France find for his dust so sublime and appropriate a monument as the huge rock of St. Helena. Its dark walls of a thousand yards, compared with which the walls of great Babylon were as hillocks raised by the mole, — the unceasing surge that idly frets itself against its base, — the vast surrounding sea, with its dim and distant line, — the sublime o'erarching canopy, — the minute and speck-like tomb rising towards the clouds on a pedestal not its own, — where else will it be felt with such soul-stirring effect that man is so very little, and God so very great? Not among the mingled palaces and hovels of Paris, or amid the half-infidel, half-idolatrous veneration of its frivolous and theatrical people.

"Change grows too changeable," says Byron, when referring to the state of matters twenty years ago. "In what circumstances, think you," says Dr. M'Crie, in addressing a correspondent, "if you and I were to retire for two years to some sequestered island, would we find our native country on our return?" The amount of vicissitude and revolution spread over centuries in the past has been concentrated in the present within the compass of a single lifetime; and there are perhaps few things more interesting than those tide-marks, if we may so express ourselves, which, like the measure of Thiers, show the ebbs and flows of circumstance and opinion, and the wonderful suddenness of their rise and fall. Who would have said twelve years ago that a minister of France would have set himself to court popularity and to strengthen the kingly authority by finding a tomb for the Emperor in Paris? And who that remembers that the remains of Henry IV. — *Henri Quatre* — the beloved of the people, the theme of their tales and their songs, the hero of their only epic — were torn by these very people from the sepulchre, and cast ignominiously into a ditch, will venture to say that another and very different chapter may not yet be added to the posthumous history of Napoleon? The current that sets in so powerfully in one direction to-day, may flow as powerfully in a different direction to-morrow; and the half-idolatrous respect that more than canonizes the memory and the remains of a great warrior and statesman now, may be soon exchanged by a fickle and varying people, ever in extremes, for a detestation equally strong, and surely not less rational, of the despotic subverter of popular rights, — the destroyer of a million and a half of creatures with souls as undying as his own, — the cold-hearted and selfish calculator, who made human lives the coin with which he bought and sold, and who could reckon out his tale of these, and pay them down, as coolly, for some definite extent of wall or trench, or some certain amount of territory, as the land-agent or the merchant could the common circulating medium, when employed in their respective professions. We are afraid there still awaits a discipline of despotism, suffering, and blood for the people whose admiration can rise no higher than the greatness of a Napoleon.

V.

JEAN D'ACRE.

The fortress of Jean d'Acre, the main stronghold of the Levant, is now connected a second time, within the course of forty years, with the history of Great Britain. And in both instances the national success has been very signal, and the objects attained of a strikingly similar character. Europe was first taught before the ramparts of Jean d'Acre that the greatest of modern conquerors was not invincible. The history of Napoleon, until he took up his position in front of this eastern fortress, was summed up in a series of victories; nor could one so familiar with conquest have anticipated defeat here. The garrison consisted mainly of a semi-barbarous and half-disciplined soldiery, who were fast losing their ancient military character; the fortifications of the place belonged at the time to that obsolete and less approved school of defence whose peculiar defects had been first detected, more than an age before, by the countrymen of the assailants, — of all military men the most skilful in carrying on such warfare. The walls yielded to an incessant storm of shot and shell; and the best troops of France, under the command of by far her ablest general, were led repeatedly to the attack. But they returned time after time baffled and broken. Jean d'Acre, tottering apparently to its fall, and half-dismantled and half-garrisoned, resisted their utmost efforts. A mere handful of Englishmen fought in the breach; and He who can save by many or by few had willed that their efforts in every instance should be crowned with complete success. The siege was raised; the key of Palestine passed not into the hands of Napoleon; the besiegers fell back upon Egypt; disaster followed disaster; Nelson annihilated their fleet at the Nile; their leader slunk away from them; the army of Abercromby cooped up their forces amid the sand-hills of Alexandria; and their scheme of eastern conquest finally terminated in so inglorious an abandonment of the enterprise that they owed their very safety mainly to the sufferance of the British. Napoleon afterwards opened his trenches before stronger fortresses, and *they* fell. A single campaign threw open the cities of Prussia to him; he gave law to the armed millions of Spain, Austria, and Holland: but on the solitary wastes of Judea there awaited another destiny; and the stars in their courses fought against him and prevailed, when his scheme of conquest led him there. His enterprise, and the apparently inadequate means through which it was defeated, remind one of the old Grecian story of the open space left by the countrymen of Ajax in the forefront of their armies, long after the hero's death, but in which they believed he still continued to take his stand. A famous warrior of the enemy, it is said, espying the opening in the heat of battle, rushed into it to make his way through, but he was instantly felled to the ground by some invisible antagonist, and then dashed back upon his friends. Judea, so long trodden under foot by every enemy, however mean or contemptible, seems in the present century to represent that open space.

Forty years have passed since the discomfiture of Napoleon before the walls of Jean d'Acre. A new scene of things has arisen. The conqueror, after performing the part to which he had been appointed, was conquered in turn, and died in captivity and exile; and another conqueror has arisen, — a man of far inferior power, but with immensely inferior powers to contend with. Hall of Leicester, in one of the most sublime of his compositions, has compared the terrible Napoleon to an eagle burying its beak and talons in the quivering flesh of living victims, — tearing the still sentient nerves asunder, and drinking the warm blood. Mehemet Ali may be regarded rather as a vulture, who attacks but the dead and dying. He has been dissevering the limbs of a victim somewhat less than half alive. A great empire seems passing into extinction; and the Pasha of Egypt — sagacious, energetic, brave — is exactly one of those characters, so frequent in history, that become at such periods the monarchs of the minor states which spring up in the room of the great departed power, just as the place of a mighty oak or chestnut comes to be occupied, when it had sunk into decay, by whole thickets of inferior growth. He had appropriated Egypt, and the claim of the successful soldier had been fully recognized by at least all possessed of power enough to challenge it. Syria lies adjacent; and of Syria, by far the most interesting portion is comprised in that land, so peculiarly a land of promise, of which prophecy is so full, and which has been the scene of events compared with which all in the course of human affairs that have taken place in the other lands of the globe sink into utter insignificance. And Syria, apparently as defenceless as the blank space in the ancient Grecian army, seemed to lie even more open to Mehemet Ali than to Napoleon. He possessed himself of Jean d'Acre, the key of the country, — the identical fortress which a few dozen Englishmen, assisted by a half-disciplined horde of Turks, had maintained against the greatest general and the best soldiers of France. He strengthened the fortifications on the most approved principles; he surmounted them with nearly two hundred cannon; he stored the magazines of the place with supplies for at least a six months' siege; he furnished it with a garrison of six thousand veteran troops. A few British vessels cast anchor before it, under the fire of at least a hundred and twenty cannon and twenty huge mortars, and bombarded it for three hours. What has been the result? An unseen hand was raised in the conflict, and the fate of Syria decided at a single blow. In the heat of the engagement a terrible explosion took place within the fortress, that shook the earth and the walls like an earthquake; a huge cloud shot up over the place, fold beyond fold, till it seemed to reach the central heavens, and then passed slowly and heavily away; and when it had cleared off, it was found that one third of the city had been utterly destroyed, as if by the earthquake predicted in the Apocalypse, and nearly one third the garrison buried in the ruins. There was scarcely a house left habitable within the walls. The principal magazine had caught fire; and thus the ruin of the fortress has been signal in proportion to the means taken for its defence. The firing slacked immediately after, and then finally ceased; and at midnight the surviving portion of the garrison stole silently out of the place, which was taken possession of about daybreak by a party of the besiegers. They found it a terrific scene of devastation, covered with shattered ruins, sprinkled with blood, and strewed with dissevered limbs and dismembered carcasses. An immense hollow, like the crater of a volcano, occupies the place where the magazine lately stood; and for the space of a mile around nothing appears but the broken fragments of what were once buildings, scathed and blackened by fire, and the mangled bodies of men and animals piled in heaps upon cinders and rubbish. In the course of the day, a portion of the garrison, amounting to about seven hundred infantry, intimidated apparently by the mountaineers, marched back to the place, and, delivering up their arms, surrendered themselves prisoners of war. There are more than two thousand prisoners besides; and of the whole garrison, not one sixth part is said to have escaped. The loss of the assailants amounts to but twenty-three killed, and about fifty wounded.

One inevitable result of this important and very remarkable event will be the signal diminution, perhaps the thorough annihilation, of the aggressive power of Mehemet Ali. If possessed of sagacity enough rightly to estimate his position, — a position not unlike that of Napoleon when he rejected the terms of peace proffered him after the disaster of Moscow, — he may still retain the sovereignty of Egypt; but it is impossible that he can now become the conqueror of Syria. He has entered the gap, like the old warrior, and been struck down, just as Napoleon, when he attempted entering upon it, was forced back. The space, for that great purpose at which the finger of Revelation has been pointed for nearly the last three thousand years, has been kept clear, and the time of the accomplishment of this purpose seems fast approaching. Is there nothing remarkable in the fact that the two great conquerors of the nineteenth century should have made their attempts upon it, — the one backed by the identical means which had been employed against the other, — and that what the one found so powerful in resisting him should have proved of no avail in the other's defence? Napoleon was to be resisted, and Jean d'Acre became impregnable; Mehemet Ali was to be dispossessed and turned back, and Jean d'Acre fell in three hours. And yet it seems to be a mere gap among the nations, a solitary and empty space, that has been thus defended, — a few skeleton cities, a few depopulated villages, a few sandy plains, a few barren hills, the long valley of the Jordan, Gennesaret, Galilee, Jerusalem, and the rocky eminence of Calvary. A great nation seems dying away from amid the wastes of this more than classic country, leaving the place well-nigh tenantless. Others have arisen to take possession in their room, but they have been violently held back. The land still waits unoccupied for the appointed inhabitants.

VI.

THE CROMWELL CONTROVERSY.

Our readers must have remarked, with some degree of amusement, the progress of the controversy still raging regarding an important clause in the Marble History of England now in the course of being chiselled, at the national expense, in the new House of Commons. Every one agrees that, in order to impart to the record any degree of truth at all, it must contain a vast number of clauses that will do no honor to the marble, and that will be unable to receive honor from it. It will contain the Marian clause, in the form of a grim, ugly female, smelling horribly of blood and fire ; and the Henry VIII. clause, in the shape of a puffy-cheeked, truculent bully, surrounded by a group of skeleton wives, some of them bearing their dissevered heads under their elbows; and the Charles II. and James VII. clauses ; and a great many other disreputable clauses besides, some of them of more modern, some of them of more ancient date, on the insertion of which all are agreed. The entire dispute hinges on the singularly brilliant clause Oliver Cromwell, respecting the insertion of which there are, it would seem, many diverse opinions. Some assert that the clause Oliver should, like the clause William the Conqueror, or the clause Richard III., be introduced in full; others maintain, on the contrary, that it should not be introduced in full, nor introduced at all, and that there should be even no hiatus left to indicate its existence, but that the flat, moody clause Charles I. should *run in* without break, as printers say, with the miserable clause Charles II.; while a third class, content to halve the difference, recommend that the clause Oliver should not be inserted, but that its place should be represented by a wide blank, suited to serve the purpose of a line of asterisks in a piece of abridged narrative. Now, doubtful as the fact may seem, there is actually some meaning in this controversy. In its ostensible relation to a bit of marble, it merely involves the not very important question, whether the new House of Commons is to be adorned by some sixty statues or so, or by only fifty-nine; but in its true relation to principle it involves a question of somewhat greater magnitude, — the existing amount of liberal opinion; and its producing springs lie deep among the great parties of the country.

One very important party in the transaction is the Book of Common Prayer. Among the Presbyterians of Scotland, as with the better English historians, Charles I. does not stand high. Such was the character of his government, that they had as one man to take up arms against it; and it is known that, save for their success on that occasion, the Star Chamber would have become as permanent an institution in England as the Bastile did on the opposite side of the Channel; that the new mode of raising ship-money would have formed the model for levying all the other taxes; and that the English House of Commons would have shared exactly the same fate as that of the nearly contemporary French Chamber, the States-General, under Louis XIII. The British Government would have ceased to be representative; the religion of Laud would have become for a time that of the two kingdoms, and then have merged into the Romanism of the third; and the state officers, assisted by the bishops, would have meanwhile carried on the agreeable amusement of shutting up honest men for life in dungeons, confiscating their properties, and cutting out their ears, or, if the ears had been previously cut out, of grubbing up the stumps. Nor do we estimate more highly the personal character of the man than his principles of government. He was a kind husband, and amiably suffered his popish wife to influence the national councils, which was, of course, something in his favor; and, when unfortunate, he had a profound sympathy for himself and his family, — the true way of eliciting the sympathy of others; and this was doubtless something in his favor too. But we decidedly demur to the titles of Saint and Martyr, in their ordinary and unqualified meanings. We must at least be permitted to regard him as the unique saint, who, according to the old Scotch chronicler, "swaire terribly," and played golf on the Sabbath; and as the extraordinary martyr, whose head was cut off because his word could not be believed. Such, pretty generally, is the Presbyterian estimate of Charles; but in the estimate of the Book of Common Prayer there are no such qualifications. He is there the glorious martyr and the blessed king; and Episcopacy still fasts once a year in all her churches, to avert the judgments that may be still impending over the land for his death. Much, of course, depends on being used to a thing; and there are, we doubt not, devout men who can join in the hymn which she sings on the occasion with much earnestness; but to us it has ever appeared to be considerably more akin to the parodies of Hone and Carlile than to the greater part of the compositions which Hone and Carlile parodied. There can be no mistake regarding the slain man to which it refers; the title fixes that: it is a hymn "to be used yearly upon the thirtieth of January, being the day of the martyrdom of the blessed King Charles the First, to implore the mercy of God" "that the guilt of that sacred and innocent blood may not at any time hereafter be visited upon" the people or their children. The slain man is unequivocally the man Charles; and yet it is thus we find him spoken of in the hymn, — a bizarre piece of mosaic, it may be mentioned in the passing, composed of a curious mixture of gems filched from the Scriptures, and of bits of paste broken from off the Apocrypha: —

"O my soul! come not thou into their secret; unto their assembly, mine honor, be not thou united; for in their anger they slew a man. — *Gen.* xlix. 6.

"*Even the man of thy right hand: the Son of man, whom thou hadst made so strong for thine own self.* — *Ps.* lxxx. 17.

"In the sight of the unwise he seemed to die; and his departure was taken for misery. — *Wisd.* iii. 2.

"They fools counted his life madness, and his end to be without honor; but he is in peace! — *Wisd.* v. 4 and iii. 3.

"How is he numbered with the children of God; and his lot is among the saints! — *Wisd.* v. 5.

"But, O Lord God, to whom vengeance belongeth, thou God to whom vengeance belongeth, be favorable and gracious to Sion. — *Ps.* xciv. 1 and li. 18.

"Be merciful, O Lord, unto thy people whom thou hast redeemed, and lay not innocent blood to our charge. — *Deut.* xxi. 8."

Charles I., the man of God's right hand! the Son of man, whom God made strong for himself! It is not wonderful that the church which can thus continue to appropriate to the wretched Charles the glory of the adorable Redeemer, should exert some little influence in preventing the apotheosis of the Pilate who put him to death.[1] The revival, too, of the old Canterburian party in England, — true representatives of Charles's infatuated advisers, — who, amid the light of the present day, can regard the revolution to which her Majesty owes her crown as simply the Rebellion of 1688, has of course its effect on the controversy. The special admirers of the "Blessed Charles the Martyr" still muster stronger within the pale of the English Church — though they seem fast quitting it for a more congenial communion — than they have done for at least a century previous; and we may be well assured that

[1] The fast of the royal martyr is no longer celebrated.

in their hands the "form of prayer and fasting for the thirtieth of January" will be no dead-letter. In no other churches will the hymn from which we have quoted be sung with half such energy as in the churches that have got their crucifix-mounted altars perched up under the east window, and in which the priest prays with his back to the people. There is a story told by Franklin of the good old Puritans of New England, which the more rational members of the English Church might perhaps do well to ponder. The poor people, forced from their homes by the fierce intolerance of the blessed martyr, whose martyrdom led to the blessings of toleration, felt at first exceedingly melancholy in the savage country of solitary wilds and deep forests in which they were compelled to sojourn, and for a series of years kept the anniversary of their arrival as a fast; and the oftener they fasted, the more melancholy they became. At length, at one of their meetings on the eve of an anniversary, when the usual fast had been proposed, an honest farmer rose and moved an amendment. They were all getting very comfortable, he said, if they could but see it. Their farms were improving and extending, their crops becoming every year more weighty and the country less wild; they were living in peace, too, and enjoying liberty of conscience; and he moved, therefore, that, instead of holding their anniversary as a fast, they should forthwith convert it into a day of thanksgiving. The suggestion approved itself to the judgment of the meeting; the fast was suffered to drop, and the day of thanksgiving substituted in its place; and from that day forward the colony continued to prosper. Now, we are of opinion that the Church of England has fasted quite long enough for the martyrdom of Charles. It was an event of an exceedingly mixed character; it had party-colored sides, like the gold and silver shield in the story; and the church, regarding it on merely the unfavorable one, has now been fasting for it nearly a hundred and ninety years.

She should now by all means try to get a glimpse of it on the other side, and, like the worthy New England Puritans, convert her fast into a thanksgiving. We are pretty much assured the country would be none the worse for the change. There are weightier national sins for which to fast than the sin of the martyrdom; and were we but grateful enough for its benefits, we might avoid, among other perils, all danger of committing the great national folly of excluding from our marble records the name of our greatest ruler.

The question at issue in this case is unquestionably a British one, — Scotch as well as English; but it strikes us that the Scotch are in more favorable circumstances for arriving at an impartial decision regarding it than their neighbors in the south. In England the two great parties still exist, with many of their old predilections and antipathies undiluted and unchanged; the one of which Cromwell led on to victory, and the other of which he defeated and threw down. The question regarding him is still a party question there, argued on the one side in many a goodly volume, and sung once every year in their churches by the other, in music set to the organ. Scotland, on the other hand, dealt more with Crmowell as a nation: the Protesters stood widely aloof, and did not take up arms; but the great bulk of the nation — all its Resolutioners and all its Cavaliers — joined issue against him on behalf of Charles II., and got heartily drubbed for their pains. We are nearly in such circumstances as the English themselves would be were the question, not whether Cromwell should have a statue in the British House of Commons among the other supreme rulers of England, but whether Napoleon should have a statue in the French Chamber of Deputies, among the other supreme rulers of France. True, Cromwell beat us, — and we don't much like the memory of our defeats; but we flatter ourselves that it was only because our "Committee of Church and State," contrary to the judgment of Leslie, was a little too eager to beat him.

We solace the national vanity, too, by remembering that he himself was half a Scotchman: we can still point out, from the burgh of Queensferry, the old house on the opposite side of the Frith in which his mother, Elizabeth Stewart, first saw the light; and, further, we call to mind that the blood of the Bruce flowed as purely in the veins of the plebeian Cromwell as in those of the royal martyr himself, and that he represented the indomitable hero of Bannockburn immensely better. Above all, we remember how very different the treatment which we received from the man we fought against, from that which we received from the man we fought for. And so we at least deem ourselves impartial, and marvel how there should live Englishmen in the present age who could so much as dream of excluding the record of the Protector from the general record of the country, as exhibited in its house of representatives. Save for Puseyism, High Churchism, and the rather equivocal service in "our most excellent Prayer Book," the question could never have been mooted. We have seen it virtually decided in children's toy-books that were written half a century ago. Some thirty years since, when we kept our library in a chip-box six inches square by five inches deep, we were in the proud possession of two tall volumes, four inches high by three inches across, the one of which, for the use of good boys and girls, contained notices and woodcuts of all the Scottish monarchs, from the Davids down to James VI.; and the other, notices and woodcuts of all the English ones, from William the Conqueror down to George III. And each little book, we well remember, had its single uncrowned figure, and its single notice, pretaining to a great monarch that wanted the kingly title. The figure in the one case was that of a mailed warrior trampling on a lion, and the figure in the other that of a warrior, also in mail, with a marshal's truncheon in his hand. The legend affixed to the one was "Sir William Wallace, Protector of Scotland," and that borne by the other, "Oliver Cromwell, Lord Protector of England;" and such was the interest attached to the prints and the notices, that the little books at length learned to open of themselves at the pages which exhibited the uncrowned warriors; for the one, with scarce a single exception, was the greatest and noblest of the Scotchmen, as the Bruce, though of a heroic nature, was less disinterestedly a patriot; and the other, with scarce a single exception, was, as far as we know, the greatest and noblest of the Englishmen; for, though the figure of Alfred looms large in the distance, the exaggerating mists of the past close thick around him, and we fail to ascertain his true proportions through the cloud. Here, we contend, in the child's books, and by the child, the grave question at issue, of statue or no statue, was fairly decided. The child's books found fitting space in their pages for the effigies of the two Protectors; and the child soon learned to give unwitting evidence that the effigies of none of the others had at least a better right to be there.

But Cromwell, it is urged, was not a *king:* he said *no*, though he might have said *yes*, when offered the crown; and his statue ought to be excluded on the strength of the monosyllable. We would be inclined to sustain the objection had it been proposed to erect Cromwell's statue, not in the British House of Commons, but in the Herald's Office. But history is a thing of veritable facts, not of heraldic quibbles. *King* is a simple word of four letters, and *Lord Protector* a compound word of thirteen; but, translated into their historic meaning, their import is exactly the same. They just mean, and no more, the supreme governor of the country. The only real difference between Cromwell and the Charleses on either side is, that he was a great and good supreme governor, and that they were little and bad ones. Ah! but Cromwell, it is urged further, has no legal existence in our chronicles. A statute, still enforced, effaced his name from the constitutional annals, by giving his years and his acts to his successor. History, we reply, is no more a thing of legal fictions than of heraldic quibbles. By a legal fiction Cromwell may be merely a bit of Charles II., and we know that by a legal fiction husband and wife are but one person; but we also know that the historian who should represent George IV. and his wife Caroline as merely the two halves of a single individual, would make sadly perplexed work of the "Queen's Trial." If Charles II. was also Cromwell, he was assuredly the most extraordinary character that ever lived — much more emphatically than Bacon, as described by Pope, —

"The wisest, greatest, *meanest* of mankind;"

and his statue, if that of the Protector is to be excluded, should by all means indicate the fact. Let him be represented as an eastern sept represent one of their gods, — the "man-lion," as a compound monster, half-brute, half-man, with double fore-arms articulated at his elbows; or let his effigy be placed astride that of a tall figure in a cloak, like the Old Man of the Sea astride the shoulders of Sinbad; or, to render the allegory complete, let there be no human form placed on the pedestal at all, but simply a good representation in stone of Æsop's live ass and dying lion. For the sake of truth, however, the lion would require to be exhibited, not as dying, but dead. Cromwell was dead, and, as if to make all sure, cold, for considerably more than a twelvemonth ere a monarch or lawyer dared to raise the assinine heel against him. "They hung your father, lady," was the ungenerous taunt dealt, many years after the event, to one of his daughters. "Yes," was the proud reply, "but he was *dead* first."

We do not think the statue of Cromwell should be assigned its proper niche, were it but for the sake of the associations which it is fitted to awaken, and the lesson in behalf of supreme governors in general which it is suited to teach. Quivedo, in one of his Visions of Hell, as quoted by Cowper, requested his black conductor to show him the jail in which they kept their kings. "*There*," said the guide, "*there* you have the whole group full before you."

"Indeed!" exclaimed Quivedo, "they seem but few!" "Few, fellow!" replied the indignant guide, "few! — they are all that ever reigned, though." Cowper objects to the undiscriminating severity of the wit, and names one or two kings, such as Alfred and Edward VI., who could hardly be regarded as inmates of Quivedo's prison; but certainly, were all kings of the type of the royal martyr, his father, and his two sons, — the British kings of nearly an entire century, be it remembered, — the objection would scarce have been lodged. It would be of importance, surely, as suited to produce the moral effect of Cowper's exception, to have inserted full in the middle of the line one supreme governor who was not a scoundrel, and who was not a fool. Very different indeed would be the associations that would hang on the central effigy, from those which the two effigies on either side must of necessity suggest. The smell of blood rises rank from these miserable Stuarts, and it is invariably the blood of the best of their land, — the blood of honest patriots and of godly men. We find the insensate marbles associated with a dark record of crime, and cruelty, and monstrous infatuation; they are suggestive of the melancholy of protracted exile, the gloom of dungeons, the agonies of torture, and the pangs of death, — of the blood of God's saints shed on the hills like water, or flooding the public scaffolds, — of Scottish maidens tied to stakes under floodmark, to perish amid the rising waters, — and of venerable English matrons burnt alive. It speaks of national degradation and impotency; of ever-recurring defeats, and inefficient, disastrous wars; of unavenged insults to the British flag; of English fleets chased into the Thames by the victorious enemy; and of English towns burnt unavenged on its shores. Surely it were well to have some means of relief at hand from such thick-coming forces. The antidote of the central marble is imperatively required. It opens up, amid the darkness on either hand, a vista of surpassing glory. We see England throned in the midst of the nations, — her armies victorious in all their battles; her navies sweeping the seas invincible; her voice of thunder resounding all over the world in behalf of religious liberty and the rights of man, and all over the world feared, respected, and obeyed; good men everywhere living in peace, however little friendly to the magnanimous Cromwell; and the sword of persecution dropping from the terror-palsied hand of the Papacy.

VII.

THE THIRD FRENCH REVOLUTION.

WAS there ever an age of the world like the present! The painted scenes in a theatre do not shift before the eyes of the spectators more suddenly, or apparently more on that principle of strong contrast on which the poet and the artist rely for their most striking effects, than dynasties and forms of government in the times in which we live. The biography of Louis Philippe could belong to but one period in the history of the species.

It will be eighteen years, first July, since the writer was employed, on a clear beautiful evening, in the immediate neighborhood of a busy seaport of the north of Scotland, all alive at the time with the turmoil and bustle of the herring-fishery; and a few neighbors, whose labors for the day had closed, were lounging beside him. There were two French luggers in the harbor furnished with crews of stout English-looking seamen from Normandy, — crews at least thrice as numerous as the herring speculation in which they were engaged could ever pay; but the Government of their country, still as anxious as in the days of Napoleon to create a navy, made up, by what was nominally a very extravagant bounty on fish, but in reality a bounty on sea-faring men, the amount necessary to render their undertaking remunerative. And so there, in the middle of a group of fishing-boats and small craft of the British type, lay the two hulking-looking foreigners, one of which rejoiced in the august name of "*Le Charles Dix*," with their bare brown masts and their dark, half-unfurled sails, and crowded with seamen attired in dirty Guernsey frocks and red nightcaps. The post came in, and a newspaper, still damp from the press, was handed to a neighbor. He opened it, and repeated, with an air of mingled astonishment and incredulity, a few magical words, —"Revolution in France! — Three days' fighting!—Flight of Charles X.!" We were sensible, as the words were pronounced, of a thrilling sensation similar to that produced by an electric shock. Nothing could be more evident than that the consequences of an event so truly great could not be restricted to France. A new epoch had arrived in the history of civilization and of man; but what was to be its character? The curtain had arisen literally at the ringing of a bell; and the stage, at the opening of the piece, as at the close of some tragedies, red with blood and cumbered with dead bodies, presented the imposing spectacle of a falling dynasty. But who could predicate regarding the nature of the plot on which the general drama was to turn, or anticipate with aught of confidence the outlines of even its next scene? The poor Frenchmen of the two luggers, with just enough of bad English to bargain for herrings, but not enough to understand the details of a revolution, were sadly perplexed by the intelligence, of which the townspeople present, in as plentiful a lack of French as they manifested of English, could but communicate to them vaguely enough the general result. They got hold of the newspaper, and scanned it with all the eager excitability of their nation, though apparently to little purpose. They could merely here and there pick out a few Norman words which the conquest of William had served to naturalize in our language, and pronounce them with tremendous emphasis after the French mode ; but all they succeeded in picking out of the broad sheet seemed to be summed up in the emphatic heading of the editor's article, — *Revolution Francaise* — *Trois Jours de Combat* — *La Fuite de Charles Dix!* They learned quite enough, however, to exhibit in a small way how slight a hold French kings have in these latter times on the affections of the French people. One of the masters, seizing a lump of chalk, stepped to the stern of his vessel, and, with great coolness, blotted out from the board the name of *Charles Dix.* He did in the harbor of Cromarty, on a minute scale, what his countrymen had just done in Paris.

And now Paris has witnessed yet another revolution. The bell has rung ; the scene has shifted ; drama the second has come as suddenly to a close as drama the first ; and the after-piece begins, like its predecessor, with fighting and bloodshed, and the masque-like pageant — picturesquely symbolical of the whole event — of an empty throne paraded through the streets, and then dashed down and burned at the foot of the "column of July." The effacing chalk has been applied, and the name of another monarch blotted out. And amid the general thrill of undefinable electric interest and restless anxiety, there obtains exactly the same uncertainty regarding what is to come next. It is not unworthy of notice, that the three French revolutions have in reality all turned on one pivot, and that some of the shrewdest of our contemporaries have been led egregiously into error in their calculations on the present occasion, simply by losing sight of it. Nay, a similar disregard of this hinging-point, and of its controlling principle, seems to have been the fatal error of Louis Philippe himself. "It will require a most extraordinary and unforeseen combination of circumstances," said the *Times*, in an article on the Parisian outburst, "before any government, supported by an army of one hundred thousand men, under the command of Marshal Bugeaud, quartered with great skill in the outskirts of

Paris, perfectly prepared for action, and backed by eighteen fortresses, will be compelled to capitulate to a popular insurrection. We suspect, however, it will turn out that no serious popular insurrection is even probable. The people have been stirred, but not inflamed. They are shaken, but not irritated; they are unarmed, and no preparation for insurrection had been made. Under such circumstances the result is certain. But if lives are lost in this miserable brawl, the reckoning will be heavy, not only on those who inconsiderately commenced an agitation which they had no power to bring to a successful termination, but on those whose obstinate resistance to a well-founded demand rendered such an appeal to the populace successful." Such were the anticipations of the *Times;* and not a few of our other contemporaries followed in its wake. Had they taken into account in their calculations the principle to which we refer, — a principle first pointed out at a time when there had been but *one* French Revolution from which the necessary data could be derived, — they would have reckoned less securely on the hundred thousand soldiers and the eighteen fortresses. France is emphatically the great military nation of Europe. But though it possesses what are in reality the sinews of war, that is, great military ardor and many people, — for to regard money as such is an idle unsolidity, which, while it has the disadvantage of being commonplace, wants the balancing advantage of being true, — while France possesses, we say, a warlike people, it is wanting now, as in the days of Napoleon, and at every former period of its history, in the wealth necessary to *purchase* their service. Its rulers, therefore, in order to raise those great armies on which the power and character of the nation depend, must always appeal to its warlike sympathies; and the armies thus formed are, in consequence, what armies, in at least the same degree, are nowhere else in Europe, — merely armed portions of the people, — most formidable, as all modern history has shown, for purposes of foreign aggression, but in the hands of a despot, unless like Napoleon, the idol of the soldiery, dangerous chiefly to himself. This apparently simple, but in reality profound principle on which all the French revolutions have hinged, and which Louis Philippe, untaught by experience, so entirely forgot, was enunciated for the first time by Sir James Mackintosh, when the seventy thousand soldiers brought by Louis XVI. to invest the "Legislature and capital of France, felt that they were citizens, and the fabric of despotism fell to the ground." "It was the apprehension of Montesquieu," said the philosopher, "that the spirit of increasing armies would terminate in converting Europe into an immense camp, in changing our artisans and cultivators into military savages, and reviving the age of Attila and Genghis. Events are our preceptors, and France has taught us that this evil contains in itself its own remedy and limit. A domestic army cannot be increased without increasing the number of its ties with the people, and of the channels by which popular sentiment may enter it. Every man that is added to the army is a new link that unites it to the nation. If all citizens were compelled to become soldiers, all soldiers must of necessity adopt the feelings of citizens; and the despots cannot increase their army without admitting into it a greater number of men interested to destroy them. A small army may have sentiments different from the great body of the people, and no interest in common with them; but a numerous soldiery cannot. This is the barrier which nature has opposed to the increase of armies. They cannot be numerous enough to enslave the people, without becoming the people itself." It was on the unseen rock so skilfully marked out here that Louis XVI., Charles X., and Louis Philippe made shipwreck in turn, and that led to the error of our contemporaries. They took note of the hundred thousand men and the eighteen fortresses, but not of the all-influential principle which, in the revolution of last week, rendered them of no avail.

Events have exhibited the influence of the second French revolution on this country as, in the main, wholesome. It furnished the moving power through which parliamentary reform was carried, and the representation of the empire placed on a broader and firmer basis than at any former period. It formed the primary cause of the abolition of slavery in our colonies; destroyed monopoly in the East Indies; reorganized our municipal corporations; and, above all, gave to the people a standing-room virtually, though not nominally, legislative, through which, in the character of a league such as that which carried the great free-trade question, they can constitute themselves into a kind of outer chamber, whose decisions, if there be in reality a clamant case to give union and energy to their exertions, the *two inner* chambers must ultimately be content to register. And if, after all, it did not do more, it is only because all merely external reforms, whether political or personal, are in their nature unsatisfactory, and because men can only be made happier by being made wiser and better. It was through the inherent justice of the second French revolution, be it remembered, and the great moderation manifested in turning it to account, that this amount of good was produced. Never, on the other hand, was there an event less friendly to the progress of civilization, and to the true rights of man, than the first French revolution. Its atrocities, through the violent reaction to which they led, served to prop up every existing abuse, by rendering whatever professed to be the cause of reform suspected and unpopular. It was Robespierre and his colleagues, more than any set of men the world ever saw, that imparted to the cause of a blind, undiscriminating conservatism, not merely the character of sound policy, but also of justice. They arrayed the moral sense of mankind against their measures in the mass; and hence many an antagonist abuse was suffered to exist, which would otherwise have been singled out and swept away. The general war, too, in which the revolution terminated, and which was so peculiarly marked by the rise of one of the greatest military despots the world ever saw, militated against the progress of the species, and nowhere more powerfully than in Britain. The general effect of the first French revolution was as disastrous as that of the second was favorable. But what is to be the character and tendency of the third? We have our serious misgivings and fears. It is no doubt well for our country that, since the revolutionists have been successful, Louis Philippe should have been so decidedly in the wrong. Had he fallen five years ago by an assassin, and had Paris, in the distraction consequent on the event, been overmastered by the mob, the case would have been different; the sympathies of the British people would have been with the king and his family; Toryism would have profited in consequence, and Tory councils would have acquired a dangerous ascendancy. But there will be, in the existing state of the case, little British sympathy on the side of Louis Philippe. The policy of the later years of his reign has belied the promise of its opening, and he falls enveloped in the weakness inherent in whatever is palpably selfish and unjust. Still there is much cause for fear. There may be yet a reaction in France in favor of wiser heads and more moderate measures; but, for the present at least, the destinies of the country and the peace of Europe seem to be in the hands of an unthinking and reckless mob.

To what are we to attribute the singularly mistaken policy of Louis Philippe during the last few years, — so unlike, in at least the degree of sagacity which it evinced, that of the earlier portion of his reign? "Forget," said Napoleon, in urging one of his generals to exert all the energy of his more vigorous days, — "forget that you are fifty." Has the ex-king of the French been unable to *forget* that he is considerably turned of seventy? Has that peculiarly solid understanding for which in his more vigorous years the man was so remarkable, been gradually giving way during the last few years of his life; and are we to recognize in the gross imprudence — to give it no harsher name — which led to the present catastrophe, as in his shameless attempts to aggrandize his family in Spain, and his homologation as national of the revoltingly unjust assault on Tahiti, the signs of a decaying intellect, no longer able to control, as formerly, the selfish instincts of his nature, constitutionally very strong? And is this wise and brave man to be regarded as forming one illustrious example more of that class of the wise and brave so well described by Johnson? —

> "In life's last scene what prodigies surprise! —
> Fears of the brave and follies of the wise.
> From Marlborough's eyes the streams of dotage flow,
> And Swift expires a driveller and a show."

Certainly the latter scenes of the drama of his reign, to whatever they owe their peculiarity of character, read a fearful lesson. By virtually ceasing to be — what the title conferred on him exclusively recognized — "King of the French," and by setting himself, on the *effete* principles of the ancient *regime*, to be a king on but his own behalf and that of his family, he has ceased to be a king at all. It is noticeable, too, that he should have fallen a victim to a spirit evoked, indirectly at least, by that second French revolution to which he owed his throne. Save for that revolution, and its more immediate consequences, the Anti-Corn-Law League of Richard Cobden would have been altogether an impracticability, even in Britain. It was in order to prevent any such quiet but powerful combination of the British merchants from thwarting his plans in France, that the monarch's ill-judged stand against the reform banquet was so uncompromisingly taken. He resolved that no French league formed on the model of that of Britain should give law to him; and to that rash determination the third French revolution owes its origin.

VIII.

THE DUKE OF WELLINGTON.

"TRUST not," says an ancient English writer, "to the haleness of an old man's appearance, however stout and hearty he seem. He is a goodly tree, but hollow within, and decayed at the roots, and ready to fall with the first blast of wind." The country has received a startling illustration of that enhanced uncertainty of tenure by which men hold their lives when they have passed the indicated term, and "fallen due to nature," in the death of one of the most extraordinary men of modern times. A goodly but ancient tree has suddenly yielded to internal decay when no one looked for its fall; and the echoes of the unexpected crash resound mournfully, far and wide, through the forest. The consideration will, we are afraid, form but a doubtful solace to Britons of the present generation, that they will scarce again witness the fall of aught so goodly or so great.

The Duke of Wellington was the last, and at least *one* of the greatest, of that group of men whose histories we find specially connected with the history of the first French revolution. He pertained to a type of man so rare that we can enumerate only two other examples in the great Teutonic family to which he belonged, — George Washington and Oliver Cromwell. Of spare and meagre imagination, and of intellect not at all cast in the literary or oratorical mould, they yet excelled all their fellows in the possession of a gigantic common sense, — rarer, we had almost said, than genius itself, but which, in truth, constituted genius of a homely and peculiar, but not the less high order, and which better fitted them to be leaders of men than the more brilliant and versatile genius of a Shakpeare or of a Milton would have done. The ability of seeing what in all circumstances was best to be done, and an indomitable resolution and power of will which enabled them to do it, constituted the peculiar faculties in which they surpassed all their contemporaries. With more imagination they would have perhaps attempted more, and, in consequence, have accomplished less. Napoleon possessed powers which in Cromwell, or in Napoleon's great rival and ultimate conqueror the Duke, had no place. Neither the Lord Protector nor Wellington could have gloated over the overwrought sentiment and vivid description of an Ossian; nor yet could they have entranced, by their extempore tales, brilliant parties of thoroughly cultivated taste, and familiar with the best literary models of the age. But then, neither Cromwell nor the Duke would have sealed their ruin by a Russian campaign. Had the Lord Protector been in Napoleon's place, misled by no high imaginings, and infinitely less selfish than his great antitype, he would have restored their ancient independence to the Poles, erected their kingdom into a powerful barrier against the Czar, taken his revenge on Russia, not by attempting to dictate to it from its ancient capital, but by undoing all that Peter the Great had done, and shutting it up from the rest of Europe. Whatever he attempted he would have performed; and, instead of dying in exile, a solitary prisoner at St. Helena, he would have expired at Paris, Emperor of the French, and his son would have quietly succeeded him. The three great military *doers* of the Anglo-Saxon race were all alike remarkable for their sobriety of mind and spareness of imagination, and for exactly knowing — much in consequence of that sobriety and of that spareness — what could and what could not be accomplished. And so, unlike many of the great men of antiquity, or of the more volatile races of the world in modern times, they rose to eminence and glory by comparatively slow degrees, and finished their course without experiencing great reverses. There is a still rarer type of greatness, of which the entire history of man furnishes only some one or two examples, in which the imagination was vigorous, but the judgment fully adequate to restrain and control it; and we would instance Julius Cæsar as one of these. By far the greatest man of action of the age in which he lived, he was also one of the greatest of its orators — second, indeed, only to perhaps the greatest orator the world ever saw; while as an author his work takes its place in literature as one of the ever-enduring classics. By the way, has the reader ever remarked how thoroughly the features of Wellington, Washington, and Cæsar were cast in one type? Had they all been brethren, the family likeness could not have been more strong. There is the same firm, hard, *mathematical* cast of face, the same thin cheeks and prominent cheek-bones, the same sharply-defined nether jaws, the same bold nose, — in each case an indented aquiline, — and the same quietly keen eye. And in the countenance of Cromwell, though more overcharged, as perhaps became his larger structure of bone and more masculine frame, we find exactly the lineaments, united to a massiveness of forehead possessed by neither Washington nor Wellington, and only equally by that of Cæsar. Chateaubriand's graphic summary of the character of the Protector is in singular harmony with his physiognomy. "To whom among us," he says, in drawing a parallel between the first French revolution and that which in England led to the execution of Charles, "can we compare Cromwell, who concealed under a coarse exterior all that is great in human nature, — a man who was profound, vast, and secret as an abyss, — who hid in his soul the ambition of a Cæsar, and hid it in so superior a manner, that not one of his colleagues, except Hampden, could dive into his thoughts and views?"

Wellington, like the other great men with whose names we associate his, was remarkable for seeing, in his own especial province, what even the ablest and shrewdest of his contemporaries could not see. Jeffrey and Brougham were both able men, talkers of the first water, and, even as judges and reviewers, not beneath the highest average found among men; and yet we have but to take up those earlier numbers of the "Edinburgh Review," in which these accomplished judicial critics discussed the Peninsula campaigns, to find how utterly ignorant they both were — and, with them, all the party which they represented — of that simple but really great idea which formed the basis of Wellington's operations, and which ultimately led him to results so brilliant and successful. Nor was the *mediocre* ministry of the day, though they lent him from time to time their driblets of support, at first most meagrely and unwillingly, until compelled to liberality by his successes, less in the dark regarding it than their opponents. Once and again unable to make out a case for him, and gravelled by what seemed the unanswerable arguments of their antagonists, they had to throw the entire responsibility on their indomitable general; and Wellington was content to bear it. Nor was it in the least wonderful that they should have found the case of the Peninsula a peculiarly hard one. Appearances, as all ordinary, and even almost all superior observers, were able to remark them, seemed sadly against the British. The brilliancy of Napoleon's military tactics — above all his splendid powers of combination — had astonished the world. His marshals had learned in his school almost to rival himself; they were, besides, under his direct guidance; and they had three hundred thousand French soldiers in the Peninsula. The British there at no time amounted to sixty thousand. They had allies, it is true, in the Portuguese and Spaniards, but allies on which they could reckon but little; and yet, such was the apparently inadequate force with which Wellington had determined to clear the Peninsula. What could the man mean? Was he possessed of the vulgar belief that "one Englishman is a match for five Frenchmen at any time?" No: Wellington was perfectly sober-minded; and with a confidence in the native prowess of the well-disciplined Briton such as that which Nelson possessed, — a confidence that, if opposed, man to man, on equal terms of position and weapons, the Englishman would beat the Frenchman, just as a stronger mechanical force bears down a weaker, — he was particularly chary of risking his men against overpowering odds. On what, then, was his confidence founded? He saw better than any one else the true circumstances of the Peninsula, and the true difficulties of the French. Spain, and especially Portugal, had their strongly defensible lines, which a weaker force, if through neglect it gave the enemy no undue advantage, and if liberally supplied with the munitions of war, might defend forever against a stronger. The successes of the British navy under Nelson had given it the complete command of the sea; and so to a British army these indispensable munitions could be supplied. On the other hand, the base-line from which the French had to carry on their operations was distant. The wild Pyrenees, and with them wide tracts of rough and hostile country, stretched between the French armies and their native France. They could not be supported in consequence by munitions drawn from their own country; and the hostile country in which they encamped was by much too poor to enable them to realize that part of Napoleon's policy through which he made hostile countries support the war which wasted them, and to which he had given such effect on the fertile plains of Germany. Spain could not support great armies; and so great combinations could be maintained within its territories for only a few days at a time, and then fall apart again. Wellington, from behind his lines, marched out now upon one separate army, anon upon another; now upon one strong fortress, anon upon another; never opposed himself to overpowering odds; and, when the odds were not overpowering, or the fortress not impregnable, always carried the siege or gained the battle. He broke up in detail the armies of France. When they effected one of their great combinations against him, he fell coolly back on his lines; sometimes, as he saw opportunity, stopping by the way, as at Busaco, to gain a battle, and to convince the enemy that he was merely retreating, not running away. And then, when the combination fell to pieces, as fall to pieces he saw it could not fail, he again began to beat piecemeal the armies of which it had been composed. Time after time were the best troops of France poured across the Pyrenees to bear down the modern Fabius, and time after time did they sink under the peculiar difficulties of their circumstances and the tactics of Wellington. At length a day came when France could spare no more troops to the Peninsula; all its armies were required for the defence of its northern frontier, for the army of Napoleon had been broken in his disastrous Russian campaign, and the allies were pressing upon their lines. And then Wellington, taking off his hat, and rising in his stirrups, — for he saw that his time had at length come, — bade farewell to Portugal. He broke the power of the French in Spain in one great battle; repressed and beat back Soult, who had rushed across the Pyrenees to oppose him; and finally terminated the war at Toulouse, far within the frontiers of France. He had wrought out his apparently unsolvable problem by sweeping the Peninsula of three hundred thousand French troops that had held it; and, though once so inexplicable, it now seems in the main an exceedingly simple problem after all. But Christopher Columbus was the only man in a certain company who could make an egg stand on end; and the only man of the age who could have swept out of Spain, with his handful of troops, the three hundred thousand Frenchmen, was Arthur Wellesley. At least none of the others who attempted the feat — including even Sir John Moore — had got any hold whatever of the master idea through which it was done; and we know that some of our ablest men at home held that there was no master idea in the case, and that the feat was wholly impracticable.

As a statesman, the Duke of Wellington held a considerably lower place than as a warrior. With bodies of men regarded simply as physical forces, no man could deal more skilfully; with bodies of men regarded as combinations of faculties, rational and intellectual, he frequently failed. He could calculate to a nicety on the power of an armed battalion, but much less nicely on the power of an armed opinion. And all the graver mistakes of his career we find in this latter department. Latterly he is said to have taken, sensible of his own defect, his opinions and judgments in this walk from the late Sir Robert Peel; and it has been frequently stated that he intermeddled but little with politics since the death of his adviser and friend. But, though immeasurably inferior in this department to Cromwell, and even to Washington, — for to these great men pertains the praise of having been not only warriors, but also statesmen, of the first class, — few indeed of the countrymen, and scarce any of the party, of the deceased Duke equalled him in the shrewdness of the judgments which he ultimately came to form on the questions brought before him. Even some of his sayings, spoken in bootless opposition, and regarded at the time as mere instances of the testiness natural to a period of life considerably advanced, have had shrewd comments read upon them by the subsequent course of events. It seemed to be in mere fretfulness that he remarked, a good many years since, in opposition to some new scheme for extending the popular power, that he saw not how in such circumstances "the Queen's Government could be carried on." But that strange balance of parties in the country which leaves at present scarce any preponderating power on any side to operate as the moving force of " the Executive," has, we dare say, led many to think that the old man saw more clearly at the time than most of his critics or opponents. Though of an indomitable will, too, he was in reality too strong-minded a man to be an obstinate one. He *could* yield; and the part which he took in emancipating the Roman Catholics, and in abolishing the corn laws, are evidences of the fact. Further, it is not unimportant to know that had the advice of the Duke of Wellington been acted upon in our ecclesiastical controversy, no disruption would have taken place in the Scottish Church, and the Scottish Establishment would have survived in all its integrity, as the strongest in Britain. Wellington's ability of yielding more readily was based on his ability of seeing more clearly than most of the other members of his party; *they* resembled the captains of Captain Sword, in Hunt's well-known poem; but he was the great Captain Sword himself. When the peaceable Captain Pen threatens to bring a "world of men" at his back, and to disarm the old warrior, the poet tells us that —

> "Out laughed the captains of Captain Sword,
> But their chief looked vexed, and said not a word;
> For thought and trouble had touched his ears,
> Beyond the bullet-like sense of theirs;
> And wherever he went he was 'ware of a sound,
> Now heard in the distance, now gathering round,
> Which irked him to know what the issue might be,
> For the soul of the cause of it well guessed he."

In his moral character the Duke was eminently an honest and truthful man, — one of the most devoted and loyal of subjects, and one of the most patriotic of citizens. His name has been often coupled with that of the great military captain of England in the last century, — Marborough; but, save in the one item of great military ability, they had nothing in common. Wellington was frank to a fault. One of the gravest blunders of his political life, his open declaration in Parliament that the country's system of representation possessed the country's full and entire confidence, and that he would resist any measure of reform so long as he held any station in the Government, was certainly egregiously impolitic; but who can deny that it was candid and frank? Marlborough, on the other hand, was one of the most tortuous and secret of men. Wellington was emphatically truthful; Marlborough a consummate liar. Wellington would have laid down life and property in the cause of his sovereign; Marlborough was one of the first egregiously to deceive and betray his royal master, who, however great his faults and errors, was at least ever kind to him. Wellington was, in fine, a thoroughly honest man; Marlborough, a brilliant scoundrel.

There seemed to be but little of the soft green of humanity about the recently departed warrior. He was, in appearance at least, a hard man, who always did his own duty, and exacted from others the full tale of theirs. He had seen, too, in his first and only disastrous campaign, — that of the Duke of York in the Netherlands, — the direful effects of unrestrained license in an army. Enraged by numerous petty acts of violence and plunder, the people of the country became at length undisguisedly hostile to their nominal allies, and greatly enhanced the dangers and difficulties of their frequent retreats. And Wellington, taught, it is said, by the lesson, was ever after a stern disciplinarian, and visited at times with what was deemed undue severity the liberties taken by his soldiery with the property of an allied people. And so he possessed much less of the *love* of the men who served under him than not only the weaker but tender-hearted Nelson, but than also the genial and good-humored Duke of York, — a prince whom no soldier ever trusted as a general, or ever disliked as a man. But never did general possess more thoroughly the *confidence* of his soldiers than Wellington. Wherever he led, they were prepared to follow. We have been told by an old campaigner, who had fought under him in one of our Highland regiments in all the battles of the Peninsula, that on one occasion, in a retreat, the corps to which he belonged had been left far behind in the rear of their fellows, and began to express some anxiety regarding the near proximity of the enemy. "I wish," said one, "I saw ten thousand of our countryfolk beside us." "I wish, rather," rejoined another, "that I saw the long nose of the Duke of Wellington." A few minutes after, however, the Duke was actually seen riding past, and from that moment confidence was restored in the regiment. They felt that the eye of Wellington was upon them, and that all was necessarily right. Nor, with all his seeming hardness, was Wellington in any degree a cruel or inhuman man. He was, on the contrary, essentially kind and benevolent. The same old campaigner to whom we owe the anecdote, — a gallant and kind-hearted, but, like many soldiers, thoughtless man, — had, notwithstanding a tolerably adequate income for his condition, fallen into straits; and he at length bethought him, in his difficulties, of availing himself of that arrangement made by the Whigs about twenty years ago, when they first came into office, through which he might sell his pension. The proposed terms, however, were hard; and poor Johnston, wholly unconscious of the politics of the day, wrote to his old general, to see whether he could not procure for him better ones from his Majesty's ministers, recounting, in his letter, his services and his wounds, and stating that it was his intention, with the money which he was desirous of raising, to emigrate to British America. And prompt by return of post came the Duke's reply, written in the Duke's own hand. Never was there sounder advice more briefly expressed. "The Duke of Wellington," said his Grace, "has received William Johnston's letter; and he earnestly recommends him, first, not to seek for a provision in the colonies of North America, if he be not able-bodied, and in a situation to provide for himself in circumstances of extreme difficulty; secondly, not to sell or mortgage his pension. *The Duke of Wellington has no relation whatever with the King's ministers.* He recommends William Johnston to apply to the adjutant-general of the army. (London, March 7th, 1832.)" The old pensioner did not take the Duke's advice; for he *did* sell his pension, and, though, in consequence of his wounds, not very able-bodied, he *did* emigrate to America, and, we fear, suffered in consequence; but it was not the less true humanity on the part of his Grace to counsel so promptly and so wisely the poor humble soldier. But, alas! his last advice has been given, and his last account rendered; and it will be well for our country should the sovereign never miss his honored voice at the Council Board, nor — to borrow from ancient story — the soldier ever "vehemently desire him in the day of battle."

IX.

EARL GREY.

On Saturday last, the body of Charles, Earl Grey, was committed to the tomb of his ancestors; and his lordship's existence in relation to the present scene of being ranks but among the things that were. His political life extends over the long term of sixty years. Its beginnings pertain to the annals of the last age. History has long since pronounced judgment on the illustrious group of his earlier friends and opponents, — on Pitt and Fox and Burke, Windham, Sheridan, Erskine, and Dundas, — and the portion of our literature in which they are celebrated, or which we owe to them, is a literature that has descended to us from our fathers. Were William Pitt still living, he would be but four years older than the late Earl Grey. It may seem fanciful; but the prolonged existence of this veteran statesman, so influential in the councils of his country at a period when his years had well-nigh reached the full tale indicated by the psalmist, taken in connection with the imperishable associations of his early history, has served to remind us of what we have sometimes witnessed beside the waters of a petrifying spring: we have seen tufts of vegetation, with their upper sprigs green and flourishing, and the lower converted into solid stone; the vital influences vigorous in the newer portion of the plant, while the older were imperishably fixed in marble.

There are various deeply-interesting aspects in which the political career of his lordship may be viewed. When he first entered public life, the dissolute court and infidel literature of France were busily engaged in sowing the seeds which germinated and bore fruit as the first French revolution. It was a gay winter in Paris, that of 1786, when the Earl — then Mr. Grey — was first returned to Parliament for his native county, Northumberland. The Chevalier de Boufflers was engaged in making charming songs on the new fashions; the Queen had just pensioned her milliner, and had got nine hundred thousand livres of the public money to pay some of her own "small debts;" the courtiers, who had been inconsolable for some time, — for the most accomplished opera dancer in the world had sprained her ankle, — had recovered their spirits again, for the ankle had also recovered; and, though thousands of the industrious poor were starving, and speaking ominously, in their distress, of America and its revolution, and though even the ladies had begun to wear bonnets *à la Rodney*, no one could see how trifles such as these should bear with sinister effect on the general hilarity. Nor could the young representative of Northumberland have possibly seen aught in them with which he, as a public man, had anything to do. Nothing more certain, however, than that the emphatically important portion of the history of Earl Grey which so peculiarly belongs to that of his country is entwined with the history of France. We could not better illustrate the influence which, in these times of advanced civilization, the destinies of one great European country exert on those of another, than by instancing what his lordship at one period of his life attempted, but signally failed to perform, and so completely accomplished at another. The special work of the life of

Earl Grey — that which now gives him a distinguished niche in British history — was the work of parliamentary reform. In 1793 he first introduced into Parliament his celebrated motion on this subject, and found, in a house of two hundred and twenty-three members, only forty-one supporters. The revolutionary tornado in France had reached its extreme height at the time, and had prostrated, in its fury, the king, the aristocracy, and the church; French principles were spreading among ourselves; some of the more infidel writings of Paine had just appeared, and were circulating among the people by thousands and tens of thousands; many of the more timid Whigs, alarmed at the very appearance of change, hung back from their old allies; with this timid class the great bulk of the more sober portion of the British people made common cause; and so the motion of Earl Grey was negatived by a majority that served not only to extinguish the measure for the time, but to leave scarce any hope of its ultimate success. The terrible storm raised in France blew full against it, and bore it down; and it was not until a more salutary storm arose in the same country nearly forty years after, that it fairly righted again, and, under the influence of the now auspicious gale, bore into harbor. His lordship held the helm in both cases; and the tempest that so signally baffled him in the one, and the gale that carried his bark so directly into port in the other, blew from off the same land.

When Earl Grey introduced into the House his first unsuccessful motion for parliamentary reform, he was in his twenty-ninth year. Thirty-eight years passed ere he originated the motion on the subject which was to be ultimately successful, and he was now in his sixty-seventh. In the long intervening period, the change so common to the mind of man, which modifies the Whiggism natural to youth into the semi-Toryism natural to age, seems to have taken place to some extent in the mind of Earl Grey; and his second measure was much less sweeping and extensive than his first. The first was based on the principle of household suffrage, and involved a return to the original scheme of triennial parliaments. The means, too, which he originated to give the cause a popularity and strength outside the walls of Parliament that might find it favor and secure it attention within, partook of a boldness characteristic of an early stage of vigorous and sanguine manhood. He took a prominent part in the formation of the "Association of Friends of the People," with associates such as Whitbread, Erskine, Cartwright, and Macintosh, — men almost all of whom lived long enough considerably to modify their views; and it was in the character of the leading organ of this body in the House of Commons that he brought forward his first motion on reform. There were, however, some few points in which his earlier scheme excelled that which he lived to transfer to the statute-book as part and parcel of the Constitution of the country. It gave single votes to individual electors, and single votes only; and provided that the elections, on a dissolution of Parliament, should take place simultaneously all over the empire. In order rightly to estimate the value of these provisions, we have but to look at what is perhaps the greatest defect, both in principle and practice, in the scheme of parliamentary reform which he afterwards carried. A single individual may at the present moment hold votes in at once every represented county in Scotland, and in every burgh or district of burghs that returns a member. On this principle, it is not the holders of property that vote, — property being regarded, as it ought, as a mere qualification that fixes the status of the individual and establishes his stake in the country, — but the property itself. It is the voice of the house, field, or farm that votes, — the same voice serving for several houses, fields, or farms, —just as the same voice in a puppet-show serves for Punch, Judy, and the Constable. And thus it is not men, but things, that select the lawmakers of the country. Such seems to be the objection, in point of principle, to a provision in Earl Grey's second scheme of reform, which his first scheme repudiated ; and in practice we find this provision more objectionable still. It forms the basis of the whole corrupt machinery of fictitious votes, and these, in turn, the support of not a little of the profligacy in public life that can indulge in the eye of day in its true colors, despising the wholesome restraints of general opinion, because altogether independent of them. It is at once a copious source of corruption among the representatives of the country, and of legalized perjuries among the represented. We know of no defect in the measure at all deserving of being placed in the same class with this grand one, save, perhaps, the provision that extends the political franchise to tenants-at-will. Legislation cannot give independence to the mind of a voter ; but it should at least provide, in every possible case, that independence should be communicated to his circumstances.

There is another interesting point illustrated by the long political life of Earl Grey. His lordship was unquestionably a very able man, but he did not possess one of those gigantic minds which mould and fashion the destinies of nations. He resembled rather an index-hand attached to the great political machine, than its moving power. No one can say how the civil war would have terminated in England in the seventeenth century had there been no Cromwell, or what complexion the present politics of France and the Continent generally would wear had there been no Napoleon. Had the one great man never been called into existence, it is probable that on the death of Hampden prerogative would have triumphed, and Britain have sunk to the level of the contemporary despotisms of the Continent. It is possible that, had the other great man never lived, an allied army would have marched to Paris ere the present century began, and that humbled France, restored to the despotic sway of the Bourbons, and with no proud recollections of victory to reinvigorate her, would have witnessed no second revolution. Cromwell and Napoleon belonged to the class of men to whom the destinies of their age seem intrusted; but in the career of Earl Grey we see rather the movements of an intelligent index of the course of things than the operations of a power originating and setting them in motion. And hence an interest of a particular kind in contemplating his history. We see in it the growth of popular opinion, like that of vegetation in a backward spring, now shooting forth in green vigor, now checked and prostrated by the chilling influence of great political storms, now yet again recovering itself, now again thrown back, and finally reaching, in the decline of the year, a late and somewhat blighted maturity. First come the terrors of the French Revolution; then the untoward influences of the long French war; then an intermediate period, in which the power acquired during the two previous seasons by the antagonists of all political change is employed in depressing their opponents; and then, when opinion, long cherished in its growth, and often thrown back, has arrived at the necessary degree of ripeness, a reaping-time arrives, and Earl Grey, as little able previously to control the heats and chills of the political atmosphere as the husbandman to control the weather, on which all his interests depend, reaps the harvest of his political life. It is not our present purpose to speak of the great measure which will be ever associated with his name in the history of our country. With all its defects, it indisputably did much for the *letter* of the Constitution, and nowhere so much as in Scotland. It everywhere extended the basis on which the liberty of the subject rests; and nowhere else had that basis been so narrow as in this northern kingdom, and nowhere else had it been so unsound. It is, however, the "spirit," not the "letter," that "maketh alive;" and it is not from statesmen, however enlightened or honest, that the spirit can come. They can construct the framework of constitutions; they can mould them out of the humble materials of which laws are made, as the body of Adam was moulded from the dust; but virtue in the people is that alone breath of life without which they cannot become "living souls." The people of Scotland had scarce any political standing in the sixteenth and seventeenth centuries, and yet, animated by a right spirit, they accomplished much. One of these centuries witnessed the Reformation, and the other the Revolution. During the last twelve years our people have possessed, on the contrary, ample political standing; but it would not be quite so easy to say what great work they have accomplished.

The personal character of the nobleman over whom the grave has so lately closed seems to have been truly excellent. He was a Whig of a high type; and we certainly think none the less of him from the circumstance that, while he struggled to extend the privileges of the people, his leanings were aristocratic, and that he stood determinedly by his order. He exerted himself with a life-long exertion to do what he deemed justice to one class of the community, while his feelings and predilections were mainly with another. There are incidents not a few in his biography that tell remarkably well. On the character of Fox there rests the unhappy stain left by his public denial of the marriage of the Prince Regent, afterwards George IV., to Mrs. Fitzherbert, though of that marriage Fox himself is said to have been a witness. Earl (then Mr.) Grey is known to have been exposed, in the case, to a similar temptation to that in which his leader was wanting, but he stood it vastly better. "Mr. Fox," says one of the Earl's biographers, "being authorized by the prince, had denied the marriage with Mrs. Fitzherbert. The lady was naturally offended, and, to appease her, the prince tried to restore the matter to the uncertainty which had previously hung over it. He wished, therefore, to have some ambiguous or equivocating remark made, as if from authority, in the House of Commons;

and, with singular want of discrimination, Mr. Grey was applied to for the purpose. But the unaccommodating young senator spurned the dishonorable office, and gave offence which was never forgotten or forgiven." It is further to the honor of Earl Grey, that though by no means indifferent to the pleasures of place or the possession of power, he held office, during his long political life of more than half a century, for little more than five years. He had many opportunities of being in place presented to him, had he chosen to sacrifice principle for its sake; but he did not choose it. Very different indeed would be the present position of the party to which he belonged, had they but imitated their leader in this important respect. In these times of reaction on old Toryism, none the wiser or better for all the experience of the past, they would be by far the strongest, — not what they now are, one of the weakest parties in the country.

X.

LORD JEFFREY.

The most eminent of our Edinburgh literati — a man who for nearly half a century has enjoyed European celebrity as first in the realms of criticism, and a reputation at least coextensive with his native country as a politician and a lawyer — has passed from off the stage of mortal existence, and now lives but in the unseen world. On the evening of Saturday last, Francis Jeffrey, the philosophic and tasteful reviewer, the accomplished advocate, and judicious and honest judge, died, after a few days' illness, at his Edinburgh residence in Moray Place, in the seventy-eighth year of his age.

Lord Jeffrey may be properly regarded as the last Scottish survivor of that group of eminent men, contemporary with Napoleon, to which Chalmers and Sir Walter Scott, Wordsworth and Wellington, Goëthe, Cuvier, Humboldt, and Chateaubriand belong. Professor Wilson, though fast descending into the vale of years, we regard as the member of a somewhat later group, — that of Lockhart, Carlyle, and Macaulay, Lamartine, Arago, and Sir David Brewster. It was the last Scotchman of that elder group of distinguished men who achieved celebrity or influenced opinion as early as the beginning of the present century, or nearly so, that quitted this scene of things on the evening of Saturday. And he has left to the biographer, in the story of his life, much that is of signal interest and importance in the legal and political history of our country, and much in the history of its literature that is better represented by his career than by that of any other individual. He represents a mighty revolution in letters, which has perhaps considerably lessened the number of *books*, but increased, beyond all calculation, the number of brilliant *articles*. Not a few superior men have passed away in consequence, and left no permanent mark behind them; but that literature of the periodic press which forms, perhaps too exclusively, the staple reading of the age, — which occupies men's minds and influences their opinions to-day, but which is in great part forgotten ere to-morrow, and which, in this as in other respects, forms that daily bread of the republic of letters which cannot be wanted, and which, once used up, is never more thought of, — has been immensely heightened in its tone and power, and become a great engine, without whose potent assistance no cause can succeed, and no party prosper. Previous to the appearance of perhaps the only "Edinburgh Review" known to the great bulk of our readers, there had been men, who, in calibre and literary attainment, at least equalled the ablest of its contributors engaged in writing for periodicals. We do not refer to those diurnal, or hebdomadal, or semi-hebdomadal publications of the last century, which may be regarded as commencing with the "Tatler" and Sir Richard Steele, and terminating with the "Lounger" and Henry M'Kenzie, — works which contain some of the finest writings in the language, — but simply to the newspapers and magazines. For these, compelled by stern necessity, Goldsmith wrote for several years. His "Citizen of the World"— one of the most exquisitely written books in any tongue — first appeared as a series of essays in the "Public Ledger;" and he wrote criticisms for the "Monthly Review," and articles for the "British Magazine." Smollett conducted for about seven years the "Critical Review;" Burke wrote for the "Annual Register;" and Johnson labored for years for the "Literary Magazine," the "Gentleman's Magazine," and the "Universal Visitor." And about half a century previous to the appearance of that second "Edinburgh Review" with which the name of Jeffrey must be forever associated in the history of letters, there existed for about a twelvemonth a first "Edinburgh Review," conducted by Blair, Robertson and Adam Smith. But there were no periodicals of sustained effort, or (with perhaps the exception of this first "Edinburgh Review") all of whose contributors were men of nearly equal standing and power. Burke, Johnson, and Goldsmith were associated in their compelled labors with dull amateurs, or the scribblers of Grub Street; and Smollett, in his description in "Humphrey Clinker" of a dinner of authors, is known to have drawn, in the hair-brained mediocritists which he portrays, some of the nameless contributors associated with him in his periodical. Even when, as in the Edinburgh instance, all the writers were superior, they seem to have given but half their mind to their work of article-writing. The first "Edinburgh Review" is a respectable, but not a very brilliant production. Its writers were engaged at the time on works which still live: Robertson on his "History of Scotland," Smith on his "Theory of Moral Sentiments," and Blair in maturing the thinking of his "Lectures on Criticism and the Belles Lettres;" and they could spare for their occasional critiques merely the odds and ends of their cogitations. "No man ever did anything well," says Johnson, "to which he did not apply the whole bent of his mind;" and it was reserved for Jeffrey and his associates at once to render, by their equality of talent, a periodical all of a piece, and, in generous rivalry, to do for it the very best which they were capable of performing. Robertson and Adam Smith could, and did, immensely exceed themselves in all they had done for their "Review;" whereas Jeffrey and Sydney Smith did all they were capable of doing for theirs; and so on no other occasion or form did they exceed what they had accomplished as periodical reviewers. And hence the great revolution in periodical literature which they effected. Without once designing any such thing, they succeeded in raising its platform from the level of Grub Street to very nearly that of the standard of literature of the country.

We say, without once designing any such thing. Chateaubriand shrewdly remarks of Napoleon, that, "by leading on France to the attack,"— that is, by bringing armies into the field some five or six times more numerous than had wont to be employed under the old school of strategy, — "he taught Europe also to march: the chief point which has since been considered is to multiply *means;* masses have been balanced by masses. Instead of a hundred thousand men, six hundred thousand have been brought into the field; instead of a hundred pieces of cannon, five hundred have been employed." And such was the effect produced by that introduction of first-class talent into the field of periodic literature with which we associate the name of Jeffrey. The "Edinburgh Review" was a Whig periodical; and the interests of the opposite party imperatively demanded that its park of artillery five hundred strong should be met by an antagonist park in which the guns should be as numerous and their calibre as great. And hence the origination of the "Quarterly Review," edited by Gifford, and to which men such as Southey and Sir Walter Scott contributed. And then the magazines caught the high tone communicated by the Review; and in this race, as in the other, Scotland assumed the lead. The "Christian Instructor," edited by Dr. Andrew Thompson, and supported by Dr. M'Crie, Dr. Chalmers, and Dr. Somerville, started first on the new table-land of elevation; though its theological character, and its restriction to the old Presbyterianism of Scotland, served greatly to limit both its influence and its fame. "Blackwood" followed, and took at once a place in literature which no magazine, at least as a whole, had ever taken before. It was supported by the contributions of Lockhart, Galt, DeQuincey, Moir, and Alison, and conducted, it was understood, for many years by Professor Wilson. The "New Monthly" followed, with Thomas Campbell at its head; and about much the same time,

Byron, Shelly, and Leigh Hunt originated their short-lived periodical the "Liberal." The newspapers had partaken at even an earlier period of the induced elevation. Like the magazines and reviews, they had been the occasional vehicles of very powerful writing at a comparatively earlier period. The "Letters of Junius" had appeared in the "General Advertiser." Coleridge had, for a short time about the beginning of the present century, conducted the "Morning Post." Sir James Macintosh had, at a rather earlier date, written copiously for several of the liberal papers of the day. But it was not until the "Edinburgh Review" had fairly entered on its career, that that general elevation of the newspaper platform took place which is now one of the marked characteristics of periodic literature. Edinburgh has been in this respect far behind London; but a very great change has taken place during the last forty years even in Edinburgh. There are men still connected with our newspaper printing-offices who remember when papers by the management of which fortunes were realized were conducted either without an editor at all, or by some printer or mere man of business, who would be unfitted in the present time to perform the duties of even a sub-editor or reporter. It was mainly through that indirect influence of the labors of Lord Jeffrey and his friends to which we refer that Edinburgh has reckoned among its newspaper editors, during the last thirty years, writers such as M'Cullock, M'Laren, Buchanan, Dr. James Brown, Alexander Sutherland, and John Malcolm. The provincial newspaper press has also caught the general tone. Had there been no "Edinburgh Review," newspapers such as the "Dumfries Courier" and the "Inverness Courier" would have been prodigies. No later than the day on which Lord Jeffrey died, a gentleman of business habits, who had been for some time unsuccessfully engaged in looking out for an editor to conduct a weekly paper established in a large town, remarked to us, that of all men an efficient newspaper editor was perhaps the most difficult to find.

It occurred to us not long after, on hearing of his lordship's death, that, in all probability, had *he* never lived, the difficulty would not have existed.

This indirect influence exercised on periodic literature by Lord Jeffrey was perhaps more important in the main than that which he wielded as a political writer or a critic. And yet in both departments he stood very high. His influence as a politician is of course mixed up with that of his associates, and must be regarded generally as that of the "Review" which he conducted. For about thirty years, as we had once before occasion to remark, the "Edinburgh Review" labored indefatigably with various political objects in view, mainly, however, to repress the dreaded growth of despotism, and to assert the cause of constitutional reform. And for at least the latter half of that period its exertions were accompanied by very marked success. During the war with France, the current ran strongly against it. It was thrown out in its calculations, both by that infatuation of Napoleon which led to the Russian campaign, and by the military genius of Wellington. The consequent issue of the great revolutionary struggle was a struggle which it had not foreseen. There was, besides, a principle elicited in our state of war which ran counter in its influence to that of the "Review." The resentments of the people were so enraged with their enemies abroad, that they had comparatively little indignation to spare for their rulers at home. But a period of peace told powerfully in its favor. Men found leisure to look through the spectacles which it furnished at the defects of existing institutions; its politics spread and gathered strength; a second French revolution, achieved under immensely more favorable circumstances than the first, wrought as decidedly in favor of the Liberal cause in Britain as the first French revolution had wrought against it; and Whiggism at length saw its favorite scheme of political reform embodied into a bill, and passed into a law. And in producing this result the "Edinburgh Review" had a large and sensible share. But then, Jeffrey was simply one of several powerful-minded men to whom the periodical owed its political potency. Regarded, however, in its purely critical character, and as a leader of the public taste in poetry and the belles lettres, the case was otherwise. Though Sir James Macintosh occasionally contributed a paper, — such as his critique on the Poems of Rogers, which in this department fully sustained the general character of the periodical, — Jeffrey to all intents and purposes was the Edinburgh Review." And in this his peculiar province he took his place, we have no hesitation in saying, as the first British critic of the age. He had his prejudices and his deficiencies, and occasionally — put out in his reckoning by what the poet beautifully describes as "glorious faults, which critics dare not mend" — he committed, as in the case of Wordsworth, grave mistakes; but, take him all in all, where, we ask, is the critic of the present century who is to be placed in the scale against Francis Jeffrey? His peculiar fitness for his task resulted mainly from the exquisiteness of his taste, his fearless honesty, and the integrity of his judgment. His few mistakes arose chiefly from certain partial defects in faculty. These, however, were important enough to prevent him, if not from taking his place as the first of contemporary critics, from at least entering those highest walks of British criticism in which a very few of the master minds of the past were qualified to expatiate, and but these few exclusively. There are snatches of criticism in the prefaces and dedications of Dryden, in Burke's "Treatise on the Sublime and Beautiful," and even in Johnson's "British Poets" (though there were important faculties which Johnson also lacked), which Jeffrey has not equalled. But that man rises high in an intellectual department, who, though not equal to some of the more illustrious dead, is first among his compeers. We know not at once a better illustration of what Jeffrey could do, and what he failed in doing, than that furnished by his article on the Sense of the Beautiful. There is scarce a finer piece of writing in the language; and yet it embodies, as part of its very essence, the great sophism that, apart from the influence of the associative faculty, there is no beauty in color. We know of but one other sophism in the language that at all approaches it in the elegance and delicacy of its form, and which resembles it, too, in its perfect honesty and good faith; for both authors wrote as they felt, and failed in producing more than partial truth, which is always tantamount to error, simply because they both lacked a faculty all-essential to the separate inquiries which they conducted. Both were fully sensible of the immense power of association in eliciting images of delight; but the one, insensible to the beauty of simple sounds, from the want of a musical ear, attributed all the power of music to association alone; and the other, insensible to the beauty of simple colors, attributed, from a similar want of appreciating faculty, all their power of gratifying the eye to a similar cause. All our readers are acquainted with the article on the Beautiful; but the following fine stanzas, the production of John Finlay, a Scottish poet, who died early in the present century, when he had but mastered his powers, may be new to most of them: —

"Why does the melting voice, the tuneful string,
A sigh of woe, a tear of pleasure bring?
Can simple sounds or joy or grief inspire,
Or wake the soul responsive to the wire?
Ah, no! some other charm to rapture draws,
More than the finger's skill, the artist's laws;
Some latent feeling at the string awakes,
Starts to new life, and through the fibres shakes;
Some cottage-home, where first the strain was heard,
By many a tie of former days endeared;
Some lovely maid who on thy bosom hung,
And breathed the note all tearful as she sung;
Some youth who first awoke the pensive lay,
Friend of thy infant years, now far away;

Some scene that patriot blood embalms in song;
Some brook that winds thy native vales among, —
All steal into the soul, in witching train,
Till home, the maid, the friend, the scene, return again.
'Twas thus the wanderer 'mid the Syrian wild
Wept at the strain he caroll'd when a child.
O'er many a weary waste the traveller passed,
And hoped to find some resting-place at last,
Beneath some branchy shade, his journey done,
To shelter from the desert and the sun;
And haply some green spot the pilgrim found,
And hailed and blessed the stream's delicious sound.
When on his ear the well-known ditty stole,
That, as it melted, passed into his soul, —
'O, Bothwell bank!' — each thrilling sound conveyed
The Scottish landscape to the palm-tree shade;
No more Damascus' streams his spirit held,
No more its minarets his eye beheld:
Pharpar and Abana unheeded glide,
He hears in dreams the music of the Clyde;
And Bothwell's bank, amid o'er-arching trees,
Echoes the bleat of flocks, the hum of bees.
With less keen rapture on the Syrian shore,
Beneath the shadow of the sycamore,
His eye had turned, amid the burst of day,
Tadmor's gigantic columns to survey,
That sullenly their length of shadows throw
On sons of earth, who, trembling, gaze below.
'Twas thus when to Quebec's proud heights afar
Wolfe's chivalry rolled on the tide of war,
The hardy Highlander, so fierce before,
Languidly lifted up the huge claymore;
To him the bugle's mellow notes were dumb,
And even the rousing thunders of the drum,
Till the loud *pibroch* sounded in the van,
And led to battle forth each dauntless clan.
On rush the brave, the plaided chiefs advance;
The line resounds, 'Lochiel's awa' to France!'
With vigorous arm the falchion lift on high,
Fight as their fathers fought, and like their fathers die."

Long as our extract is, there are, we suppose, few of our readers who will deem it too long. Independently, too, of its exquisite vein, it illustrates better both the merits and the defects of Lord Jeffrey's theory of beauty than any other passage in the round of our literature with which we are acquainted. For there are scores whose degree of musical taste compels them to hold that there is a beauty in "simple sounds" altogether independent of association, for the single individuals whose sense of the beauty of "simple colors" is sufficiently strong to convince them that it, like the other sense, has an underived existence wholly its own.

We have left ourselves but little space to speak of the distinguished man, so recently lost to us, as a lawyer, a statesman, and a judge. He will be long remembered in Edinburgh as one of the most accomplished and effective pleaders that ever appeared at the Scottish bar. It has become common to allude to his appearances in the House of Commons as failures. We know not how his speeches may have sounded in the old chapel of St. Stephen's; but this we know, that of all the speeches in both Houses of which the Reform Bill proved the fruitful occasion, we remember only his: we can ever recall some of its happy phrases; as when, for instance, he described the important measure which he advocated as a firmament which was to separate the purer waters above from the fouler and more turbulent waters below, — the solid worth of the country, zealous for reform, from its wild, unprincipled licentiousness, bent on subversion; and, founding mainly on this selective instinct of our memory, we conclude that the speech, which is said to have disappointed friends and gratified opponents, must have been really one of the best delivered at the time, — perhaps the very best. As a judge, the character of Jeffrey may be summed up in the vigorous stanza of Dryden: —

"In Israel's courts ne'er sat an Abethdin
With more discerning eyes, or hands more clean,
Unbribed, unsought, the wretched to redress,
Swift of despatch, and easy of access."

All accounts agree in representing him as in private life one of the kindest and gentlest of mortals, ever surrounded by the aroma of a delicate sense of honor and a transparent truthfulness, equable in temper, in conversation full of a playful ease, and, with even his ordinary talk, ever glittering in an unpremeditated wit, "that loved to play, not wound." Never was there a man more thoroughly beloved by his friends. Though his term of life exceeded the allotted threescore and ten years, his fine intellect, like that of the great Chalmers, whom he sincerely loved and respected, and by whom he was much loved and respected in turn, was to the last untouched by decay. Only four days previous to that of his death he sat upon the bench; only a few months ago he furnished an article for his old "Review," distinguished by all the nice discernment and acumen of his most vigorous days. It is further gratifying to know, that though infected in youth and middle age by the wide-spread infidelity of the first French revolution, he was for at least the last few years of his life of a different spirit: he read much and often in his Bible; and he is said to have studied especially, and with much solicitude, the writings of St. Paul.

XI.

FIRE AT THE TOWER OF LONDON.

THREE of the most interesting ancient buildings of Britain destroyed by fire within less than ten years! "Are such calamities as these really unavoidable?" asks a writer in the *Times*, "and ought we to make up our minds to hear of the conflagration of some great national treasure every five or ten years as a thing that must be?" Treasures of at least equal value still survive to England, — Windsor, Hampton Court, the British Museum, and the great University Libraries. How are *they* to be protected? Increased vigilance and care are recommended by this writer. Fires smoulder for hours ere they burst forth so as to be detected by the watchmen outside; and they have then, in most cases, become too formidable to be got under. But by stationing careful persons within our more valuable buildings, instructed to visit every apartment and passage once every hour, might not the mischief be detected at a stage when it could be easily overmastered? Statistical fact, however, comes in to show that the suggestion is less wise than obvious; buildings so watched are found more liable to destruction from fire than those for whose safety no such precautions are taken. The private watchman has to use a light in his rounds; in cold weather he requires a fire; though essential that he be of steady character, there is a liability to be deceived, on the part of the employer, considerable enough to tell in the statistical table as an element of accident. Even when there is no unsteadiness, inattention is apt to creep on men watching against an enemy that has just a chance of visiting what they guard once in five hundred years.

In short, the result of the matter is, that insurance offices, founding on their tables, demand a higher premium for houses guarded in this manner than for houses left altogether unprotected. To meet with the evil thus indicated, the writer in the *Times* suggests that the watchmen, in order to keep up their vigilance, should be changed once every two years; that each at the end of his term should have to look forward to some certain promotion as a reward of his diligence and care; and that none but active, prudent, trustworthy men should be chosen for the office. The scheme, of course, lies open to the objection just hinted at; — the inevitable liability of employers to be deceived in character would in not a few cases render the precaution useless. We question, too, whether the attention of a watchman who visited every part of a large building some ten or twelve times each night for two years together, could be so continually kept up that more than a balance would be struck between the dangers he introduced and those he prevented. It is doubtful, we say, whether, even by a scheme thus improved, the statistician would find that the watchman did more than neutralize himself.

One suggestion, however, may be made on the subject, which we are convinced the practical man will at once recognize as sound. The causes of the three great fires which within the last seven years have inflicted three great calamities on the country, seem, so far as they can be ascertained, to have been all pretty much alike: they all appear to have been connected with the overheating of flues. The buildings were all *ancient ones*, — none of them at least less so than the times of William III.; and they have all been destroyed by accidents originating in the *modern mode* of heating houses by stoves and metal flues. Any one practically acquainted with the subject must see that in every such case the liability to accidents of this nature is inevitably great. In *building* a house, the workman can take the necessary precautions as he proceeds. He can take care, for instance, that no beam or joist, or other piece of wood, approach any flue nearer than a foot, — the distance specified by act of Parliament; but in *altering* a house, he can, in striking out his flues, take no such precautions. In cutting through the hard walls, there may be wood within an inch of him, of which he can know nothing, — wood covered up at times by a mere film of mortar; and no possible care can guarantee him against accidents. He is of necessity a worker in the dark; nor, in the circumstances, can it be otherwise. Still, however, one very effectual kind of precaution *may* be taken. A medium for heating such a flue may be employed through which fire cannot be communicated. A metal flue, heated in the ordinary manner, becomes not unfrequently red hot, and sets fire to whatever wood may be in contact with it; and hence, we doubt not, the destruction of both Houses of Parliament, the Royal Exchange, and the National Armory. But steam, when employed as the heating medium, is restricted to a certain temperature, above which it cannot rise, and which cannot set fire to wood or any other substance employed in architecture. We would therefore suggest it should be laid down as a rule, that in all ancient buildings heated by metal flues, the heating medium should be steam, and that the furnance should always be in a fire-proof outhouse, disconnected from every other building. Simple as the precaution may seem, we are certain it would diminish the chances of accident from fire by full two thirds of their present amount.

It is melancholy enough that in so brief a period three of the most interesting public buildings of England or the world should have thus perished. Each of the three has been associated for centuries with the history of Britain, in all for which Britain is most famous. Her emporium of trade is still a heap of blackened ruins, — the noble and venerable pile that served to connect her commerce of the present day, spread over every land and every sea, with her commerce of three hundred years ago, when a few adventurous traders struck out in quest of yet undiscovered shores, into oceans still undefined by the geographer, and whose remoter skirts seemed as if bounded by lines of darkness! Her halls of legislation perished next, — erections the history of which is that of civil liberty, not in Britain only, but over half the world, — places suggestive of every great English name that mingles in the history of the lengthened contest between *right* and *prerogative*, from the days of Pryne and Hampden down to those of Chatham and Fox. And now the national magazine of trophies and arms has fallen a prey to the devouring element. The building representative of the wars and victories of Britain has shared the same fate with her halls of commerce and legislation; and much has perished, as in the other cases, which cannot be estimated at a money value, and which money cannot replace; — the relics of Blenheim and of Waterloo, the remains of the two rebellions in Scotland, the arms of Tippoo Saib, the bows employed at Cressy and Agincourt, the spoils of the Armada and of Trafalgar, — much that linked together the names and triumphs of many of our greatest warriors, by exhibiting their exploits, if we may so express ourselves, on one platform, — that grouped together the memories, as well as the trophies, of Blake and of Nelson, — that associated Henry the Fifth with William of Orange, and brought into close juxtaposition the names and histories of Marlborough and of Wellington. The loss is a national one, and we fear we would but lay ourselves open to a charge of extravagance were we to say at how great a rate we estimate it. Some of our readers must remember the instance given by Thomas Brown of the force with which distant existences or events are sometimes impressed on the mind through the medium of objects in themselves trivial and uninteresting. He relates the case of some English sailors moved to sudden tears by thoughts of home and their friends, on finding on the bleak coast of Labrador a metal spoon with the name "London" stamped on the handle. Such is the constitution of the mind, that the seen and the tangible impart to whatever we associate with them impressiveness and reality. The armor worn by an ancient king sets him much more vividly before us than the chronicles of his reign, however minute; the trophies of a battle enable us better to realize it than the most graphic descriptions of the historian, — or, rather, they give to the descriptions a new sense of truth, by rendering them in some degree evident to the senses; — they are the stone and earth by which we enfeoff ourselves in them as matters of solid belief. There is an interest, too, in such relics regarded in their connection with classical literature, as a sort of goods and chattles of cultivated minds. Who acquainted with letters, whether in our own country or abroad, did not regret, in the destruction of both Houses of Parliament, the loss of the old and faded tapestry which suggested to Chatham his eloquent and impressive appeal? Or who interested in Shakspeare does not feel that England was richer for possessing what it possessed only a week ago, — the identical apartment in which Clarence was smothered in his Malmsey? Whatever is intimately associated with the great names of a nation forms a portion of the national wealth. The feeling that it does so, says an eminent writer of the last age, is a feeling implanted by nature; "and when I find Tully confessing of himself, that he could not forbear, at Athens, to visit the walks and houses which the old philosophers had frequented or inhabited, and recollect the reverence which every nation, civil and barbarous, has paid to the ground where merit has been buried, I am afraid to declare against the general voice of mankind, and am inclined to believe that this regard which we involuntarily pay to the meanest relic of a man great and illustrious, is intended as an incitement to labor, and an encouragement to expect the same renown, if it be sought by the same virtues."

XII.

THE CENTENARY OF "THE FORTY-FIVE."

THE General Assembly of the Free Church of Scotland held its first meeting at Inverness on Thursday the 21st ult.; and on Tuesday the 19th, just two days before, a party of gentlemen and ladies, accompanied by half a dozen pipers, visited Glenfinnon in rather showery weather, and called their visit the "Centennial Commemoration of the Gathering of the Clans." A great reality, and the meagre ghost of what had been a great reality a hundred years ago, entered upon the stage at nearly the same place and time, but with a very different result from that which almost always takes place in the ghost scene in Hamlet. Hamlet the living — a thing, as he himself informs us, of "too, too solid flesh" — attracts but a small share of attention compared with that excited by the unsolid spectre of Hamlet the dead; the shadow fairly eclipses the substance. But here, on the contrary, it was the substance that fairly eclipsed the shadow. The solid reality so occupied the mind of the Highlands that it had not a thought to spare on the unsolid ghost; and so the ghost, all drooping and disconsolate, passed off the stage unapplauded and unseen. We could find no room at the time for the paragraph that formed the sole record of its entrance and exit: our columns were occupied to the full with matters which the "clans" deemed of more serious concernment than the centenary of their gathering in Glenfinnon, — among the rest, with the very grave fact that not a few of their present chieftains are grossly outraging their rights of conscience, and chasing them, when they meet to worship God on the brown moors and bleak hillsides of their country, to its exposed cross-roads and its wild sea-beaches. But we have found room for it now, not as a piece of news, — for, after the lapse of a month, it has become somewhat stale, — but as the record of an event which, though but a trifle in itself, is at least interesting in what it indicates. A feather has been held to the lips of dead Jacobitism, to ascertain whether there was breath enough left within to stir the fibres, and not a single fibre has moved; and the paragraph on the "Centennial Commemoration" records the experiment and its result.

There are curious mental phenomena connected with the history of the decay of Jacobitism in Scotland. Like the matter of decomposing bodies, it passed, at a certain stage in its progress, from the solid to the gaseous form, and found entrance in the more subtle state into a class of minds from which, in its grosser and more tangible condition, it had been excluded. We are introduced in the letters of Burns to an ancient lady, stately and solemn, and much a Jacobite, who boasted that she had the blood of the Bruce in her veins, and who conferred, in virtue of her descent, the dignity of knighthood on the poet. We learn further, that the poet and the ancient lady, during the evening they spent together, agreed remarkably well: she would scarce have knighted him otherwise. She proposed toasts so full of loyalty to the exiled family that they were gross treason against the reigning one; but, notwithstanding their extremeness, the poet cordially drank to them, and, in short, seemed in every respect as zealous a Jacobite as herself. But there was a wide difference between the Jacobitism of Burns and that of the ancient lady. Hers was of the solid, his of the gaseous cast. Her mind was of the order in which *effête* opinions and dying beliefs are cherished to the last; his of the salient order, that are the first to receive new impressions and to take up new views. She would undoubtedly have died a Jacobite of the old grim type, that were content to forfeit land and life in the cause of a shadowy loyalty; he, on the other hand, only a few years after, incurred the suspicion and displeasure of Government by sending a present of artillery to the French Convention, to assist in defending a people who had deposed their king, against all other kings, and the *Jacobites* of their own country. The *Jacobite* of one year, who addressed enthusiastic verses to the "revered defenders of beauteous Stuart," and composed the "Chevalier's Lament," had become in the next the uncompromising *Jacobin*, who wrote "A man's a man for a' that." Now, through the very opposite classes of minds represented by the old lady and the poet has Jacobitism passed in Scotland, in its progress to extinction. The class of true Jacobites — the men in whom Jacobitism was a solid principle — died with the generation that fought at Culloden, and they were succeeded by the class to whom Jacobitism formed merely a sort of laughing-gas, that agreeably excited the feelings. These last bore exactly the same sort of relation to the race that preceded them, that our admirers of earnestness in the present day bear to the earnest men of a bygone time whom they admire. Their principle was ineffective as a principle of action: it was purely a thing of excited imaginations, and of feelings strung by the aspirations of romance; and died away, even when elevated to its highest pitch, in tones of sweet music, or the wild cadences of ballad poetry.

But this Jacobitism of the middle stage of decay had at least the merit of being a reflection of the real Jacobitism that had gone before. It was Jacobitism mirrored in poetry. Not such, however, the character of yet a third species of Jacobitism, that exists at the present in a few calculating minds wretchedly unfitted for the work of calculation. We have heard of an English divine of the last century, who, having grafted on his theology the philosophy of Bolingbroke and Pope, used to assert in his discourses that whatever *was* was right, and who was urged after sermon, on one occasion, by an individual of his congregation, — a little, thin man, formed somewhat like the letter S, with one shoulder greatly higher and one leg greatly shorter than the other, — to say whether *he* was all right. "Oh yes, all right," was the unhesitating reply of the reverend doctor; "you are quite right *for a cripple.*" Now, the middle stage of Scotch Jacobitism was in like manner quite a right thing of its kind: its legs and shoulders were not equal; it stumped about on a Jacobitical leg to-day, and sometimes, as in the case of Burns, stood on a Jacobinical leg to-morrow; but then it was all quite right for a cripple, and, if it could do nothing more, produced at least some pretty music and some exquisite song. The existing Jacobitism, or, rather, the Jacobitism not existing, but merely supposed to exist, — a shadow of a shade, — a cripple a thousand times more lame than the Jacobitism its immediate predecessor, for it has got no legs at all; and not only no legs, but it can neither sing nor make poetry, — is rendered ridiculous by being represented as all right absolutely, and not as a cripple, — as one of, not the fantasies, but the forces, of the country, — as one, not of its mere night-dreams, but of its waking-day realities, — as not a phantom, but a power. The grand mistake of the *Times* on this subject must still be fresh in the minds of our readers, as it took place little more than three years ago, during the time of her Majesty's first progress through Scotland. The Scotch Lowlanders, said this journal, — usually so sagacious in its estimates, but sorely bemuddled in these days by its Puseyism, — were no doubt a narrow-minded, fanatical, puritanical, selfish set, all agog about non-intrusion and the independence of the Kirk; but very different was the spirit of the Highlands. There the old generous loyalty still existed entire; the long-derived devotion to hereditary claims, and the ancient implicit subjection to divine right. There, in short, ambitious Puseyism, eager to fling its shoe over Scotland, was to find in existing Jacobitism such a friend and ally as the "king over the water" had found in it a century ago. The *Times* has since been undeceived. But there still exist quarters in which Highland Jacobitism continues to be fondly clung to as an actual power, and a religious party that regard it as a *bona fide* ally. We found, when in the Western Highlands last summer, that the approaching commemoration was regarded as a popish movement at bottom; and it would be certainly not uninteresting to know what proportion of the some three or four hundred Highlanders that are said to have turned out on the occasion belonged to the Romish communion. Certainly, if Rome wished, by masquerading at the Centenary in the romance of "The Forty-five," to make an impression on the more active imaginations of the country, she has not been very successful. There is vastly more of the bizarre than of the solemn in the trappings of the Jacobite domino, as accident and pretension have conspired to trim it. It has got bells to its cap. We see it championed by "Young Scotland," — a personage recognized by the half-dozen that ever heard of him as *very young* indeed, — and headed by a Percie Shafton, the undoubted descendant of the royal Stuarts, that edits tartan patterns, the strips of which had been preserved in manuscript in the library of the Scotch Church at Douay, and trembles, meanwhile, lest some unlucky bodkin should establish the maternal relation of old Overstitch the tailor. Happy modern Jacobitism! It is no more a great-grandson of the Pretender that you can boast of as the central figure in your picturesque group, but the Pretender himself, whole and entire.

Yes; the river, with its deep pools and eddying currents, has turned into a different channel from that in which it flowed a century ago; and it is but idle work to be wandering along the deserted course, with its few stagnant shallows, where a handful of landlocked minnows await the droughts that are to lay them dry, as if the water and the great fish were still there. The tide of Highland devotion has long since set in, in a direction entirely opposite. The meeting at Glenfinnon was a meaningless pageant, and, it would seem, a miserably poor pageant to boot. Its enthusiasm, warmed up specially for the occasion, and but lukewarm after all, had no more truth or reality in it than that of the ancient Pistol in the play. The heart of the Highlands was to be found beating elsewhere. It was at the Assembly at Inverness, to which from distant valley and solitary hillside the earnest-minded Celtæ had congregated by thousands, that the enthusiasm was spontaneous and the devotion true. There beat, with all its old truth and warmth, the heart of the Highlands. But alas for the poor Highlanders! It seems to be their destiny as a people to give evidence of their earnest and truthful natures by endurance and suffering. Such was the evidence they had to tender of old of their devotion to the Stuarts, and such the evidence which they have to tender now of their devotion to the cause of evangelical religion and a preached gospel. We saw the stalwart Camerons of Lochiel, whose country a century ago had been wasted by fire and sword, and themselves chased to the rocks and hills, for a loyalty to a hereditary king, again chased from the tombs of their fathers and their little holdings to the oozy sea-beach, and there worshipping God under the tide-line; and the Grants of Strathspey, — of all our Highland clans the clan that last manifested, after the old type, its devotion to its hereditary lord; for, little more than twenty years ago, on learning that his person was endangered in some electioneering contest in the Lowlands, five hundred of its fighting men marched down from their hills to protect him; — these poor clansmen, over a wide and exposed district, denied a place of shelter, have to worship in the open air. And in both cases the persecutor of the clan was its chief, anxious, apparently, that his hereditary followers should be his followers no longer, nor run any further risk of getting into awkward collisions with the law for his sake. We have heard wonder expressed that a single century should be sufficient to effect in the Highland mind so great a change as the revolution indicated by the opposite aspects of the "Centenary of the Forty-five" and the Inverness Assembly. We do not see that there is much cause for wonder. The Presbyterian Highlander of the present day is removed further, by some ten or twelve years, from his popish ancestor who fought at Culloden, than the Presbyterian Covenanter of 1638 was removed from his popish ancestor who fought at Pinkie. It does not require centuries to effect the change in opinion and character which evangelism, when once introduced into a country, is sure always to induce. One peculiarity, however, of the Highlander's position, in reference to the comparatively late introduction of evangelism among his hills, seems not unworthy of mention. Unlike the Southern Scot, who recognizes the old Covenanter as his ancestor, and is, in some instances, a Free Churchman in virtue of the fact, the Highlander of at least the Western and Midland Highlands has no hereditary associations on the side of his beliefs. His hereditary associations, on the contrary, are ranged on the side of Jacobitism. But he is not the less, but the more earnest in his Free Churchism in consequence. His feelings are more fresh, direct, and simple. He is no mere admirer of the Covenanters; he is what the Covenanters themselves were.

Alas! how the short-lived children of men press on to the tomb! A century has now passed since the clans mustered in Glenfinnon; and there are few Scotchmen in middle life to whom that event does not stand as a sort of beacon in the tide of time, to indicate how wave after wave of the generations of the past has broken on the silent shores of eternity, and disappeared from the world for ever. The writer of these remarks was born within the present century, and yet even he can look back on some three or four several generations of men, peculiarly marked in their neighborhood by the epoch of the rebellion, who have passed in succession from this visible scene of things, lighted up by the sun, to the dark land of forgetfulness.

First, we remember a few broken vestiges of a generation that had been engaged in the active business of life when the field of Culloden was stricken. We attended, when a mere boy, the funeral of an old Highlander, a Stuart, who had fought in it on the side of the Prince. We knew another old man, who had been a ship-boy at the time in a vessel with some government stores aboard, that, shortly before the battle, was seized by the rebels; and have heard him tell how, when joking with them, — for they were by no means a band of cut-throat-looking men, — he ventured to speak of their Prince as the Pretender, and was cautioned by one of them to use a more civil word for the future. We remember, too, being brought by two grown-up relatives to visit an old man on his death-bed, who, like the first, had fought at Culloden, but on the side of Hanover. He had been settled in life at the time as the head gardener of a northern proprietor, and little dreamed of being engaged in war; but the rebellion broke out; his master, a kindly man, and a great Whig, volunteered in behalf of his principles under Duke William, and his attached gardener went with him. At the time of our visit, when stretched on the bed from which he never afterward rose, he had outlived his century. He had been an extremely powerful man in his day; and the large wrinkled hand, and huge structure of bone, and deep, full voice, still remained, to testify, amid the general wreck, to what he had once been. His memory for all the later events of his life was gone, so that the preceding forty years of it seemed a blank; but well did he remember the battle, and still more vividly, and with deep execration, the succeeding atrocities of Cumberland. These vestiges of the age of Culloden passed away, and the generation immediately behind them fell into the front ranks, — ancient men and women, who had been mere boys and girls at the time of the "fight," but who vividly remembered some of its details. We knew one of these, an aged woman, who, on the day of the battle, had been tending some sheep on a solitary moor, separated from that of Culloden by an arm of the sea, and screened by a lofty hill, and who had sat listening in terror to the boom of the cannon and the rattle of the musketry, scared as much by the continuous howling of her dog, which she regarded as coupled with some supernatural cause, as by the deadly "thunders in the moors." We intimately knew another who witnessed the battle, though in no very favorable circumstances for minute observation, from the Hill of Cromarty. The day, he has told us, was drizzly and thick; and on reaching the brow of the hill, where he found a vast group of the townsfolk already assembled, he could scarce see the opposite land. But the fog gradually cleared away; first one hill-top came into view, and then another, till at length the long range of coast, from the opening of the great Caledonian Valley to the promontory of Brugh-head, was dimly visible through the haze. A little after noon there arose a sudden burst of round white cloud from the moor of Culloden, and then a second burst beside it, and then they mingled together, and went rolling slantways on the wind towards the west; and he could hear the rattle of the smaller firearms mingling with the roar of the artillery. And then, in what seemed a wonderfully short space of time, the cloud dissipated and disappeared, and the boom of the greater guns ceased, and a sharp intermittent patter of musketry passed on towards Inverness. Such was the battle of Culloden, as witnessed by the writer's maternal grandfather, then a boy in his fourteenth year. The years passed by, and he and the generation to which he belonged followed the generation that had gone before; and then the front rank in the general march to the tomb came to be occupied by those so long known in Scotland as the Culloden-year people, — a class of persons who stood in no need of consulting records and registers for the date of their birth, for the battle had drawn, as if with the sword-edge, its deep score athwart the time, so that all took note of it. But the Culloden-year people passed from the stage also; every season in its flight left them fewer and feebler; and we now see the front rank composed of their children, — a gray-haired generation, drooping earthwards, who have already spent in their sojourn the term so long since fixed by the psalmist. And thus — as wave succeeds wave, storm-impelled, from the ocean, to break upon the shore — pass away and disappear the generations of man. It were well, since our turn must come next, to be distinguishing in time between the solid and the evanescent, — the things which wear out like the old Jacobitism of the past, and become sorry shows and idle mockeries, and the things immortal in their natures, which contumely cannot degrade nor persecution put down.

XIII.

THE HALF-CENTURY.

The first fifty years of the nineteenth century terminated a few hours ago, and we have now entered upon the second fifty. As last night's clock struck twelve, the most important half-century of modern history came to its close, and a half-century which threatens to be scarce less eventful began its course. The general progress made by Great Britain during the lapsed period has been great beyond all former precedent; but there is one special department in which it is ominously, fearfully great; and should the same ratio of increase continue throughout the succeeding fifty years, there will be problems for our country to solve, compared with which those of the present day, difficult as they may seem, may be regarded as the tasks of children. At the commencement of the half-century just closed, the population of England and Scotland united did not much exceed eight millions of souls; in 1841 it considerably exceeded eighteen millions; and, as the census of the present year will by and by show, it now exceeds twenty millions. For every *two* Britons that existed on their native soil when the century began, there now exist *five*, and in fifty years there has taken place in the population an increase of a hundred and fifty per cent.; and at the close of the nineteenth century, should the same rate of increase continue, the soil of Great Britain will be encumbered by fifty millions of human creatures. How the privileges of proprietors, as now defined, are to be made good in such a state of things — should such a state of things ever arrive — against the pressing claims of the crowded masses, it is at present difficult to see; but in this element of increase alone — an element which the inadequate expedient of emigration, that, when most active, sends only *one* abroad for every additional *three* born at home, may in vain expect to counterbalance — we recognize a disturbing agent, suited, even did it stand alone, to give more than employment enough to the philanthropists and statesmen of the future. Since the death of Chalmers it has not been customary to press much on this topic; but considerably less than half a century will serve to show how entirely he was in the right regarding it.

Fifty years form a large proportion of the period assigned to man; and those whose powers of observation were active at the beginning of the present century, and their opportunities of exercising them considerable, must now be far advanced in life. We, however, reckon among our readers individuals who can compare from personal observation the Scotland of 1801 with Scotland in the present day, and who can tell how, over wide areas, the face of the country has changed. We ourselves, though born within the half-century, are acquainted with extensive localities in which, within our recollection, the breadth of corn-land has fully doubled. We have seen it slowly advancing over moory waste and brown hillside, till, where only heath and ling and unproductive brushwood used to grow, every autumn mottles over the landscape with shocks of corn. In proportion as the population was increasing were the means of their support in these localities increasing also. But it was chiefly in lowland districts, or in districts which merely bordered on the Highlands, that we witnessed this change for the better taking place. Much of the Highlands themselves has been the subject of a reverse process. During the last half-century many a sheltered glen and fertile valley have given their cultivated patches back to waste; and where human habitations once stood, and happy communities once lived, we find but moss-covered ruins and the solitude of a desert. And it would seem as if this state of management had already produced its crisis. Where the corn-land has more than doubled its area, or, what amounts to the same thing, more than doubled its produce, there is food and employment for the more than doubled population; whereas in the Highlands, on the contrary, famine stares the unhappy inhabitants full in the face, and Lowland Scotland is told, that, unless it exert itself greatly in their behalf, thousands of them must perish. It will be a question for the next half-century practically to determine whether, as the population is growing, and seems destined to grow, the Highlands must not be compelled in the general behalf to sustain their own portion of it. There is another question which this continued increase in the numbers of the people will at length render all potent. Men have wondered how, in a country such as China, where the tone of morailty is low and the government is corrupt, education should have such honors and privileges attached to it, that it forms the sole means of rising into place and affluence. The true secret of the matter is to be read in the fact that China, with its three hundred millions of inhabitants, is the most populous country on the face of the earth. Ignorance, therefore, cannot be tolerated in China; and knowledge, including, as a matter of course, a thorough acquaintance with the arts by which men live, is at a premium there. However unacquainted with what most ennobles man, the Chinese cannot be left ignorant of how — to use their own homely phrase — "men are to get their rice." Were the case otherwise, they would of necessity have to eat one another; and so in this vast nation, still in some respects a semibarbarous one, a certain measure of education is universal; and its cheap literature, notwithstanding its block-printing and its difficult character, is the most immense in the world. And, on a similar principle, the growing population of Britain will force upon the country the question of an adequate education for the people. It is difficult to overpeople any nation with a taught and industrious race of men. China is not overpeopled with its three hundred millions. Ireland, that has not half the number of inhabitants to the square mile, and the Highlands of Scotland, that have not the one-fortieth part the number to the square mile, are, on the contrary, greatly overpeopled; and the difference consists mainly in this, that whereas the Chinese have, with all their many faults, been taught how to "get their rice," the poor Highlanders and the Irish have not. But, in this special department at least, the extreme limits of the "let-alone system" have been well-nigh reached; and the next half-century will see knowledge more largely spread abroad, as a matter of necessity in which the very existence of the nation is involved, than any former age of the world. The time has at length come when "many shall run to and fro, and knowledge shall be increased."

But though knowledge during the last half-century *did* greatly increase, so that there are now single periodicals that possess a larger circle of readers than composed in the previous half-century, according to the estimate of Burke, the whole reading public of Great Britain, there is another, and, as has been generally supposed, antagonistic principle, that has increased in a still greater ratio. Popery reckons, at the close of the first half of the nineteenth century, about ten times the number of adherents within the two kingdoms that it reckoned when the century began. In producing a result so disastrous, Puseyism has no doubt had its share. There are but two elements in the religious world of Europe, — the Popish and the Puritanic; and when, some fifteen years ago, a zealous and influential section of English Episcopalians set themselves to reinvigorate their Church by revivingthe ceremonies and doctrines of a Christianity absolutely *ancient*, but comparatively modern, — for it dates at least three hundred years later than the age of the New Testament, — they had inevitably committed themselves, little as they might be aware of the fact at the time, to the popish element. And we now see the fruit of the committal in the perversions which are taking place almost every day in the English Church. But these, though of mighty importance to Rome, have done comparatively little to swell her numbers. She owes the vast increase which has filled the dingier dwellings and poorer lanes of our larger towns with her votaries, to the overflowings of the miserable population of Ireland. The Romish Church has been no doubt much encouraged by the revival of the ancient Christianity within the pale of the English one; and, save for this encouragement, it is not in the least likely that the aggression of the past year would have taken place; but there can be as little doubt that it is the poor, neglected Irish, sacrificed generation after generation to the Erastian secularities of Protestant Episcopacy, and latterly expatriated by the potato disease, that popery owes its increase in Britain. There will be work enough in this department for all the Protestant churches of the country for the coming half-century, if they would escape defeat and disgrace at their own doors. The last half-century has shown how difficult it is to calculate on the strength of churches. Its first decade witnessed the dethronement of the Pope by Napoleon; its terminating decade, his flight from Rome under the terror of his revolutionary subjects.

And yet popery possesses at the present time a vast empire in the minds of men; and it has just dared to perpetrate, in consequence, one of its boldest aggressions on the most powerful empire in the world. And that aggression has brought out the great strength of another church, which, about the time of the passing of the Reform Bill, was deemed so far from strong that statesmen of no inconsiderable calibre held that almost any sort of liberty might be taken with the status of her dignitaries, or with her property. It seems unquestionably true, that the present powerful anti-popish movement, which has done what the zeal of Dissent could never do, — stirred the nation to its very depths, — has arisen among the English Episcopalians, and has been a direct consequence of what the Dissent of England and the Presbyterians of Scotland regard as a very inconsiderable element in the matter, — the encroachment on the domains of the English bishops. We recognize in the fact the correctness of the impression made upon us when residing for a short time in England a few years ago. We crossed the borders in the belief, pretty general, we are disposed to think, among Scotchmen, that the active power of nonconformity in the southern kingdom was not much less than a match for the mere passive power of its Established Episcopacy: we came away full under the conviction that the two powers are so very unequal that it is scarce wise to name them together. Established Episcopacy in England represents the soldiers of a vast army leaning silently on their arms; whereas Dissent may be rather likened to the handful led by Gideon, making great show and much noise, but, unless miracles be wrought in their behalf, not destined to make a very considerable impression on the country. And so evangelism in Scotland has a much larger stake in the doctrinal soundness of the English Church than it seems to be aware of. Judging from present appearances, the religion of the English Church, whatever that may come to be, bids fair to be also the religion of the English Constitution; and therefore, though we respect many of the honest and good men who seem determined at the present crisis to do battle both with popery and Established Episcopacy, we cannot think they have by any means fallen on the best way of dealing with the emergency. They will, we are afraid, find either opponent quite a match for them; and should they set themselves to fight against both at once, neither Protestantism nor themselves will gain anything by their coming into the field.

Another mighty increase has taken place during the lapsed half-century in the numbers of the poor. It is generally, and, we think, justly held, that that enormous amount of pauperism in Scotland which, at the time of the Revolution, Fletcher of Saltoun could deem so formidable, was, in great part at least, a result of the previous persecution. There can be at least as little doubt that it was the termination of the church controversy, not in an equitable adjustment, suited to place under the control of our civil courts all the temporalities of the church, and under her courts ecclesiastical all her spiritualities, but in the Disruption, — an event gilded by the glory of conscientious sacrifice, but not the less, but rather the more, on that account a calamity to the country, — that brought the pauper question to a crisis, and saddled upon Scotland a crushing poor-law. It is a surely not uninstructive fact, that the proprietors of the country have paid for the support of the poor, since this event, a sum as large as would have purchased all their patronages three times over,— a sum which previous to the collision they had not to pay, and which, had they urged the question to a different issue, they would not have to pay now. The settlement which the controversy received has been, economically at least, a very bad settlement for them. But there is no party that need triumph in such a result. Free Churchmen, as certainly as Established Churchmen, suffer in consequence; and the hard problem subjected to the country through the event it may take the whole of the next half-century to solve. It is something, however, that it is already compelling attention, and that Carlyle's "Condition of the People Question" is recognized as the great question of the day. These are but desultory remarks, and, withal, sufficiently prosaic; but the magnitude of the subject oppresses us; nor dare we attempt condensing into an article what, could we devote a whole volume to the survey, would require to be even then greatly condensed.

XIV.

THE ECHOES OF THE WORLD.

DR. CHALMERS.

HAS the reader ever heard a piece of heavy ordnance fired amid the mountains of our country? First there is the ear-stunning report of the piece itself, — the prime mover of those airy undulations that travel outwards, circle beyond circle, towards the far horizon; then some hoary precipice, that rises tall and solemn in the immediate neighborhood, takes up the sound, and it comes rolling back from its rough front in thunder, like a giant wave flung far seaward from the rock against which it has broken; then some more distant hill becomes vocal, and then another, and another, and anon another; and then there is a slight pause, as if all were over, — the undulations are travelling unbroken along some flat moor, or across some expansive lake, or over some deep valley, filled, haply, by some long withdrawing arm of the sea; and then the more remote mountains lift up their voices in mysterious mutterings, now lower, now louder, now more abrupt, anon more prolonged, each, as it recedes, taking up the tale in closer succession to the one that had previously spoken, till at length their distinct utterances are lost in one low, continuous sound, that at last dies out amid the shattered peaks of the desert wilderness, and unbroken stillness settles over the scene, as at first. Through a scarce voluntary exercise of that faculty of analogy and comparison so natural to the human mind, that it converts all the existences of the physical world into forms and expressions of the world moral and intellectual, we have oftener than once thought of the phenomenon, and its attendant results, as strikingly representative of effects produced by the death of Chalmers. It is an event which has, we find, rendered vocal the echoes of the world; and they are still returning upon us, after measured intervals, according to the distances. First, as if from the nearer rocks and precipices, they arose from the various towns and cities of Scotland that possess their periodicals; then from the great southern metropolis, and the other towns and cities of England, as if from the hills immediately beyond; from Ireland next; and next from France and Geneva, and the European Continent generally. And then there was a slight pause. The tidings were passing in silence, without meeting an intelligent ear on which to fall, across the wide expanse of the Atlantic. And then, as if from more distant mountains, came the voices of the States, and the colonies, and the West Indian Islands. It was no uninteresting task to unrobe from their close brown covers, that spake in color and form of a foreign country, the Transatlantic journals, and read tribute after tribute to the worth and intellectual greatness of the departed; and to hear of the funeral sermons preached far away, on the very verge of the civilized world, amid half-open clearings in the vast forest, or in hastily-erected towns and villages that but a few twelvemonths before had no existence. Nor have all the echoes of the event returned to us even yet. They have still to arise from, if we may so express ourselves, the more distant peaks of the landscape, — from the Eastern Indies,

Australia, and the antipodes. Every more remote echo, while it indicates how great the distance which the original undulations have traversed, and how wide the area which they fill, serves also of necessity to demonstrate the far-piercing character and greatness of the event which first set them in motion. Dryden, in describing the grief occasioned by the death of some august and "gracious monarch," describes it as bounded, with all its greatness and extent, by his own dominions: —

"Thus, when some great and gracious monarch dies,
Soft whispers first and mournful murmurs rise
Among the sad attendants; then the sound
Soon gathers voice, and spreads the news around
Through town and country, till the dreadful blast
Is blown to distant colonies at last."

There have been no such limitations to the sorrow for Chalmers. The United States and the Continent have sympathizingly responded — of one mind in this matter, as of one blood, with ourselves — to the regrets of Britain and the colonies. We have few men left whose names so completely fill the world as that of Chalmers.

The group of great men to which Thomas Chalmers belonged has now well-nigh disappeared. Goldsmith has written an ingenious essay to show that the "rise or decline of literature is little dependent on man, but results rather from the vicissitudes of nature." The larger minds, he remarks, are not unfrequently ushered into the world in groups; and after they have passed away, there intervene wide periods of repose, in which there are only minds of a lower order produced. "Some ages have been remarkable," he says, "for the production of men of extraordinary stature; others for producing particular animals in great abundance; some for excessive plenty; others, again, for seemingly causeless famine. Nature, which shows herself so very different in her visible productions, must surely differ also from herself in the production of minds; and, while she astonishes one age with the strength and stature of a Milo or a Maximian, may bless another with the wisdom of a Plato or the goodness of an Antonine." In glancing over the history of modern Europe, and more especially that of the British empire, civil and literary, one can scarce fail to mark a cycle of production of this character, which now seems far advanced in its second revolution. The seventeenth century was in this country peculiarly a period of great men. Cromwell and Shakspeare were so far contemporary, that when, little turned of fifty, the poet lay on his deathbed, the future Lord Protector, then a lad of seventeen, was riding beside his father, to enter as a student the University of Cambridge; and the precocious Milton, though still younger, was, we find, quite mature enough to estimate the real stature of the giant that had fallen, and to deplore his premature death in stanzas destined to live forever. And when, in after life, the one great man sat writing, to the dictation of the other, the well-known noble letter to Louis in behalf of Continental Protestantism, the mathematician, Isaac Newton, sat ensconced among his old books in the garret at Grantham; the metaphysician, John Locke, was engaged at Oxford in his profound cogitation on the nature and faculties of mind; John Bunyan was a soldier of the Commonwealth; Cowley was studying botany in Kent; Butler was pouring forth his vast profusion of idea in the dwelling of Sir Samuel Luke; Dryden, at the ripe age of twenty-seven, was making his first rude efforts in composition in Trinity College; Sir Matthew Hale was administering justice in London, and planning his great law works; and, though Hampden and Selden were both in their graves at the time, the former, had he escaped the fatal shot, would still have been in but middle life, and the latter was but four years dead. The group was assuredly a very marvellous one. It passed away, however, like all that is of earth; and there arose that other group of men, admirable in their proportions, but of decidedly lower stature, that all in any degree acquainted with English literature recognize as the wits of Queen Anne. To this lower but very exquisite group the Popes, Swifts, and Addisons, the Gays, Parnells, and Priors belong. It also passed; and a still lower group arose, with, it is true, a solitary Johnson and Burke raising their head and shoulders above the crowd, but attaining not, at least in the mass, to the stature of their immediate predecessors. And they themselves were well aware of their inferiority. Is the reader possessed of a copy of Anderson's "Poets?" From its chronological arrangement, it illustrates very completely the progress of that first great cycle of production from the higher to the lower minds to which we refer; and with the works of the Jenyns, the Whiteheads, the Cottons, and the Blacklocks, the collection closes. And then the cycle, as if the moving spring had been suddenly wound up to its original rigidity, begins anew. The gigantic figure of Napoleon appears as the centre of a great historic group; and we see ranged around him the tall figures of statesmen such as Pitt and Fox; of soldiers such as Soult, Ney, and Wellington; of popular agitators such as Cobbett and O'Connell; of theological writers and leaders such as Hall, Foster, and Andrew Thomson; and of literary men such as Goethe, Chateaubriand, Sir Walter Scott, and Wordsworth. The group is very decidedly one of men large and massy of stature; and to this group, great among the greatest, Thomas Chalmers belonged. It has, we repeat, nearly passed away. Wellington, Wordsworth, and Chateaubriand — all well stricken in years, — turned very considerably, the youngest of them, of the threescore and ten — alone survive. Immediately beneath these, and bearing to them a relation very similar to that which the wits and statesmen of Queen Anne bore to the Miltons and Cromwells, their predecessors, stands a group, the largest of their day, including as politicians the Peels and Russells, and as literary men the Lockharts and Macaulays, of the present time. Happily the Free Church, though its great leader be removed, does not lack at least its proportional number of these. They may be described generally, with reference to their era, as men turned of forty; and, so far as may be judged from the present appearance of things, the younger and succeeding group, just entered on the stage, are composed, as during the middle of the last century, of men of a third class, that seem well-nigh as inferior in height and muscle to those of the second, as the second are inferior in bulk, strength, and massiveness to those of the first. The third stage of the second cycle of production is, it would appear, already full in view. In the poetical department of our literature this state of things is strikingly apparent. Ere the Cowpers and Burnses arose to herald the new and great era, the latter half of the last century had its Wartons and its Langhorns, — true and sweet poets, but, it must be confessed, of somewhat minute proportions. The present time has its Moirs and its Alfred Tennysons; and they are true poets also, but poets on a not large scale, — decidedly men of the third era.

In glancing over the various tributes to the memory of Chalmers, one is struck with a grand distinction by which they may be ranged into two classes. Belonging, as he did, to two distinct worlds, — the worlds literary and religious, — we find estimates of his character and career made by representatives of both. In the one, the appreciation hinges, as on a pivot, on a certain great turning incident in his life; in the other, there is either no reference made to this incident, or the principles on which it occurred are represented as of a common and obvious, and not very important character. Is it not truly strange, that the most influential event that can possibly take place in the history of individual man — which has lain at the foundation of the greatest revolutions of which the annals of the species furnish any record, and which constitutes the main objective theme of revelation — should be scarce at all appreciated, even in its palpable character as a fact, by the great bulk of the acutest and most intelligent writers of the present age? That change in the heart and life which sent the apostles forth of old to Christianize the world, and the Reformers at a later time to re-Christianize it, — which, forming the charm of the successes of Cromwell, preserved to Britain its free Constitution, and which altered *in toto* the destinies of Chalmers, — that change, we say, is rightly appreciated, in even its obvious character as a fact, by none of our purely literary men; or, at least, if we must make one exception, by Thomas Carlyle alone. It constitutes a mighty spring of action, — by far the mightiest in this world, — of which the rest are ignorant. Regarded in this point of view, the following extract from the "People's Journal"— a periodical conducted chiefly, it is understood, by Unitarians — is not uninstructive. It refers to the conversion of Chalmers, and describes that event as occurring on a few obvious commonplace principles: —

"A new era in the development of Chalmers' mind commences with his engagement upon the article 'Christianity.' The powerful devotional tendency of his mind had hitherto, to all appearance, lain dormant. The protracted and unintermitting attention to religious questions which, in the compilation of that essay, he was compelled to bestow, was favorable to the formation of a devotional habit of mind in one who, like all men of poetical temperament, was eminently liable to take the tone and color of his mind from the element in which he lived. The Leslie controversy, too, had bridged over the gulf which had hitherto intervened between the higher orders of minds among the *literati* and the orthodox clergy of Scotland. The Dugald Stewarts and the Jeffreys on the one hand, the Moncreiffs and Thomsons on the other, had, while acting in concert, learned to know and appreciate each other's peculiar merits. The sentiment of political independence, and that liberal tolerance, the most uniform feature of superior minds, had infused permanent feelings of mutual good-will into minds which by their organization were irreconcilably different. Chalmers, who had been thrown among the purely intellectual class in a great measure by the accident of position, was now attracted to the religious class, with whom his natural sympathies were, if anything, still greater. He devoted himself more exclusively to the duties of his ministerial office, and, carrying into the pulpit the same buoyant enthusiasm, the same Herculean powers, he soon became one of the most distinguished inculcators of 'evangelical' views of religion."

Among the numerous funeral sermons of which the death of Chalmers has proved the occasion, we know not a finer, abler, or better-toned than one of the Transatlantic discourses. It is from the pen of Dr. Sprague, of Albany, United States, so well known in this country by his work on revivals. His estimate of the great change which not only expanded the heart, but also in no slight degree developed the intellect, of Chalmers, differs widely, as might be expected from the general tone of his writings, from that of the Unitarian in the "People's Journal." It is strange on what analogies men ingenious in misleading themselves when great principles are at stake contrive to fall. We have lately seen Cromwell's love of the Scriptures, and his diligence, according to the divine precept, in searching them, attributed to the mere military instinct, gratified, in his case, by the warlike stories of the Old Testament, as the resembling instinct was gratified in that of Alexander the Great by the stories of the Illiad.

"He [Dr. Chalmers] removed to Kilmeny," says Dr. Sprague, "in 1803, where he labored for several years, and where occurred at least *one* of the most remarkable events of his life. It was nothing less, as he himself regarded it, than a radical change of character. Previous to that period he seems to have looked upon the duties of his profession as a mere matter of official drudgery; and not a small part of his time was devoted to science, particularly to the mathematics, to which his taste more especially inclined him. But having been requested to furnish an article for the "Edinburgh Encyclopædia" on the evidences of divine revelation, in the course of the investigation to which he was led in the prosecution of this effort he was brought into communion with Christianity in all its living and transforming power. He not only became fully satisfied of its truth, of which before he had had only some indefinite and inoperative impression, but he discovered clearly its high practical relations; he surrendered himself to its teachings with the spirit of a little child; he reposed in its gracious provisions with the confidence of a penitent sinner; and from that time to his dying hour he gloried in nothing save in the cross of the Lord Jesus Christ. He stood forth before the world strangely unlike what he had ever been before. There was a sacred fervor, an unearthly majesty, in all his utterings and all his writings. Scotland, Britain, the world, soon came to look at him with wonder, as one of the brightest luminaries of his time, — as destined to exert a controlling influence upon the age, if not to work an epoch in the world's history. It was quickly found that there was a far higher effect produced by his ministrations than mere admiration, — that the sword of the Spirit, wielded with such unwonted energy, was doing its legitimate work; for worldliness could not bear his rebuke; scepticism could not stand erect in his presence; while a pure and living Christianity was constantly reproducing itself in the hearts of some one or other of his enchained hearers."

Dr. Sprague's estimate of the intellectual character of Chalmers seems eminently just, and, formed at the distance of more than three thousand miles from the more immediate scene of Chalmers' personal labors, — for distance in space has greatly the effect in such matters of distance in time, — it may be regarded as foreshadowing the judgment of posterity.

"The intellectual character of Dr. Chalmers was distinguished chiefly by its wonderful combination of the imaginative, the profound, and the practical. If there be on earth a mind constituted with greater power of imagination than his, we know not where to look for it. And because he was so preëminent in respect to this quality, I am inclined to think that some have underrated his more strictly intellectual powers, — his ability to comprehend the more distant bearings of things, or to grapple with the subtilties of abstract philosophy; and they have reached their false conclusion on the ground that it were impossible that a mind so highly gifted in one respect should be alike distinguished in the other. But if his productions may be allowed to speak for him, I think it will be difficult to show that he was not equally at home in the depths as on the heights; and some of his works, particularly that on Natural Theology, exhibit the two qualities blended in beautiful proportions. I hesitate not to say, that any man who could reason like Chalmers and do nothing else, or any man who could soar like Chalmers and do nothing else, or any man who could contrive and execute like Chalmers, as is evinced by his connection with the whole Free Church movement, and do nothing else, would be a great man in any country or in any age; but the union of the several faculties in such proportion and such degree constitutes a character at once unparalleled and imperishable."

Among the various references to this genius of Chalmers for the practical, which, according to Sprague, would have constituted him a great man even had it been his only faculty, we know not a finer or more picturesque than that which we find in a truly admirable article in a late number of the "North British Review." The picture — for a picture it is, and a very admirable one — exhibits specially the inspiriting effect of the quality in a time of perplexity and trial. It is when dangers run high that the voice of the true leader is known: the storm in its hour of dire extremity exhibits the skill of the accomplished pilot.

"When the courts of law revoked," says the reviewer, "the liberty of the Scottish Church, much as he loved its old Establishment, much as he loved his Edinburgh professorship, and much more as he loved his two hundred churches, with a single movement of his pen he signed them all away. He had reached his grand climacteric; and many thought that, smitten down by the shock, his gray hairs would descend in sorrow to the grave: it was time for him "to break his mighty heart and die." But they little knew the man. They forgot that spirit which, like the trodden palm, had so often sprung erect and stalwart from a crushing overthrow. We saw him that November. We saw him in its Convocation, — the sublimest aspect in which we ever saw the noble man. The ship was fast aground; and as they looked over the bulwarks, through the mist and the breakers, all on board seemed anxious and sad. Never had they felt prouder of their old first-rate, and never had she ploughed a braver path, than when, contrary to all the markings in the chart, and all the experience of former voyages, she dashed on this fatal bar. The stoutest were dismayed; and many talked of taking to the fragments, and, one by one, trying for the nearest shore ; when, calmer because of the turmoil, and with the exultation of one who saw safety ahead, the voice of this dauntless veteran was heard propounding his confident scheme. Cheered by his assurance, and inspired by his example, they set to work; and that dreary winter was spent in constructing a vessel with a lighter draught and a simpler rigging, but large enough to carry every true-hearted man who ever trod the old ship's timbers. Never did he work more blithely, and never was there more of athletic ardor in his looks, than during the six months that this ark was building, though every stroke of the mallet told of blighted hopes, and defeated toil, and the unknown sea before him. And when the signal-psalm announced the new vessel launched, and leaving the old galley high and dry on the breakers, the banner unfurled, and showing the covenanting blue still spotless, and the symbolic bush still burning, few will forget the renovation of his youth, and the joyful omen of his shining countenance. It was not only the rapture of his prayers, but the radiance of his spirit, which repeated, 'God is our refuge.' It is something heart-stirring to see the old soldier take the field, or the old trader exerting every energy to retrieve his shattered fortunes; but far the finest spectacle of the moulting eagle was Chalmers, with his hoary locks, beginning life anew. But, indeed, he was not old. They who can fill their veins with every hopeful, healthful thing around them, — those who can imbibe the sunshine of the future, and transfuse life from realities not come as yet, — their blood need never freeze. And his bosom heaved with all the newness of the Church's life, and all the bigness of the Church's plans. And, best of all, those who wait upon the Lord are always young. This was the reason why on the morning of that exodus he did not totter forth from the old Establishment a blank and palsy-stricken man, but, with flashing eye, snatched up his palmer-staff, and, as he stamped it on the ground, all Scotland shook, and answered with a deep God-speed to the giant gone on pilgrimage."

Of all the tributes to the memory of Chalmers which we have yet seen, one of at once the ablest and most generous is that by Dr. Alexander of this city.[1] Belong-

[1] A Discourse on the Qualities and Worth of Thomas Chalmers, D.D. LL.D., etc. By William Lindsay Alexander, D.D.

ing to a different family of the church catholic from that whose principles the illustrious deceased maintained and defended, and at issue with him on points which neither deemed unimportant, the doctor has yet come forward, in the name of their common Christianity, to record his estimate of his character and his sorrow for his loss. It was one of the points worthy of notice in Chalmers, that none of his opponents in any controversy settled down into personal enemies. We saw, among the thousands who attended his funeral, Principal Lee, with whom he had the controversy regarding the Moderatorship; Dr. Wardlaw, his opponent in the great controversy on establishments; and the carriage of the Lord Provost, as representative of the Provost himself, with whom he had the controversy regarding the Edinburgh churches and their amount of accommodation; and who was on business in London at the time. And to this trait, and to what it indicated, Dr. Alexander finely refers. The doctor was one of Chalmers' St. Andrew's pupils; and his opportunities of acquaintanceship at that period furnish one or two singularly interesting anecdotes illustrative of the character of the man: —

" Sometimes it was my lot to be his companion," says the doctor, "to some wretched hovel, where I have seen him take his seat by the side of some poor child of want and weakness, and patiently, affectionately, and earnestly strive to convey into his darkened mind some ray of truth that might guide him to safety and to God. On such occasions it was marvellous to observe with what simplicity of speech that great mind would utter truth. One instance of this I must be allowed to mention. The scene was a low, dirty hovel, over whose damp and uneven floor it was difficult to walk without stumbling, and into which a small window, coated with dust, admitted hardly enough of light to enable an eye unaccustomed to the gloom to discern a single object. A poor old woman, bedridden, and almost blind, who occupied a miserable bed opposite the fire-place, was the object of the doctor's visit. Seating himself by her side, he entered at once, after a few general inquiries as to her health, etc., into religious conversation with her. Alas! it seemed all in vain. The mind which he strove to enlighten had been so long closed and dark that it appeared impossible to thrust into it a single ray of light. Still, on the part of the woman there was an evident anxiety to lay hold upon something of what he was telling her; and, encouraged by this, he persevered, plying her, to use his own expression, with the offers of the gospel, and urging her to trust in Christ. At length she said, 'Ah, Sir, I would fain do as you bid me, but I dinna ken how: how can I trust in Christ?' 'Oh, woman,' was his expressive answer, in the dialect of the district, 'just *lippen* to Him.' 'Eh, Sir,' was her reply, 'and is that a'?' 'Yes, yes,' was his gratified response; 'just lippen to Him, and lean on Him, and you'll never perish.' To some, perhaps, this language may be obscure; but to that poor, blind, dying woman it was as light from heaven; it guided her to the knowledge of the Saviour; and there is good reason to believe it was the instrument of ultimately conducting her to heaven."

We had marked for quotation various passages in this admirable discourse, unequalled, we hold, by aught that has yet appeared, as an analysis of the mental and moral constitution of him whom Dr. Alexander at once eloquently and justly describes as "a man of brilliant genius, of lovely character, of sincere devotion, of dignified and untiring activity, the most eminent preacher our country has produced, the greatest Scotchmen the nineteenth century has yet seen." We have, however, much more than exhausted our space, and so must be content for the present with recommending to our readers an attentive perusal of the whole. One passage, however, we cannot deny ourselves the pleasure of extracting. It meets, we think, very completely, a frequent criticism on one of the peculiarities of Chalmers, and shows that what has been often instanced as a defect was in reality a rarely attainable excellence:—

"In handling his subjects Dr. Chalmers displayed vast oratorical power. He usually selected one great truth or one great practical duty for consideration at a time. This he would place distinctly before his hearers, and then illustrate, defend, and enforce it throughout his discourse, again and again bringing it up before them, and urging it upon them. By some this has been regarded as a defect rather than a merit in his pulpit addresses; and it has been ascribed to some peculiarity of his mind, in virtue of which he has been supposed incapable of turning away from a subject when once he had hold on it, or, rather, it had laid hold on him. I believe this criticism to have been quite erroneous. That his practice in this respect was not an accidental result of some mental peculiarity, but was purposely and designedly followed by him, I know from his own assurance; indeed, he used publicly to recommend it to his students as a practice sanctioned by some of the greatest masters in oratory, especially the great parliamentary orator Charles James Fox; and the only reason, I believe, why it is not more frequently adopted, is, that it is immeasurably more difficult to engage the minds of an audience by a discourse upon one theme, than by a discourse upon several. That it constitutes the highest grade of discourse, all writers on oratory, from Aristotle downward, are agreed. But to occupy it successfully requires genius and large powers of illustration. When the speaker has to keep to one theme, he must be able to wield all the weapons of address. He must be skilled to argue, to explain, persuade, to apply, and, by a fusion of all the elements of oratory, to carry his point whether his audience will or no. Now these requisites Dr. Chalmers possessed in a high degree. He could reason broadly and powerfully; he could explain and illustrate with exhaustless profusion; he could persuade by all the earnestness of entreaty, all the pathos of affection, and all the terrors of threatening; he could apply, with great skill and knowledge of men's ways, the truth he would inculcate; and he could pour, in a torrent of the most impassioned fervor, the whole molten mass of thought, feeling, description, and appeal, upon the hearts and consciences of his hearers. Thus singularly endowed, and thus wisely using his endowments, he arrived at a place of the highest eminence in the highest walk of popular oratory."

XV.

GLEN TILT TABOOED.

A RENCOUNTER of a somewhat singular character has taken place in Glen Tilt between the Duke of Atholl, backed by a body of his gillies, and a party of naturalists headed by a learned professor from Edinburgh. The general question regarding right of way in Scotland seems fast drawing to issue between the people and the exclusives among the aristocracy, and this in a form, we should fain hope, rather unfavorable to the latter, seeing that the popular cause represents very generally, as in this case, that of the sentiment and intellect of the country, while the cause of the exclusives represents merely the country's brute force, — luckily a considerably smaller portion of even that than falls to the share of even our physical force Chartists. Should thews and muscles come to sway among us, the *regime* must prove a very miserable one for Dukes of Leeds and of Atholl.

From time immemorial the public road between Blair-Athole and Braemar had lain through Glen Tilt. In most questions regarding right of roadway witnesses have to be examined; the line of communication at issue is of too local and obscure a character to be generally known; and so the claim respecting it has to be decided on the evidence of people who live in the immediate neighborhood. Not such, however, the case with Glen Tilt. There is scarce in the kingdom a better-known piece of roadway than that which runs through the glen; and all our ampler Guide-Books and Traveller's Companions assume the character of witnesses in its behalf. Here, for instance, is the Guide-Book of the Messrs. Anderson of Inverness, — at once one of the most minute and most correct in its details with which we are acquainted, and which has the merit of being derived almost exclusively from original sources. It does not indicate a single route which the writers had not travelled over, nor describe an object which they had not seen and examined. And in it, as in all the other works of its class, we find the road running through Glen Tilt which connects Blair-Athole and Braemar laid down as open to the tourists, equally with all the other public roads of the country. The reader will find it marked, too, in every better map of Scotland. In the "National Atlas"—a work worthy of its name — it may be seen striking off, on the authority of the geographer to the Queen, Mr. A. K. Johnston, at an acute angle from the highway at Blair-Athole; then running on for some twelve or thirteen miles parallel to the Tilt; and then, after scaling the heights of the upper part of the glen, deflecting into the valley of the Dee, and terminating at Castleton of Braemar. The track which it lays open is peculiarly a favorite one with the botanist, for the many interesting plants which it furnishes; and so much so with the geologist, that what may be termed the classic literature of the science might, with the guide-books of the country, be brought as evidence into court in the case. Playfair's admirably-written "Illustrations of Hutton" take part against the Duke and his gillies. That curious junction of the granite and stratified schists in which Hutton recognized the first really solid ground for his theory, and of which, as forming the great post of vantage in the battle between his followers and those of Werner, a representation may be found in almost every geological treatise since published, occurs in Glen Tilt, and possesses a more than European celebrity. There is not a man of science in the world who has not heard of it. The history of its discovery, and of what it establishes, as given in a few sentences by one of the most popular of modern geologists, we must present to the reader.

"The absence of stratification in granite," says Mr. Lyell, "and its analogy in mineral character to rocks deemed of igneous origin, led Hutton to conclude that granite must also have been formed from matter in fusion; and this inference, he felt, could not be fully confirmed, unless he discovered, at the contact of granite and other strata, a repetition of the phenomena exhibited so constantly by the trap rocks. Resolved to try his theory by this test, he went to the Grampians, and surveyed the line of junction of the granite and superincumbent stratified masses, and found in Glen Tilt, in 1785, the most clear and unequivocal proofs in support of his views. Veins of red granite are there seen branching out from the principal mass, and traversing the black micaceous schists and primary limestones. The intersected stratified rocks are so distinct in color and appearance as to render the example in that locality most striking; and the alteration of the limestone in contact is very analogous to that produced by trap veins on calcareous strata. This verification of his system filled him with delight, and called forth such marks of joy and exultation that the guides who accompanied him, says his biographer, were convinced that he must have discovered a vein of silver or gold."

There are various other objects interesting to the geologist on this track through the property of the Duke of Atholl. We understand that when Agassiz was last in this country, he accompanied to the locality an Edinburgh professor, well known both in the worlds of letters and of science, with the intention of visiting a quarry on the grounds of his Grace; but, on addressing his Grace for permission, there was no answer returned to his letter, and the distinguished foreigner had to turn back disappointed, to say how much more liberally he had been dealt with elsewhere, and to contrast, not very favorably for our country, the portion of *liberty* doled out to even the learned and celebrated among the Scottish people, with that enjoyed under the comparably free and kindly despotisms of the Continent. The incident happily illustrates the taste and understanding of his Grace the Duke of Atholl, and intimates the kind of measures which the public should keep with such a man. If the Scottish people yield up to his Grace their right of way through Glen Tilt, they will richly deserve to be shut out of their country altogether; and be it remarked, that to this state of things matters are fast coming with regard to the Scottish Highlands. It is said of one of the Queens of England, that in a moment of irritation she threatened to make Scotland a hunting-park; and we know that the tyranny of the Norman Conqueror did actually produce such a result over extensive tracts of England. The formation of the New Forest is instanced by all our historians as one of the most despotic acts of a foreign conqueror. William, in order to indulge his tastes as a huntsman, depopulated the country, and barred out the human foot from an extent, says Hume, of more than thirty miles. It is to this act of despotism, and its consequences, that the master-poet of the times of Queen Anne refers in his exquisite description:

> "The land appeared in ages past
> A dreary desert and a gloomy waste,
> To savage beasts and savage laws a prey,
> And kings more furious and severe than they,
> Who claimed the skies, dispeopled air and floods,—
> The lonely lords of empty wilds and woods."

The pleasures of the chase are necessarily jealous and unsocial. The shepherd can carry on his useful profession without quarrel with the chance traveller; the agriculturist in an open country has merely to fence against the encroachments of the vagrant foot the patches actually under cultivation at the time; whereas it is the tendency of the huntsman possessed of the necessary power, to "empty" the "wilds and woods." of their human inhabitants. The traveller he regards as a rival or an enemy: he looks upon him as come to lessen his sport, either by sharing in it or by disturbing it; and so, when he can, he reigns, according to the poet, a "lonely lord," and the country spreads out around him, as in the days of the Conqueror, "a dreary desert and a gloomy waste." And into this state of savage nature and jealous appropriation — characteristic, in the sister kingdom, of the times of the Conquest — many districts in the Highlands of Scotland are fast passing. The great sheep-farms were permitted, in the first instance, to swallow up the old agricultural holdings; and now the let shootings and game-parks are fast swallowing up the great sheep-farms. The ancient inhabitants were cleared off, in the first process, to make way for the sheep; and now the people of Scotland generally are to be shut out from these vast tracts, lest they should disturb the game. There is no exception to be made by cat-witted dukes and illiterate lords in favor of the man of letters, however elegant his tastes and pursuits; or the man of science, however profound his talents and acquirements, or however important the objects to which he is applying them. The Duke of Leeds has already shut up the Grampians, and the Duke of Atholl has *tabooed* Glen Tilt. The gentleman and scholar who, in quest of knowledge, and on the strength of the prescriptive right enjoyed from time immemorial by even the humblest of the people, enters these districts, finds himself subjected to insult and injury; and should the evil be suffered to go on unchecked, we shall by and by see the most interesting portions of our country barred up against us by parishes and counties. If one proprietor shut up Glen Tilt, why may not a combination of proprietors shut up Perthshire? Or if one sporting tenant bar against us the Grampians, why, when the system of shooting-farms and game-parks has become completed, might not the sporting tenants united shut up against us the entire Highlands? They would be prevented, it may be said, by certain rights of roadway. No; these rights of roadway as certainly exist in the case of Glen Tilt and the Grampians as over the Highlands generally.

Regretting, as we do, that a gentleman and scholar, with his friends, of character resembling his own, should have been subjected to unworthy treatment, we yet deem it fortunate that it should have fallen rather on men such as he and they than on some party of humble individuals, possessed of no adequate means of making their case known, or of attracting for it any general sympathy even if they had. Were, however, the party of humble men to be very numerous, — some such pleasure party as occasionally, in these days, sets out from Edinburgh for Berwick, Glasgow, or the land of Burns, — we could afford to wish them substituted for the naturalist and the professor. There is, we repeat, a right of roadway through Glen Tilt: the Duke of Atholl is quite at liberty to challenge that privilege in a court of law; but he has no right whatever violently to arrest travellers on the public way; and all good subjects, when the policeman or the soldier is not at hand to protect them, in the name and authority of the civil magistrate, from illegal violence, have a right to protect themselves. And we are pretty sure a few scores of our working men could defend themselves very admirably amid the solitudes of Glen Tilt, even though assailed by the Knight of the Gael and all his esquires. As the case chanced, however, it is well that a learned professor and a party of amateur naturalists should have been the sufferers. We may just mention in the passing, as a curious coincidence, that the professor in question is one of the nearest living relatives of the philosophic Hutton, who sixty-two years ago rendered Glen Tilt so famous; the professor's father is, we understand, the philosopher's *nearest* living relative. We trust to see the country roused all the sooner and the more widely in consequence of the character of the outrage, to assert for the people a right to walk over the country's area, — to share in that cheap enjoyment of the beauties of its scenery which softens and humanizes the heart, — and to trace unchallenged, amid its wild moors, on its lonely hilltops, or in the rigid folds of its strata, those revelations of the All-wise Designer which serve both to expand the imagination and to exercise the understanding. Not merely the rights of the poor man, but the privileges of the man of literature, and the interests of the man of science, are involved in this question, — those rights, interests, and privileges which the true aristocracy of the country have ever been the first to recognize. Our better proprietors have often admitted where they might have excluded, — never excluded where they ought to have admitted; and the experience of our men of literature and of science, save in those singularly rare instances in which they come in contact with men of the peculiar mental cast of the Duke of Atholl, has been invariably that of Cowper's in the park of the Throckmortons: —

> "The folded gates would bar my progress now,
> But that the lord of this enclosed demesne,
> Communicative of the good he owns,
> Admits me to a share: the guiltless eye
> Commits no wrong, nor wastes what it enjoys."

We will not instance the case of Sir Walter Scott, nor say how, in his kindliness of heart, he flung open the grounds of Abbotsford to his humbler neighbors, without ever finding occasion to repent his liberality; for Sir Walter was no ordinary proprietor, nor, perhaps, was he dealt with in this matter according to the ordinary experience of the class. The passage, however, in which, in his "Letters of Malachi Malagrowther," he sums up some of the balancing advantages which make up to the poor Highlander for the general hardness of his lot, seems so entirely to our purpose that we cannot forbear reference at least to it. Taken in connection with the shutting up of Glen Tilt and the Grampians, it forms a piece of peculiarly exquisite irony: —

"The inhabitants of the wilder districts in Scotland," says Sir Walter, "have actually some enjoyments, both moral and physical, which compensate for the want of better subsistence and more comfortable lodging. In a word, they have more liberty than the inhabitants of the richer soil. Englishmen will start at this as a paradox; but it is very true, notwithstanding, that if one great privilege of liberty be the power of going where a man pleases, the Scottish peasant enjoys it much more than the English. The pleasure of viewing 'fair nature's face,' and a great many other primitive enjoyments, for which a better diet and lodging are but indifferent substitutes, are more within the power of the poor man in Scotland than in the sister country. A Scottish gentleman in the wilder districts is seldom severe in excluding his poor neighbors from his grounds; and I have known many that have voluntarily thrown them open to all quiet and decent persons who wish to enjoy them. The game of such liberal proprietors, their plantations, their fences, and all that is apt to suffer from intruders, have, I have observed, been better protected than when severe measures of general seclusion were adopted. But in many districts the part of the soil which, with the utmost stretch of appropriation, the first-born of Egypt can set apart for his own exclusive use, bears a small proportion indeed to the uncultivated wastes. The step of the mountaineer on his wild heath, solitary mountains, and beside his far-spread lake, is more free than that which is confined to a dusty turnpike, and warned from casual deviations by advertisements, which menace the summary vindication of the proprietor's monoply of his extensive park by spring-guns or man-traps, or the more protracted, yet scarce less formidable, denunciation of what is often, and scarce unjustly, spelled '*persecution* according to law.' Above all, the peasant lives and dies, as his father did, in the cot where he was born, without ever experiencing the horrors of a workhouse. This may compensate for the want of much beef, beer, and pudding, in those to whom habit has not made this diet indispensable."

"Give us a good trespass act," say some of our proprietors, "and we care not though you abolish the game-laws to-morrow." The country sees in the affair of Gen Tilt and the Grampians what a good trespass act means, and has fair warning to avoid effecting the work of abolition — for effected it will be — in a careless and slovenly style, that might result ultimately in but shutting the Scotch out of Scotland. We trust, meanwhile, that the rencounter of the Duke of Atholl with the Edinburgh professor will not be unproductive of consequences. The general question could not be fought on more advantageous ground; and at least nineteen twentieths of the population of the kingdom have an interest in taking part in it, and fighting it out. There already exists in Edinburgh a "Footpath Society;" and we think the country could not do better than make the Society the nucleus of a great league, and, in the case of the professor, bring his Grace the Duke into court. By scarce any other means, in times like the present, can the rights of the people be asserted. Combination and a general fund formed the policy of Cobden and of O'Connell, and of a greater than either, — Thomas Chalmers; and only through combination and a common fund can our country be now preserved to its people from the ungenerous and narrow-minded aggressions of Dukes of Atholl and of Leeds.

XVI.

EDINBURGH AN AGE AGO.

Edinburgh for about a hundred and thirty years after the Union continued to be in effect, and not in name merely, the capital of a kingdom, and occupied a place in the eye of the world scarcely second to that of London. In population and wealth it stood not higher than the third-class towns of England ; it had no commerce, and very little trade, nor did it form a great agricultural centre ; and as for the few members of the national aristocracy that continued to make it their home after the disappearance of its parliament, they were not rich, and they were not influential, and added to neither its importance nor its celebrity. The high place which Edinburgh held among the cities of the earth it owed exclusively to the intellectual standing and high literary ability of a few distinguished citizens, who were able to do for it greatly more in the eye of Europe than had been done by its court and parliament, or than could have been done through any other agency, by the capital of a small and poor country, peopled by but a handful of men. Ireland produced many famous orators, shrewd statesmen, and great authors ; but they did comparatively little for Dublin, even previous to the Union. With the writings of Swift, Goldsmith, Burke, Sheridan, and Thomas Moore before us, we can point to only one work which continues to live in English literature, — "The Draper's Letters," — that issued originally from the Dublin press. London drew to itself the literary ability of Ireland, and absorbed and assimilated it just as it did a portion of that of Scotland represented by the Burnets, Thomsons, Armstrongs, Arbuthnots, Meikles, and Smolletts of the three last ages; and in London the Irish became simply Britons, and served to swell the general stream of British literature. But Scotland retained not a few of her most characteristic authors; and her capital — in many respects less considerable than Dublin — formed a great literary mart, second at one time, in the importance and enduring character of the works it produced, to no other in the world. Nothing, however, can be more evident than that this state of things is passing away. During the last quarter of a century one distinguished name after another has been withdrawn by death from that second great constellation of Scotchmen resident in Edinburgh to which Chalmers, Sir Walter Scott, and Lord Jeffrey belonged; and with Sir William Hamilton the last of the group may be said to have disappeared. For the future, Edinburgh bids fair to take its place simply among the greater provincial towns of the empire; and it seems but natural to look upon her departing glory with a sigh, and to luxuriate in recollection over the times when she stood highest in the intellectual scale, and possessed an influence over opinion coextensive with civilized man.

We have been led into this train by the perusal of one of the most interesting volumes which has issued from the Scottish press for several years, — "Memorials of his Time; by Henry Cockburn." Lord Cockburn came into life just in time to occupy the most interesting point possible as an observer. He was born nearly a year before Chalmers, only eight years after Scott, and about fourteen years before Lockhart. The place he occupied in that second group of eminent men to which the capital of Scotland owed its glory was thus, chronologically, nearly a middle place, and the best conceivable for observation. He was in time too to see, at least as a boy, most of the earlier group. The greatest of their number, Hume, had indeed passed from off the stage, but almost all the others still lived. Home, Robertson, Blair, Henry, were flourishing in green old age, at a time when he had shot up into curious, observant boyhood; and Mackenzie and Dugald Stewart were still in but middle life. It is perhaps beyond the reach of philosophy to assign adequate reasons for the appearance at one period rather than another of groups of great men. We know not why the reign of Elizabeth should have had its family of giants, — its Shakspeare, Spencer, Raleigh, and Bacon; or why a Milton, Hampden, and Cromwell should have arisen together during the middle of the following century; and that after theirtime only men of a lower stature, though of exquisite proportions, should have come into existence, to flourish as the wits of Queen Anne. Nor can it be told why the Humes, Robertsons, and Adam Smiths should have appeared in Scotland together in one splendid group, to give place to another group scarce less brilliant, though in a different way. We only know, that among a people of such intellectual activity as the Scotch, a literary development of the national mind might have been expected much about the earlier time. The persecutions and troubles of the seventeenth century had terminated with the Revolution; the intellect of the country, overlaid for nearly a hundred years, had been set free, and required only a fitting vehicle in which to address that extended public to which the Union had taught our countrymen to look; but for more than thirty years the necessary vehicle was wanting. Scotchmen bred in Scotland had great difficulty in mastering that essentially foreign language the English; and not until the appearance of Hume's first work in 1738 was there an English book produced by a Scotchman, within the limits of the country, which Englishmen could recognize as really written in their own tongue. But the necessary mastery of the language once acquired, it was an inevitable consequence of the native mass and quality of the Scottish mind that it should make itself felt in British literature; though, of course, why it should have given to Britain at nearly the same time its two greatest historians, its first and greatest political economist, and a philosophy destined to be known as peculiarly the Scotch philosophy all over the world, cannot, of course, be so readily shown.

It is greatly easier to say why such talent should have found a permanent centre in Edinburgh. Simple as it may seem, the prescriptive right of the capital to draft to its pulpits the *elite* of the Established clergy did more for it than almost aught else. Robertson the historian had been minister of Gladsmuir; Henry the historian, minister of a Presbyterian congregation in Berwick; Hugh Blair, minister of Collessie; Finlayson, so distinguished at one time for his sermons, and a meritorious Logic Professor in the University, had been minister of Borthwick; Macknight, the harmonist of the Gospels, minister of Jedburgh; and Dr. John Erskine, minister of Kirkintilloch. But after they had succeeded in making themselves known by their writings, they were all concentrated in Edinburgh, with not a few other able and brilliant men; and in an age in which the Scottish clergy, whatever might be their merely professional merits as a class, were perhaps the most literary in Europe, such a privilege could not fail to reflect much honor on the favored city for whose special benefit it was exerted. The University, too, was singularly fortunate in its professors, in especial in its school of anatomy and medicine, long maintained in high repute by the Monroes, Cullens, and Gregories, and which reckoned among its offshoots, though they concentrated their energies rather on physical and natural than on medical science, men such as Hutton and Black. In mathematics it had boasted in succession of a David Gregory and Colin Maclaren, both friends and *protegés* of Sir Isaac Newton; and in later times, of a Matthew Stewart, John Playfair, and Sir John Leslie. Both these last, with their predecessor Robinson, had also rendered its chair of natural philosophy a very celebrated one; and of its moral science, it must be enough to say that its metaphysical chair was filled in succession by Dr. Adam Ferguson, Dugald Stewart, Thomas Brown, and latterly by the brilliant Wilson, who, if less distinguished than his predecessors in the walks of abstract thought, more than equalled them in genius, and in his influence over the general literature of the age. Such men are the gifts of Providence to a country, and cannot be produced at any given time on the ordinary principle of demand and supply. But even when they exist, they may be kept out of their proper places by an ill-exercised patronage; and it must be conceded to the old close corporation of Edinburgh, that in the main it exercised its patronage with great discrimination, and for the best interests of the city. It was of signal advantage that the established religion of the country was numerically and politically so strong at the time that the disturbing element of denominational jealousy could have no existence in the body; and, influenced and directed by the general intellect of the city, its choice fell on the best possible men, whether Episcopalian or Presbyterian, that lay within its reach. Further, the legal profession contributed largely to the earlier intellectual glory of Edinburgh. Kames was one of its first cultivators of letters on the English model. Monboddo, with all his vagaries a very superior man and very vigorous writer, belonged to the same class. Mackenzie, though in a different walk and of a later time, belonged also to the legal profession. Almost all the contributors to the two periodicals which he edited in succession — the "Mirror" and "Lounger" — were also lawyers. And in Edinburgh's second intellectual group the legal faculty greatly predominated. Scott, Wilson, Lockhart, were all, at least nominally, of the faculty; and the editor of the "Edinburgh Review," with its most vigorous contributors, were, even when they wrote most largely for its pages, busied with the toils of the bar. Such were the elements of that intellectual greatness of the Scottish capital which gave it so high a place among the cities of the world. How have they now so signally failed to keep up the old supply?

It would of course be as idle to inquire why Edinburgh

has at the present time no Scotts, Humes, or Chalmerses, as to inquire why Britain has no Shakspeares, Newtons, or Miltons. Such men always rank among the rarest productions of nature; and centuries elapse in the history of even learned and ingenious nations in which there appear none so large of calibre or so various of faculty. Further, it must be confessed that both the bar and the university have in a very considerable degree come under that law of paroxysm which leaves occasional blank spaces in the production of men of a high class, and the equally obvious law that gives to a highly cultivated age like the present great abundance everywhere of men of mere talent and accomplishment. Aberdeen, Glasgow, and the great second-classs towns of England, are all, from this double circumstance of a lack of the highest men and a great abundance of men of the subordinate class, much nearer the level of Edinburgh than they were only a quarter of a century ago, when Scott and Jeffrey might be seen every day in term-time at the Parliament House, and Chalmers, Wilson, and Sir William Hamilton lectured in the University. That change, too, which has passed over the pervading literature of the age, and given a first place to the daily newspaper, and only a second place to the bulky quarterly, has of necessity militated against the capital of a small country whose most successful newspapers must content themselves with a circulation of but from two to three thousand. For the highest periodic literature London has, of consequence, become the only true mart; and the Scotchman who would live by it must of necessity make the great metropolis his home. Yet further, the source whence Edinburgh derived so much of at least her earlier halo of glory can scarce be said any longer to exist. Edinburgh has still the old privilege of drafting to her established churches the *elite* of the body that can alone legally occupy them; but that great revolution in matters ecclesiastical which has rendered the abolition of the tests so essential to the efficient maintenance of the educational institutions of the nation, has manifested itself within the pale of the Establishment; and we suppose there is no one who will now contend that aught of the old ability is to be derived from this privilege. We have before us a bulky volume, entitled "Men of the Time," which, with its biographic notices of only the living, forms a sort of supplement to those ordinary works of biography which record the names of only the dead. All the men whose names it records have made themselves known in the worlds of thought or action. There are, no doubt, omissions of names that ought to have found a place in it, and some of the names which it records might well have been omitted; but it is an English, not a Scotch publication; it does not seem to have been got up for any party purpose, — certainly not for any party purpose of the Free Church; and its evidence, positive and negative, on a question like the present, may, we think, be safely received. And while we find in this volume at least three names of Edinburgh ministers who were brought into the place previous to the Disruption through the exercise of the old privilege, but who quitted the Establishment on the Disruption, we do not find in it the name of a single minister who now occupies any of the city churches.

In that altered state of things to which we refer, Edinburgh must of course acquiesce with the best grace it can. It seems greatly less to be wondered at that such a fate should overtake it now than that it should not have overtaken it earlier. There are two circumstances on which the great interest of Lord Cockburn's "Memorials" seems to depend, independently of the very pleasing manner in which the work is written. The recollection of two such groups of men as for a whole century gave celebrity to a nation, could scarce fail to secure perusal, from the interest which ever attaches to the slightest personal traits or peculiarities of men of fine genius or high talents. We read the lives of poets and philosophers, not for the striking points of the stories which they embody, — for striking points there may be none, — but simply for the sake of the men themselves. We also feel a natural interest in acquainting ourselves with the strongly-marked manners and broadly-defined characters of comparatively rude and simple ages, and seek to derive our amusement rather from the well drawn-portraits of men who bear all the natural lineaments than from the masked and muffled men of a more polished time. No small portion of the amusement we derive from the glowing fictions of Scott results from the well-drawn *manners* of ages a century or two in advance of our own. And in Lord Cockburn's "Memorials" we have both elements of interest united. In Scotland, as in several other countries of northern Europe, the intellectual development of the leading minds preceded the general development of even the upper classes in the politenesses and amenities. Macaulay, in describing the mental standing of Scotland at the time when the accession of James VI. to the throne of Elizabeth virtually united it to England, remarks, that, though it was the poorest kingdom in Christendom, it already vied in every branch of learning with the most favored countries. "Scotsmen," he adds, "whose dwellings and whose food were as wretched as those of the Icelanders of our time, wrote Latin verses with more than the delicacy of Vida, and made discoveries in science which would have added to the renown of Galileo." High intellectual cultivation and great simplicity, nay, rudeness, of manners, with an entire unacquaintance with what are now the common arts of life, existed in the same race, and, though the conventionalisms gained ground as the years passed by, continued to do so till at least the commencement of the present century. Not a few of the best writers and most vigorous thinkers Britain ever produced bore about them all the sharp-edged angularity of that early state of society in which every individual, instead of being smoothed down to a common mediocre standard, carries about him, like an unworn medal, the original impress stamped upon him by nature; and they were thus not only interesting as men of large calibre, but also as the curious *characters* of a primitive age. We have not only no such writers or thinkers now as Hume, Robertson, Kames, and Adam Smith, but no such *characters*. In some respects, however, society seems to have improved in well-nigh the degree in which it has become less picturesque. Lockhart remarks, in his "Life of Burns," that there was at least one class with which the poet came in contact in Edinburgh, that, unlike its clerical literati, were "shocked by his rudeness or alarmed by his wit." He adds, that among the lawyers of that age "wine-bibbing and the principle of jollity was indeed in its high and palmy state; and that the poet partook largely in those tavern scenes of audacious hilarity, which then soothed, as a matter of course, the arid labors of the northern *noblesse de la robe*." And then he goes on to show that there is too much reason to fear that Burns, who had tasted but rarely of such excesses in Ayrshire, caught harm from his new companions, and became nearly as lax in his habits, and nearly as reprehensible in his morals, as most respectable judges of the Supreme Court and influential elders of the General Assembly. And the work before us shows how very much may be involved in the remark. Certainly, if Burns ever drank half so hard as some of the leading lawyer elders, who, laudably alarmed lest the foundations of our faith should be undermined by the metaphysics of Sir John Leslie, took most decided part against the appointment of that philosopher, he must have been nearly as bad as he has been represented by his severer censors. The late Lord Hermand may be regarded as no unmeet representative of the class.

"He had acted," says Lord Cockburn, — his nephew, by the way, — "in more of the severest scenes of old Scotch drinking than any man at least living. Commonplace topers think drinking a pleasure; but with Hermand it was a virtue. It inspired the excitement by which he was elevated, and the discursive jollity which he loved to promote. But beyond these ordinary attractions, he had a sincere respect for drinking: indeed, a high moral approbation, and a serious compassion for the poor wretches who could not indulge in it; but due contempt of those who could but did not. He groaned over the gradual disappearance of the *Fineat* days of periodical festivity, and prolonged the observance, like a hero fighting amidst his fallen friends, as long as he could. The worship of Bacchus, which softened his own heart, and seemed to him to soften the hearts of his companions, was a sacred duty..... No carouse ever injured his health...... Two young gentlemen, great friends, went together to the theatre in Glasgow, supped at the lodgings of one of them, and passed a whole summer night over their punch. In the morning a kindly wrangle broke out about their separating or not separating, when, by some rashness, if not accident, one of them was stabbed, not violently, but in so vital a part that he died on the spot. The survivor was tried at Edinburgh, and was convicted of culpable homicide. It was one of the sad cases where the legal guilt was greater than the moral, and, very properly, he was sentenced to only a short imprisonment. Hermand, who felt that discredit had been brought on the cause of drinking, had no sympathy with the tenderness of his temperate brethren, and was vehement for transportation. 'We are told that there was no malace, and that the prisoner must have been in liquor. In liquor! Why, he was drunk! And yet he murdered the very man that had been drinking with him! They had been carousing the whole night; and yet he stabbed him, after drinking a whole bottle of rum with him! Good God! my laards, if he will do this when he is drunk, what will he do when he is sober!'"

As an elder this worthy representative of the old school was no less extraordinary than as a judge. The humor of Goldsmith has been described as hurrying him into mere unnatural farce when he describes his incarcerated debtor as remarking from his prison, in the prospect of a Gallican invasion,—"The greatest of my apprehensions is for our freedom!" and the profane soldier, very much a Protestant, as chiming in with the exclamation, "May the devil sink me into flames, if the French should come over, but our religion would be utterly undone." But from the real history of Lord Hermand similar examples might be gleaned —quite extreme enough to justify Goldsmith. We find

Lord Cockburn thus describing his zeal for what he deemed sound views, in the famous Sir John Leslie case: —

"Hermand was in a glorious frenzy. Spurning all unfairness, a religious doubt, entangled with mystical metaphysics, and countenanced by his party, had great attractions for his excitable head and Presbyterian taste. What a figure, as he stood on the floor declaiming and screaming amidst the divines! — the tall man, with his thin powdered locks and long pigtail, the long Court-of-Session cravat flaccid and streaming with the heat and the obtrusive linen! The published report makes him declare that the 'belief of the being and perfections of the Deity is the solace and delight of my life.' But this would not have been half intense enough for Hermand; and accordingly his words were, '*Sir, I have sucked in the being and attributes of God with my mother's milk.*' His constant and affectionate reverence for his mother exceeded the devotion of any Indian for his idol; and under the feeling, he amazed the house by maintaining (which was his real opinion) that there was no apology for infidelity, or even for religious doubt, because no good or sensible man had anything to do except to be of the religion of his mother, which, be it what it might, was always best. 'A sceptic, Sir, I hate! — with my whole heart I detest him! But, Moderator, I love a Turk.'"

Such was one of the *characters* of Edinburgh not more than half a century ago; and yet he belongs as entirely to an extinct state of things as the oldest fossils of the geologist. And there are many such in this volume, drawn with all the breadth, and in some instances all the picturesque effect, of the best days of the drama. But, though a thoroughly amusing volume, it is also something greatly better; and there is, we doubt not, a time coming when the student of history will look to it, much rather than to works professedly historic, for the true portraiture of Edinburgh society during the periods in which it maintained its place most efficiently in the worlds of literature and of science. And yet, as may be seen from the sketch just given, all was not admirable in the ages in which our capital excited admiration most; and we must just console ourselves by the reflection, that, though we live in a more mediocre time, it is in the main a more quietly respectable one.

XVII.

BURNS FESTIVAL AND HERO WORSHIP.

"The Burns Festival," writes a respected correspondent in the west, in whose veins flows the blood of Gilbert Burns, "is already well-nigh forgotten in Ayr." We are not at all sure that it ought to be forgotten so soon. Could we but look just a little below the surface of the event, with its checkered patchwork of the bizarre and the picturesque, and its, doubtless, much genuine enthusiasm, blent with at least an equal anount of overstrained and awkward simulation, we might possibly discover in it a lesson not unworthy of being remembered. Deep below the ridiculous gaud and glitter, we may find occult principles of our nature at work in this commemorative festival — principles which have been active throughout every period of the history of man, — which gave of old their hero gods to the Greek and the Roman, and the red-letter saint days to the calender of the Papist; and which in these latter times we may see scarce less active than ever in the worlds of politics and letters. We find them alike developed in the "hero-worship" of Carlyle, and the Pitt and Fox dinners and clubs of our politicians.

As a piece of mere show, the festival of Burns, like the tournament of Lord Eglinton, was singularly unhappy. Both got sadly draggled in the mud, and looked like bepowdered *beaus* who set out for the ballroom in their thin shoes and silk stockings, and are overwhelmed in a thunder-shower by the way. Serious earnest stands a ducking; mere show and make-believe becomes ridiculous in the wet. The 92d Highlanders were thoroughly respectable at Waterloo, though drenched to the skin; and we have seen from twelve to fifteen thousand of their devout countrymen gathered together amid their wild hills, in storm and rain, on a sacramental Sabbath, without appearing in the slightest degree contemptible. But alas for a draggled procession, or a festival first dressed up in gumflowers and then bespattered with mud! Processions and festivals cannot stand a wetting. Like some of the cheap stuffs — half whitening and starch — of the cotton-weaver, they want *body* for it. Their respectability is painfully dependent on the vicissitudes of the barometer. Every shower of rain converts itself into a jest at their expense, that turns the laugh against them, and every flying pellet of mud becomes a practical joke. And as the festival of Burns, like the tournament of Eglinton, got particularly wet, — wet till it streamed and smoked like a salt-pan, and the water that streamed downwards from its nape to its heels discharged the dye of its buckram inexpressibles on its white silk stockings, and flowed over the mouth of its thin-soled pumps, — it returned to its home in the evening, looking, it must be confessed, rather ludicrous than gay. It encountered the accident of being splashed and rained upon, and so turned out a failure. Nay, even previous to its mishap, there were visible in the getting-up of its scene-work certain awkward-looking strings and wires, that did not appear particularly respectable in the broad daylight. Its prepared lightning took the form of pounded rosin; and the mustard-mill destined to produce its thunder was suffered to obtrude itself all too palpably on the sight of the public. People remarked, that among the various toasts given at the banquet, there were no grateful compliments paid, no direct notice taken, of its first originators. No one thought of toasting them. They were found to compose part of the vulgar string-and-wire work, — part of the pounded rosin and mustard-mill portion of the exhibition; and so, according to the poet, —

> "What would offend the eye
> The painter threw discreetly into shade."

But it is well to remark that the Burns festival had an element of actual power and significancy in it, altogether separate and apart from the lowness of its immediate origin, the staring rawness of its rude machinery, or the woful ducking in which it made its ridiculous exit. It is significant that the mind of the country should exist in such a state in reference to the memory of the departed poet, that a few obscure men over their ale could have originated such a display. The call to celebrate by a festival the memory of Burns, seems, with reference to those from whom it first proceeded, to have been a low and vulgar call; but that it should have been responded to by thousands and tens of thousands, — that town and village should have poured forth their inhabitants to the spectacle, — that eminent men from remote parts of the country should have flocked to it, — are matters by no means vulgar or low. The surface of the pageant, like its origin, seems to have been a sufficiently poor affair; but underneath that surface there must have beat a living and vigorous heart, neither poor in its emotions nor yet uninteresting in its physiology.

We would recognize in it, first of all, the singularly powerful impression made by the character of Burns on the people of Scotland. The *man* Burns exists as a large idea in the national mind, altogether independent of his literary standing as the writer of what are preëminently the national songs. Our English neighbors, as a people at least, are much less literary than ourselves. The fame of their best writers has scarce at all reached the masses of their population. They know nothing of Addison with his exquisitely classic prose, or of Pope with his finished and pointed verse. We have been struck, however, by finding it remarked by an English writer, who lived long in London, and moved much among the common people, that he found in the popular mind well-marked though indistinct and exaggerated traces of at least one great English author. He could learn nothing, he observed, from the men who drove cabs and drays, of the wits and scholars of Queen Anne, or of the much greater literati of the previous century; nay, they seemed to know scarce anything of living genius; but they all possessed somehow an indistinct, shadowy notion of one Dr. Samuel Johnson, — a large, ill-dressed man, who was a great writer of they knew not what; and almost all of them could point out the various places in which he had lived, and the house in which he died. Altogether independently of his writings — for these are far from being of a popular cast — the doctor had made an impression by the sheer bulk and energy of his character; he loomed large and imposing simply as a man; an impression of the strange kingly power which he possessed, and before which his contemporaries, the Burkes and Reynolds and Charles James Foxes of the age, were content to bow acquiescent, had somehow reached the masses, and the lapse of two generations had failed to efface it. The only other man of whom the author of the remark found similar traces among the common people of London was not a writer at all; but he, too, far excelled his contemporaries in the kingly faculty, and stamped, not on the mind of his country alone, but on that of civilized man everywhere, the impress of his power. The men who carried about with them this curious shadowy idea of Johnson had an idea, also existing in exactly similar conditions, of one Oliver Cromwell, — an idea of some kind of undefinable greatness and power, not extrinsic and foreign to him, but inherent and self-derived, and before which all opposition was prostrated. The intrinsic weight of the two characters had sunk their impress deep into the popular memory. And on a similar principle has the popular memory been impressed in Scotland by the character of Robert Burns.

Scotland has produced many men eminent for literature and philosophy, — exquisite poets, like him who wrote the "Battle of the Baltic," and scholars of the highest reach, such as the author of the "Franciscan;" the history of Hume is still supereminently *the* English History; the novels of Scott are the most popular fictions ever produced in any age or nation; but the authors of these works, though great writers, were not properly great men. Some of them, on the contrary, were rather small men. Campbell was decidedly diminutive, maugre his fine genius and exquisite taste; Hume was merely a cold, though not ill-tempered sceptic, who enjoyed life at his leisure and grew fat; nor would Scott, though rather a happy-dispositioned, hospitable country gentleman, who made money and then lost it, have greatly shone as a hero in one of the dramas of Goethe or Shakspeare. But, altogether independently of his writings, the character of Burns, like that of Johnson, was one of great massiveness and power. There was a cast of true tragic greatness about it. There was a largeness in his heart, and a force in his passions, that corresponded with the mass of his intellect and the vigor of his genius. We receive just such an impression from reading his life as we do from perusing one of the greater tragedies of Shakspeare. Like the Othellos or Macbeths of the dramatist, — characters that fasten upon the imagination and sink into the memory from causes altogether unconnected with either literary taste or moral feeling, — we feel in him, per force, an interest which exists and grows alike independently of the excesses into which his passions betrayed him, or the trophies which his genius enabled him to erect. Burns was not merely a distinguished poet, — he was a man on a large scale; and the festival of the present month bore emphatic testimony to the fact.

It is not uninstructive to mark how this admiration of the merely great and imposing grows upon mankind, until at length, at the distance of an age or two, the departed great man reckons among his semi-worshippers individuals of not less calibre than himself. Burns — to borrow from Cowper's allusion to Garrick and Garrick's commemorative festival at Avon — "was himself a worshipper." "Man praises man!" The great hero of the poet was Robert the Bruce. He was selected by him to form the leading character in his projected drama; we find frequent allusions to him in his letters and journals; and the most spirit-stirring of all his songs is the address of the hero king to his troops at Bannockburn. Now we have seen faithful casts of the skulls of worshipper and worshipped resting side by side on the same shelf in a museum, and have been greatly struck by the fact that they should have existed in such a relation to each other. The worshipper, if there be a shadow of truth in the science that professes to draw conclusions from the material organ of mind regarding the energy and direction of the mind's immaterial workings, must have been altogether as powerful a man as the worshipped. In general size, the head of the indomitable king, who so strongly impressed his character on a rude and turbulent age, and the head of the not less indomitable peasant, who in an age of thinking men stamped the impress of *his* scarce less deeply, exactly resemble one another. They were heads of about the same bulk as the head of Dr. Chalmers. Both display great animal power. There is a towering organ of firmness in the head of the monarch which we miss in that of the poet, and larger developments of *caution* and *hope;* but in imagination, intellect, benevolence, the scale predominates greatly on the other side. In these — the *man-like* faculties — the worshipper was superior to his demi-god. And yet he was a worshipper. The felt influence of greatness, removed by distance, — that identical influence which a fortnight since drew so many thousands to the Burns festival, — had been operative on his imagination and his feelings: the departed hero loomed large and imposing through the magnifying fogs of the past; and the worshipper, though not greatly disposed to yield to contemporaries, and fully aware that he himself was no common man, never once suspected that the object of his worship was in the main not a greater man than himself, and in some respects an inferior one.

Could we but lay open the inner springs of this tendency to man-worship, they would enable us, we are convinced, to comprehend many a curious chapter in the early history of the species. Departed greatness, enveloped by its peculiar atmosphere of reverential respect and awe, and exaggerated by distance, is suffered to retain within the bright circle of its halo many an attendant littleness and impurity that contemporaries would have at least not admired. The greatness is doubtless the staple of the matter, — that which dazzles, impresses, attracts; and the littlenesses and impurities, mere accidents that have mixed with it; and yet how strange a tone do they not too frequently succeed in imparting to the worship! There was much of apology at the Burns festival for the errors of the poet; and it said at least something for the morals of the time, whatever it might for the taste of the speakers, that such should have been the case. In a remoter and more darkened age of the world, like those ages in which hero worship rose into religion, the errors would have been remembered, but the apology would have been wanting. Burns would have been deified into an Apollo, and his love passages with the nymphs Daphne, Levcothoe, and Coronis, and his drinking bouts with Admetus and Hyacinthus, would have been registered simply as incidents in his history, — incidents which in the course of time would have come to serve as precedents for his worshippers. We are afraid that, maugre regret and apology, there is too much of this as it is. His hapless errors, so fatal to himself, have been too often surveyed though the dazzling halo of his celebrity. The felt influence of his greatness has extended to his faults, as if they were part and parcel of that greatness. The atmosphere of the sun conceals the sun's spots from the unassisted eye of the observer; but the atmosphere of glory that surrounds the memory of Burns has not had a similar effect. To many at least it has the effect of making his blemishes appear less as original flaws than as a species of beauty-spots, of a fashion to be imitated. How can we marvel that the old worshippers of the offspring of Saturn or of Latona should have imitated their gods in their crimes, if in these our days of light, with the model of a perfect religion before our eyes, hero worship should be found to exert, as of old, a demoralizing tendency! But it would not be easy to say where more emphatic or more honest warning could be found on this head than in the writings of Burns himself. We stake his own deeply-mournful prediction of the fate which he saw awaiting him against all ever advanced on the opposite side: —

> "The poor inhabitant below
> Was quick to learn, and wise to know,
> And keenly felt the social glow
> And safter flame;
> But thoughtless follies laid him low,
> And stained his name."

Despite the authority of high names, we are no admirers of hero worship. We are not insensible to what we may term the natural claims of Burns on the admiration of his country, both as a writer and as a character of great bulk and power. It would be hypocrisy in us to say that we were. Were his writings to be annihilated to-morrow, we could restore from memory some of the best of them entire, and not a few of the more striking passages in many of the others. Nor are we unimpressed by the massiveness of his character as a man. We bear about with us an adequate idea of it, as developed in that deeply-mournful tragedy, his life. But we would not choose to go and worship at his festival. There was a hollowness about the ceremony, independently of the falseness of the principles on which its ritual was framed. Of the thousands who attended, how many, does the reader think, would have sympathized, had they seen the light some fifty years earlier, with the *man* Robert Burns? How many of them grappled in idea at his festival with other than a mere phantom of the imagination — a large but intangible shade, obscure and indefinable as that conjured up by the uninformed Londoner of Cromwell or of Johnson? Rather more than fifty years ago, the sinking sun shone brightly, one fine afternoon, on the stately tenements of Dumfries, and threw its slant rule of light athwart the principal street of the town. The shadows of the houses on the western side were stretched half way across the pavement; while on the side opposite, the red beam seemed as if sleeping on jutting irregular fronts and tall gables. There was a world of well-dressed company that evening in Dumfries; for the aristocracy of the adjacent country for twenty miles round had poured in to attend a county ball, and were fluttering in groups along the sunny side of the street, gay as butterflies. On the other side, in the shade, a solitary individual paced slowly along the pavement. Of the hundreds who fluttered past, no one took notice of him; no one seemed to recognize him. He was known to them all as the exciseman and poet Robert Burns; but he had offended the stately Toryism of the district by the freedom of his political creed; and so tainted by the plague of Liberalism, he lay under strict quarantine. He was shunned and neglected; for it was with the *man* Burns that these his contemporaries had to deal. Let the reader contrast with this truly melancholy scene the scene of his festival a fortnight since. Here are the speeches of the Earl of Eglinton and Sir John M'Neil, and here the toast of the Lord Justice General. Let us just imagine these gentlemen, with all their high aristocratic notions about them, carried back half a century into the past, and dropped down, on the sad evening to which we refer, in the main street of Dumfries. Which side, does the reader think, would they have chosen to walk upon?

Would they have addressed the one solitary individual in the shade, or not rather joined themselves to the gay groups in the sunshine who neglected and contemned him? They find it an easy matter to deal with the phantom idea of Burns now; how would they have dealt with the man then? How are they dealing with his poorer relatives; or how with men of kindred genius, their contemporaries? Alas! a moment's glance at such matters is sufficient to show how very unreal a thing a commemorative feast may be. Reality, even in idea, becomes a sort of Ithuriel spear to test it by. The Burns festival was but an idle show, at which players enacted their parts.

There is another score on which we dislike hero worship. We deem it a sad misapplication of an inherent disposition of the mind, imparted for the most solemnly important of purposes. "Man worships man," says Cowper. The tendency, either directly or in its effects, we find indicated in almost every page of the history of the species. We see it in every succeeding period, from its times of full development, when the men-gods of the Greek were worshipped by sacrifice and oblation, down to the times of the Shakspeare jubilee at Stratford-on-Avon, or the times of the Burns festival at Ayr. But the sentiment, thus active in expatiating in false direction, has a true direction in which to expatiate, and a worthy object on which to fix. As if to dash the dull and frigid dreams of the Socinian, the instinct of man worship may find a true man worthy the adoration of all, and who reigns over the nations as their God and king. Every other species of man worship is a robbery of Him. It is a worship that belongs of right to the man Christ Jesus alone, — the "God whose throne is for ever and ever," and whom "all the angels of God worship."

POLITICAL AND SOCIAL.

I.

OUR WORKING CLASSES.

NEVER in the history of the world have so many efforts been made to improve the condition of the working classes as at the present time. The legislator, the philanthropist, the city missionary, the theorist, who would do his best to uproot the very foundations of our social system, and the man of practice, who would spare no exertion to ameliorate its actual condition, have been at work, each in his several direction, honestly, earnestly, and unremittingly toiling to a single purpose, — the elevation of our working people. We have passed laws; we have devised model dwellings; we have sent pious men to hunt out ignorance and vice; we have schemed out theories that would mow down the institutions of ages; we have speculated in the direction of secular socialism and in the direction of Christian socialism; we have tried coöperative societies, building societies, and model lodgings; we have written, lectured, and taught; we have appointed commissions, printed acres of reports; pried into every hole and corner of society (except the convents); we have exported hundreds of thousands of what we termed, only a year or two back, our "surplus population;" we have raised wages, diminished competition, and founded magnificent colonies with those who were too many at home; we have done these and many other things; and what has been the result? Have we moved the living mass of our work-people a single step higher in the scale of moral existence? Have we taught them wisdom as well as knowledge? Have we taught them to be provident, and to manage their own affairs with prudence and discretion? Have we placed them in circumstances where they fulfil their duties as men? Have we, in fact, succeeded, after all our labors, in promoting the genuine welfare of the working population? To answer this question either with a summary affirmative or with an emphatic No, would be out of place. That all the expended labor has been wasted and thrown away, we cannot for a moment believe; but it is equally certain that the present condition of our working classes is preëminently unsatisfactory, and that no such general improvement has taken place as would entitle us to say that we had arrived at the true solution of this great social problem. Two things there are which, in every condition of life, mark the wellbeing of society; namely, the integrity of the family, and the sufficiency of the dwelling. The family is the foundation of everything, — the root out of which the social world grows. Break it up, and you have as certainly introduced a corrupting poison into the framework of the community, as if you had inoculated the human frame with a deadly and malignant agent that destroys the very issues of life. The whole of our factory system where women are employed is merely a systematic destruction of the family, — practical socialism, in fact, which prepares the way for theoretic socialism of the direst and most disastrous tendency, atheistic and material, without natural affection, without law, without order, without the thousand amenities of domestic life. It matters little whether the women are employed as married, or only as unmarried. If married women are engaged in factory works, they of course neglect their children, who, between the period of childhood and that of labor, have the education of the public streets, with its unconcealed vice, its oaths and curses, its idleness and its vagabondism. We have only to go into our streets in the lower quarters of any of our towns to be painfully assured that every one is a broad road to destruction for the young, and that no mere school-education can ever effectually compete with the force of evil habit, any more than wholesome food will effectually nourish those who dwell continually in a polluted atmosphere. We are all aware that the decent portion of our country population look with absolute horror on the habitual circumstances of a town life. And why so? Is it not because in their country dwellings they have been accustomed to the sacred integrity of the family, and that their isolated cottage was a *home*, containing father, mother, and children: God's first institute, — a family? The cottage may be small, ill-thatched, ill-ventilated, ill-floored, and smoky; it may have its dubs, its puddles, and its national *midden;* it may be high up on a hill, where winter blasts and winter snows are more familiar than blue skies or green fields; or it may be down in a glen, miles away from other mortal habitation, — so solitary that every stranger who appears is a spectacle and amazement to the children. No matter: wherever or whatever it may be, it is a home, and contains a family, every member of which would look with instinctive horror at the indiscriminate sort of existence common in many of our towns. Thanks to the bothy system, however, this feeling of family sacredness is beginning to be eradicated out of even our rural population; and perhaps in time a certain portion of our peasantry may be duly brought to believe that the family is a superfluous invention, after which they will be fit for anything, and good for nothing. The same principle pervades every rank of society, high or low. Where the family is broken up, — whether from what are termed the necessities of trade, from polygamous customs, from fashionable usages, or from particular accidents, — evil follows as a regular and constant effect. Of all the social laws that have ever been discovered, this is the most indisputably certain, that the family is an institution of nature, an organized association established immutably by God's providence for the welfare of mankind. What, we ask, is it that has made the most powerful monarch in the world the most universally and enthusiastically popular among her subjects? It is neither her power nor her possession of the imperial throne. It is the splendor of the wife and mother, beaming with a light far brighter than a kohinoor, and carrying to every subject-land and to every subject-household the royal proclamation that the family is respected by the throne, and that monarchs themselves may find their truest happiness in those institutes of God which are common to the humblest household that obeys their sway. The preservation of the family in its full integrity we regard as the first absolute requisite, without which there can be no permanent improvement, and without which all efforts to ameliorate the condition of our working classes must certainly fail.

Next to the family comes the dwelling. As dress is the clothing of the individual, so is the house the clothing of the family. It ought to be sufficient, — sufficient for all the purposes of family life, — for decency, for convenience, for warmth, for shelter, for washing and cooking, for retirement, and for the separation of the sexes. Here society has failed. It is idle to speak of sanitary reform, and almost idle to speak of moral reform, when we comtemplate the dwellings of a large portion of the working population. We can no more expect propriety of conduct in the individual if we clothe him in rags, and keep him in rags, than we can expect propriety of conduct in a family that lives habitually in the wretched lodgments which disgrace our towns and cities. For our towns, however, there is some excuse. They have increased so rapidly in population, that the supply of house-room did not, and could not, under the ordinary course of private speculation, equal the demand. When Ireland was pouring her thousands into Glasgow, and the Highlands were undergoing the process of clearing, it could not be expected that more than the very meanest accommodation should be obtained by such a class. The past must be palliated; but now that the pressure is in a great measure over, and a breathing-time is afforded by the stream of emigration setting no longer from the country to the town, but out of the kingdom to the colonies and the United States, we can conceive no object on which society may more profitably fix its attention than on the systematic improvement of the dwellings of the industrial classes. A universal crusade against every tenement that did not afford the proper requisites of domestic life would be at least one step towards the desirable result. But this would be insufficient. It would be only a negative reform, and all negative reforms are insufficient. It would be only cutting off the evil, whereas the true object is to produce the good. If we were to pull down every tenement that did not fulfil even the moderate conditions that would in all probability be fixed by the Government, we should only have rendered our working people houseless. We must devise some plan by which proper buildings shall be erected, and insure the future wellbeing of the people by a systematic scheme, that could not legally be departed from within the limits of any town containing a given number of inhabitants. What is already evil we must reform as best we may; but what is future we ought intelligently to design — to leave nothing to accident, nothing to the hazards of avaricious speculation; but, duly considering what is needed, to provide for it beforehand with a wise precaution, which in course of time would react powerfully on the whole habits and manners of the laboring community.

We believe, however, that we have reached a turning-point in our downward course, — that we have passed the worst, — and that there is, both in the legislature and in society at large, a very general desire to favor the requisite improvements, provided it could be clearly shown what the improvements should consist of, and upon what principle they should be undertaken. When we find men like the Duke of Buccleuch candidly confessing — to his honor be it spoken — that he had done wrong in so long neglecting the dwellings of the smaller tenantry, cottars, and bothy-men, — when we find Mr. Stuart of Oathlaw succeeding in banding together some of the most influential and extensive landed proprietors for the purpose of improving the dwellings in the country districts, — and when we find the Rev. Mr. Mackenzie of North Leith only stopped in his career of practical benevolence by the absurd and antiquated usages of feudal lawyerism, — we are not without ground for hope that a general movement may be made at no very distant period, and that we may see model towns not only projected, but actually erected, inhabited, and in vital operation. Without the integrity of the family and the sufficiency of the dwelling there can be no satisfactory reform, either in a sanitary or a moral aspect; and we propose in a future article to discuss some of the main causes that have led to the present condition of our working population. We propose to inquire whether, and in what circumstances, the laboring agriculturist or artisan might profitably be the proprietor of his dwelling, and how far the acquisition of real property might operate as a check on the habitual improvidence that is proven to exist. Among all the experiments that have been made, at least in this country, it is plainly evident that a vast field, and *that* certainly not the most unpromising, has been left untouched and unexplored. To promote the habit of providence in our working classes, it is not only necessary to exhibit a moral restriction which cautions them from going wrong, but to present a positive stimulus which induces them to go right, — to exhibit something good before their eyes, after which they shall strive, — and to make them act of their own free will, as if they had an object to attain. This stimulus may possibly be found in the desire to possess real property; and although no mere change of laws or circumstances may ever do more than facilitate the progress of good, it is quite possible that a change of circumstances might eminently promote a change of habits, and lead gradually but surely to a more enlightened appreciation of the advantages that might accrue if the present recklessness and extravagance were exchanged for prudence and economy.

II.

PEASANT PROPERTIES.

In our present observations on peasant properties, we do not intend to inquire into the ethics of the question. We do not ask whether it was morally right or morally wrong for England to pursue that vast system of inclosure by which the English peasantry were permanently ejected from their commons, and deprived of their prescriptive rights; or whether it was right or wrong for the legislature and the Highland proprietors to convert, by a fiction of law, what was once to all intents and purposes the property of the clans into the private domains of individual landlords, thereby disinheriting all save the chief and his family. These questions are practically settled; the facts are achieved; society has accepted them; and it is now useless to speculate on what might have been the result if a different principle had pervaded the arrangements. Within a century and a half a vast revolution has been wrought in the occupation of the lands both of England and Scotland. By the inclosure of the commons, about five thousand parishes, constituting nearly a half of the soil of England, were subjected to a legal process which severed the peasant from all direct interest in the land, and left it ultimately in the hands of large proprietors. And by the introduction of the English doctrine of property into the Highlands, the old system of customary occupation was entirely superseded, and a new system substituted which threw vast territories into the absolute control of single individuals, who had previously been only the representatives of their tribe, and who had held the lands not as their own, but in virtue of their office as chiefs or petty sovereigns, who ruled over a given district, and administered the public affairs of the clan. These measures have produced a radical change in the whole structure of society. The first, by leading to the absorption of the smaller properties, abolished the English yeoman; and the second bids fair to abolish the Highland population. Both measures had essentially the same result in one respect, — essentially a different result in another. They both left a country population composed of a very small number of great landed proprietors, surrounded by a dependent and almost subject tenantry, outside of which remained the mass of those who live by labor alone, who have been cast loose from all interest in the soil, and who are regarded as machines for the execution of work. In this respect the results have been similar in the two countries. But a very striking difference presents itself to view when we turn our attention to the soil itself, and ask how it has been affected by the change. In England the pretext for the inclosure of the commons was, that the land was uncultivated, and to a great extent unproductive. This was actually true, and, being so, it was a good and sufficient reason for the introduction of some new system by which the lands should be brought into cultivation. Still, even supposing that the produce after the inclosure was five or ten times greater than before, it was more advantageous to the peasantry — that is, to the great body of the rural population — to have only the fifth or the tenth as their own than to be deprived of it altogether, and to see ten times the produce passing into the hands of the great landlords and great agriculturists. The lands, however, *were* cultivated, and the produce was obtained; so that although the English peasant was ousted from his common rights, the land was turned to its proper agricultural use, and grew corn for the service of the nation. The landlords and farmers acquired wealth, the peasants went on the parish, and were supported by the parish rates. In Scotland the effect has been entirely of an opposite character. The lands, instead of being brought *into* cultivation, have been thrown *out* of cultivation. The cottage and the croft have been *herried* to make way for grouse and deer; and, so far as the production of food is concerned, — food available for the ordinary purposes of life, — hundreds of thousands of acres that once grew and supported soldiers second to none who ever stpeped, might as well be sunk in the bottom of the sea. Not only are they not cultivated, but in some cases they are not even to be *seen*.

What, then, is to be the termination of this course, that has been gradually but surely working an entire change in the relations of the British population to the British soil? The number of proprietors has been constantly diminishing, and the land is passing into fewer and fewer hands. If the process were to continue, a time might come when the very stability of the state itself might be endangered, and a change of system would be imperatively required for the safety of the nation. Already many parts of the country are both materially and martially much weaker than at any former period. They can neither turn out the same amount of food for the support of the nation, nor the same number of men for the national labor or the national defence. In other districts where the population is dense, the stature of the people has diminished; that is, the people are undergoing a course of physical deterioration. Great numbers of our healthiest, strongest, and most athletic sons are emigrating; for it is no longer the half-starved pauper who emigrates, but the very pick of our industrial classes. The nation, powerful as it is, and perhaps presuming a little too much on its past career, is certainly at the present time undergoing a process of debilitation, — becoming relatively weaker; increasing in wealth, but not improving, or even maintaining, the solid element of a well-arranged and well-conditioned population.

To arrest the progress of this growing evil, various remedies have been proposed. Some have asserted that a total abolition of entails would effectually prevent the accumulation of estates into the hands of a single proprietor; forgetting that the estates *have* been so accumulated simply because the large estates were entailed, and the small estates were not entailed; and that the usual purchaser, whenever land is exposed for sale, is either a great proprietor or a great capitalist. When an evil has grown to a certain point, it will perpetuate itself, like iron, which, when heated to a certain temperature, will burn of its own accord. In the present condition of Britain, the abolition of entails would be quite as likely to throw the land into fewer hands as to increase the number of landholders, because the great proprietors, who have large revenues, or almost unlimited credit, will give more for the land than its actual mercantile worth, estimated by the rate of interest that might be derived from other investments. The abolition of entails would in all probability only transfer the estates of the impoverished families to those who are already possessed of extensive domains. There would be no tendency to subdivision, because the offer of ten thousand pounds for a small property that was only worth five thousand would be no temptation to a lord or duke who has perhaps a clear income of a hundred thousand a year, and whose object is not to get money, but to get more land. That the abolition of entails would lead to the sale of land in such portions as would be convenient to the purchaser, — that a farmer, for instance, who had been saving and successful, could go to his landlord and buy his farm at a fair market-price, as he would buy a house or a ship, — we certainly do not anticipate; for if the farm lay in the centre of an estate, the proprietor would not sell it for ten times its estimated value; nay, he would not sell it at all. The mere abolition of entails, therefore, although in itself a good and proper measure, would not be calculated to work any great change for the general welfare. It might relieve some spendthrift families from the inconvenience of estates which they were unable to manage or redeem, and it might infuse new capital into the agricultural improvements of the country; but that it would materially affect the mass of the rural population to their advantage is by no means probable. At the same time, the total abolition of every remnant of the feudal system and of feudal practice in land conveyance is perhaps the first step to improvement.

Another proposed remedy is the formation of peasant properties, — a measure that has vehement advocates, and quite as vehement opponents, even among those who are supposed impartially to have investigated the subject. Mr. M'Culloch, carried away with the one idea of cultivation on a large scale, assures us that anything like peasant proprietorship would submerge us into a sea of pauperism. Mr. Joseph Kay, on the contrary, whose ability we take to be quite equal to that of Mr. M'Culloch, and whose opportunities for extensive, accurate, and personal observation we apprehend to have been even superior, assures us that the measure would tend to make our poorer classes happy, prudent, and prosperous. Mr. M'Culloch's objections we regard as a long course of special pleading, based on the fallacy of taking a small portion of the population as the index of the whole. It is quite easy to point to one of our large farms, or to our whole system of large farming, and to compare the amount of produce with the amount obtained from the labor of the same number of individuals in France, Germany, or Ireland. From such premises, however, the conclusion is a mere partial inference from insufficient data. It is quite easy to point to one of our regiments, and to admire the order, cleanliness, and seeming perfection of the military organization, just as Mr. Carlyle adduces the line-of-battle ship as an instance of indubitable success, and asks why the same system is not universally introduced into the field of labor. But human nature is neither composed of regiments nor of line-of-battle ships, nor of any select body of men from whom the very young, the very old, the halt, the lame, and the blind are sedulously and intentionally excluded. When we look at a regiment, we must ask not only what is the condition of these young men, but what is the condition of their wives, their children, and their aged parents? Muster the whole on parade; let us inspect the whole; and then we shall be able to form an opinion as to the success of the system. And so, also, when Mr. M'Culloch tells us to look at the success of our large properties and large farms, let us look at the whole population; let us look at the fact, that, at the very moment of his writing, about every tenth person in Enland was a pauper; let us look at our prisons, our poor-laws, our union workhouses, our poisonings for the sake of burial-fees, our emigration, as if our people were flying like rats, helter-skelter, from a drowning ship. Let us sum up the whole, and then perhaps we should find that our boasted system of social distribution was no more successful than the muster of one regiment, where we should find, on the one hand, order and competence; on the other, rags and tatters, wives abandoned, parents neglected, children left to the hazard of casual charity, and too often a dark shadow of vice and wretchedness following in the train of our vaunted institutions. But there is another special fallacy involved in the objections to peasant properties. We are told to compare ourselves with those countries where the great majority of the people are engaged in agriculture, and to mark their condition. We are told, with a singularly lame species of reasoning, that France is a nation of peasants; that France has peasant properties; and, consequently, that if we have peasant properties, we shall become a nation of peasants also. But, in the first place, the question is not whether

France may have run rather far in one direction, but whether we have not run incomparably further in the other; and, in the second place, France has at present no other means of employing her population except on the soil, whereas we can employ a hitherto unknown proportion of our people in manufacturing and commercial industry. No disposition of the land could ever again reduce Britain to the condition of France, because we have profitable manufactures, holding out the prospect of a higher reward than can be derived from agriculture; and consequently it is as absurd to suppose that our people should again return to mere tillage, as that they should return to the hunting and savage state of the earlier ages. The question of peasant properties does not affect the majority of our population, but only that portion actually engaged in the culture of the soil; and here we believe that the allocation of a certain portion of land to our laboring agriculturists would go a great way to restore the stability and independence of our country population, and perhaps to revive those homely virtues which were once more common than they are now, and which have waned exceedingly within the memory of those who are still alive. Of the positive advantages of having a peasantry rooted and grounded in the soil itself we say nothing, because there are at present no means by which the change from the prevailing system could be effected; but it seems evident that if our colonies and the States continue to present advantages which cannot be obtained at home, and if our people come to regard emigration, not as a matter of necessity, — not as a change which the indigent are obliged to make for the sake of the necessaries of life, — but as an attractive removal to another sphere, in which they can employ their labor much more satisfactorily than in their native country, then we must anticipate that a larger and larger portion of our best laborers will seek to establish an independent existence elsewhere, and leave to Britain only the inferior remnants of a class that has fought her battles, cultivated her fields, manned her ships, worked in her manufactories, peopled her colonies, and brought her, ungrateful as she is, up to the highest pitch of power. To those patriotic gentlemen who are about to improve the dwellings of our rural population we particularly recommend the experiment of attaching at least to some of the cottages as much land as would keep a cow, with a rood or two of croft, that would enable the cottar to instruct his children in spade husbandry, and to teach them regular and constant habits of industry from their earliest years. Let those gentlemen read in the "Quarterly Review" for July, 1829, how Thomas Rook did his hired work regularly, and yet made thirty pounds a year out of a little bit of land; and how Richard Thomson kept two pigs and a Scotch cow on an acre and a quarter — worth, when he got it, five shillings per acre of rent; and how the widow at Hasketon brought up her fourteen children, and saved them from the degradation of the parish, by being allowed to retain as much land as kept her two cows; and, above all, let them remark how poor-rates and degradation have always followed the severance of the peasantry from the soil. If they wish to improve the people as well as the dwellings, let them lay these things to heart, and let them be assured that the first thing to improve the laboring man is to hold out to him the prospect of an independent position, which he may hope to attain by prudence, economy, and honest labor.

III.

THE FRANCHISE.

One of the most remarkable sayings of which the discussion in Parliament on the Reform Bill proved the occasion, was that of Lord Jeffrey, then Lord Advocate for Scotland. "It was a measure," he said, "that would separate the waters above the firmament from the waters below." The remark embodied both a striking figure and a solid truth, — a figure which, by appealing to the imagination, has sunk deeper into the memory of the country than any other produced at the time; and a truth which recent events have served peculiarly to substantiate and elucidate. It was in consequence of this separation of the waters, that, while the revolutionary hurricane raged wide upon the Continent, dashing into one wild, weltering ocean of anarchy and confusion the dense and ponderous masses, whose inherent strength no such measure had divided into antagonistic, self-balancing forces, Britain escaped at least all the more terrible consequences of the storm. It is doubtful, however, whether we are permanently to escape. We are told by men of science, that, save for the continuous belt of ocean which girdles the globe in the southern hemisphere, we of the northern regions would have scarce any tides. In the equatorial and arctic oceans, the rise of the sea, in obedience to the attractive impulsions of the sun and moon, is checked by the great continents that stretch from north to south before the tidal wave becomes in the least considerable; but in the southern belt that wave rolls round the world without break or interruption, and then, travelling northwards laterally, in obedience to the law through which water always seeks its level, it rises and falls twice every twenty-four hours on the most northern shores of Europe, Asia, and America. It has been thus with the tidal wave of revolution. The Reform Bill in this country stretched abreast of the privileged classes like a vast continent, and would have effectually checked every rising tide of revolution that originated in the country itself. But there lay in the neighboring States great unbroken belts of the popular ocean, in which the revolutionary wave has risen high. The popular privileges have been elevated, in consequence, in these States, considerably above the British level; and it is very questionable whether this country will be long able to preserve its lower surface-line unaltered, when the flood is toppling at a higher line all around it. It would be at least well to be prepared for a steady setting in of the flood-tide on our shores; it would be wise — to return to the figure of Lord Jeffrey — to be casting about for some second firmament, through which a further modicum of bulk and volume might be subtracted from the waters below, and added to the waters above.

But does there exist, we ask, a portion of these lower waters that might be so separated with safety? We think there does. The *bona fide* property qualification we have ever regarded as peculiarly valuable, — greatly more so than the mere tenant qualification. The man who inhabits as tenant a house for which he pays a yearly rental of ten pounds, may be in many cases a man as well hafted in society, and possessed of as considerable a stake in the stability of the country and the maintenance of its institutions, as the proprietor to whom the ten pounds are paid. But the *class* are by no means so safe on the average. Their stake, as a body, is considerably less; they are a greatly more fluctuating portion of the population, and more unsteady and unbalanced in their views and opinions. There is really no comparison between the man who, in some of the close alleys of a city like Edinburgh, opens a spirit-cellar on speculation for which he pays a yearly rent of ten pounds, and the man who, after steadily adding pound to pound during the course of half a lifetime, at length invests his little capital in a house that brings him in ten pounds per annum, or, if he be his own tenant, that saves him that sum. The ten-pound tenants and the ten-pound proprietors compose, in the aggregate, bodies of men of an essentially different status and standing; and we hold that along the scale of proprietorship the franchise might safely descend a very considerable way indeed ere it corresponded with the existing level, if we may so express ourselves, on the tenant scale. We hold that the proprietor who *possessed* a house valued at *five* yearly pounds, would be on a higher, not a lower level, than the tenant who merely *occupied* a house valued at *ten* yearly pounds. His stake in the stability of the national institutions would be greater; and it might be rationally premised regarding him, if the house had been purchased out of his savings, or if, being derived to him by inheritance, he continued to preserve it unsquandered, that he was a steadier and safer man than the mere ten-pound tenant, of whom it could only be premised that present circumstances had enabled him, or hopes of future advantage had induced him, to inhabit a dwelling of a certain value. Nay, we are by no means sure whether there be not a principle in human nature through which a descent along the scale of proprietorship, very considerably beneath the five yearly pounds, might be rendered safe. We have ever found men valuing the property which they possessed, especially if of their own earning, not by an absolute, but by a comparative standard, — not by its price in pounds sterling, but with reference to their own circumstances and condition, and to the efforts which the acquirement of it had cost them. We have seen working men quite as proud of the little house, consisting of a *but* and a *ben*, a trap-stair and a loft, which the painful labor of years had secured to them, as the merchant on the little estate of some three or four hundred acres, in which he had invested the savings of a lifetime, or the master-builder or contractor of the half street or square of which his profession, long and successfully pursued, had enabled him to become the owner. We therefore do not attempt fixing the line to which, in this special direction, the franchise might be safely permitted to descend; but we do think it is a direction in which it might descend very safely; and though in our larger cities, such as Edinburgh and Glasgow, its descent would have scarce any effect in extending that basis on which the representation of the country rests, and which to a certainty must by and by be widened and enlarged, we are mistaken if in the smaller towns it would not considerably more than double its area. It would broaden the base of the social pyramid, and enable it to resist, without the danger of overthrow, the coming tempest which is so visibly darkening the heavens.

Is there any other portion of the "waters beneath the firmament" that might be separated from the general mass and made to balance against it, somewhat in the manner in which the tidal wave that enters through the Bristol Channel from the south balances and counteracts the tidal wave which enters the German Sea from the north, and in some parts of the coast, as at Great Yarmouth and the Hague, reduces, by fully one half, the average rise and fall of the tide? We would answer this query much more hesitatingly and doubtfully than the other. We, however, do not see on what principle it is that, while the tenants of houses of ten-pound-rental in the burghs are equally vested in the franchise with their proprietors, it is merely the *proprietors* of such houses that are vested in the franchise in the counties. Why not extend the privilege to the tenants also? The writer of this article inhabits a house a few hundred yards beyond the boundary line of the city, as drawn at the time of the Reform Bill, for which he pays a rent of rather more than thirty pounds yearly; and there are some of his neighbors, most respectable, intelligent men, who inhabit houses for which they pay rent of forty and forty-five pounds; but falling short of a fifty-pound rental, they do not possess a county vote. Why, we ask, should this state of things exist? As the tenants of thirty and forty-pound houses, they belong to an entirely different class of persons from the tenants of thirty and forty-pound farms; and the fact of actual residency in their dwellings places them in a category still more widely different from that to which the fictitious voters of our counties belong. We are disposed to hold that the exclusion of this class from the franchise is simply the consequence of a design to prevent the introduction, not of an element of subserviency, but of an element of independence, into our county elections, and that in this direction the franchise might be safely extended. It is a direction, however, in which extension could not very considerably affect the representative basis. With regard to further extension along the tenant line, our views are far from clear. It seems obvious, however, that a scale of rental common in its pecuniary amount to our cities and our smaller towns does not adequately represent classes. The ten-pound house-renters of the lesser towns are considerably superior in the average to the ten-pound house-renters of the larger; nay, it seems doubtful whether the five or six-pound tenants of burghs containing from fifteen hundred to three thousand inhabitants do not stand, on the average, on as high a level as the ten-pound tenants of the towns that possess a population of from eighty to a hundred thousand. The extension of the franchise to the five-pound householders of Edinburgh and Glasgow would in all probability wholly swamp the existing constituencies of these towns, and give them for their representatives mere loquacious Chartists, full of words, but infirm of judgment and devoid of principle; but we would have no such fear regarding a similar extension in burghs such as Tain and Dingwall, Cromarty and Nairn.

Our dread of universal, or even mere household suffrage, is derived chiefly from our long and intimate acquaintance with the classes into whose hands it would throw the political power of the country. "A poor man that oppresseth the poor," says Solomon, "is like a sweeping rain which leaveth no food." Alas! tyranny, as the wise man well knew, is not the exclusive characteristic of the wealthy and the powerful, nor is oppression the offence of a mere class. It is not the aristocracy, and they only, that are cruel and unjust: the poor can also override the natural liberties of the poor, and trample upon their rights; and it is according to our experience that there is more of this injustice and tyranny among that movement class now known as Chartists, but which we have closely studied under other names, when coming in contact with them in strikes, combinations, and political meetings, than in perhaps any other class in the country. It has been at least our own fate in life never personally to experience the oppression of the higher ranks, but not a little of the tyranny of the lower classes, especially that of this movement class. And we derive much of our confidence in the property qualification, not merely from the sort of ballast in the state which it furnishes, but from the fact that we never yet saw a workman who made a right use of his wages with an eye to his advancement in life, or who was in any respect a rising man, at all disposed to join in oppressing a comrade or neighbor. We have very frequently seen him made a victim of a tyrannical combination,—unmanly odds taken against him if at all formidable for native power,—but rarely, if ever, enacting the part of a tyrant himself. In a little work recently published, entitled the "Autobiography of a Working Man," we find an experience of this kind so truthfully rendered that we cannot resist submitting it to the reader. The working man's story is illustrative of a class of cases incalculably numerous, from the existence of which, too often and surely tested, we derive our chief dread of universal, or even houshold suffrage, and the abandonment of a property qualification. It was in the year 1830, when the cry for political reform in this country was so loud and general, that the following incident in the history of the working man took place: —

"A number of masons were hewing the blocks of stone, and each hewer had a laborer allotted to him to do the rougher work upon the stone with a short pick — technically, to 'scutch' it. The masons were intolerable tyrants to their laborers. I was in the quarry cutting the blocks from the rock when the tide was out; and when the tide was in I went and scutched with some of the hewers — chiefly with my friend Alick. One day, when we had been reading in the newspapers a great deal about the tyranny of the Tories, and the tyranny of the aristocracy in general, and some of the hewers had been, as usual, wordy and loud in denouncing all tyrants, and exclaiming, 'Down with them for ever!' one of them took up a long wooden straight-edge, and struck a laborer with the sharp edge of it over the shoulders. Throwing down my pick, I turned round and told him that, so long as I was about the works, I would not see a laborer struck in that manner, without questioning the masons' pretended right to domineer over laborers. 'You exclaim against tyranny,' I continued, 'and you yourselves are tyrants, if anybody is.' The hewer answered that I had no business to interfere; that he had not struck me. 'No,' said I, 'or *you* would have been in the sea by this time. But I have seen laborers who dared not speak for themselves knocked about by you, and by many others; and by every mason about these works I have seen laborers ordered to do things, and compelled to do them, which no working man should order another to do, far less have the power to compel him to do. And I tell you it shall not be done.'

"The laborers gathered around me; the masons conferred together. One of them said, speaking for the rest, that he must put a stop to this: the privileges of masons were not to be questioned by laborers; and I must either submit to that reproof or punishment which they thought fit to inflict, or leave the works; if not, *they* must all leave the works. The punishment hinted at, was, to submit to be held over one of the blocks of stone, face downward, the feet held down on one side, the head and arms held down on the other side, while the mason *apprentices* would whack the offender with their leathern aprons knotted hard. I said that, so far from submitting to reproof or punishment, I would carry my opposition a great deal further than I had done. They had all talked about parliamentary reform; we had all joined in the cry for reform, and denounced the exclusive privileges of the anti-reformers; but I would begin reform where we then stood. I would demand, and I then demanded, that if a hewer wanted his stone turned over, and called laborers together to do it, they should not put hands to it unless he assisted; that if a hewer struck a laborer at his work, none of the laborers should do anything thereafter, of any nature whatever, for that hewer. The masons laughed. 'And further,' said I, 'the masons shall not be entitled to the choice of any room they choose, if we go into a public house to be paid, to the exclusion of the laborers; nor, if there be only one room in the house, shall the laborers be sent outside the door, to give the room to the masons, as has been the case. In everything we shall be your equals, except in wages; that we have no right to expect.' The masons, on hearing these conditions, set up a shout of derisive laughter. It was against the laws of their body to hear their privileges discussed by a laborer; they could not suffer it, they said, and I must instantly submit to punishment for my contumacy. I told them that I was a quarryman, and not a mason's laborer; that, as such, they had no power over me. They scouted this plea, and said that wherever masons were at work they were superior, and their privileges were not to be questioned. I asked if the act of a mason striking a laborer with a rule was not to be questioned. They said, by their own body it might, upon a complaint from the laborer; but in this case the laborer was insolent to the mason, and the latter had a right to strike him. They demanded that I should at once cease to argue the question, and submit, before it was too late, to whatever punishment they chose to inflict. Upon hearing this, I put myself in a defensive attitude, and said, 'Let me see who shall first lay hands on me'! No one approaching, I continued, 'We have been reading in the newspaper discussions about reform, and have been told how much is to be gained by even one person sometimes making a resolute stand against oppressive power. We have only this day seen in the papers a warning to the aristocracy and the anti-reformers that another John Hampden may arise. Come on, he who dares! I shall be Hampden to the tyrannies of masons!'

"None of them offered to lay hands on me; one said they had better let the affair rest where it was, as there would only be a fight about it, and several others assented; and so we resumed our work.

"Had it been in summer, when building was going on, they would have either dismissed me from the works, or have struck, and refused to work themselves. It was only about the end of January, and they could not afford to do more than threaten me."

IV.

A FIVE-POUND QUALIFICATION.

When, owing to some deep-seated cause, the general level of a country is heightened by sudden upheaval, not only is its area extended by an apparent recession of the sea, but the outlines of its coasts are also very much changed. In places where the land is flat and low, and the water shallow, it receives accessions of great tracts of new country; whereas in other places, where high table-lands sink suddenly into the sea, and the water is deep, it is restricted to nearly its old limits. In Scotland, for instance, that last upheaval which laid dry the old coast-line added many a rich acre to the links of the Forth and the Carse of Gowrie, and gave to the country the sites of most of its seaport towns, such as Leith and Greenock, Musselburgh, Stonehaven, and Inverness; whereas, along the rocky shores of Aberdeen and Banff, and in especial Caithness and Orkney, it did little more, save here and there in a narrow inlet, than reduce by some two or three fathoms the depth of sea at the foot of the cliffs. It left the old boundaries just what they had been. The extension of area which took place in consequence of the upheaval was partial and local, though in the aggregate it added not a little to the general value of the country; and this peculiar character was altogether a result of the previous form of the surface. We have witnessed something similar in the effects of those great upheavals which occasionally take place in the political world. The Reform Bill effected a wonderful upheaval of this kind. It raised over the sea-level, in certain districts, vast tracts that had been previously submerged; while in other districts it left the old limits unchanged. The highlands of Toryism received no new accessions; while those of Liberalism it greatly enlarged. By elevating the long-buried heads of the people above the water in the character of ten-pound franchise-holders, it strengthened the trading interests, or — to carry out our parallel — gave new standing-room to the trading towns; while the agricultural interests, located, if we may so speak, on the high table-lands of the country, remained no broader or stronger than they had been before. And so, in the great struggle which ensued between the two interests, the agricultural one went down, without, however, catching any harm in the fall, and free trade won the day. Party in general was not a little affected by this great upheaval. The new accessions were chiefly accessions made to the cause of Liberalism in general; but it did quite as little for hereditary Whiggism as for hereditary Toryism; and either party feel, when in office, that it has had but the effect of making their position more precarious and less desirable than of old. Or — to carry out to a meet termination our somewhat lengthened comparison — while the upheaval has done much for those lower regions which it fairly raised over water, it has had but the effect of elevating the high official peaks on which each succeeding ministry takes its stand, into a less genial and more exposed region of the atmosphere than that which they had previously occupied. It has thrown them up nearer than of old to the chill line of perpetual ice and snow, and exposed them to the dangers of treacherous landslips and sudden avalanches.

What, let us ask, would be the effect of a still further upheaval of the political area, that would place the ten-pound franchise in the position of a second old-coast-line, by raising a widely-spread five-pound franchise outside of it? To what regions of party would such an upheaval add new breadth? In what regions would it leave the present limits unchanged? What would be its effect, for instance, on the various parties in Edinburgh, as brought out by the late election? Some of these, though of but comparatively recent appearance, must be regarded as tolerably permanent in their elements. The Forbes Mackenzie Act is a law of yesterday; but the strong reaction against the spirit traffic has been going on for some considerable time; and so long as the monstrous evil of intemperance continues to exist *it* will, we cannot doubt, continue to exist also. It will continue to form the pervading soul and spirit of a distinct party; nor will the antagonist party — the public-house one — be less permanent. The latter has in its composition that strongest, though at the same time most sordid, of all elements, a profit-and-loss one: it stands on a monetary basis, — a foundation that bids fair to remain firm till at least the millennium; and so both these parties, come of the Forbes Mackenzie Act what may, may be calculated on, in any future contest, as permanent ones. How would an extension of the franchise affect them? There are about nine hundred spirit-dealers in Edinburgh; and it has been calculated that in the late election about three hundred others voted in the spirit-dealing interest, influenced by the stake which they possess as proprietors of public-house property. A public house or tap-room in a suitable situation lets at a higher rent, by from one third to one half, than it would bring as a dwelling-house. And hence the interest of the proprietors of such, in their standing as public houses, and, of consequence, their opposition to any measure that would have the effect of either lessening the number of spirit-dealers or reducing their profits. Now, to this public-house party an extension of the franchise to the five-pound householders would bring almost no accession of strength. All the spirit-dealers pay at least ten-pound rents; all the owners of their houses are ten-pound proprietors; both classes are within the limits of the existing franchise, and certainly, as a body, exhibit great energy, and hold well and act efficiently together. The extension of the franchise would do almost nothing for them. It would do much, however, for their opponents.

The strength of the temperance cause will be found to lie chiefly among the decent five-pound house-tenants,— skilled mechanics chiefly, provident enough to meet the landlord at rent-day, and in the main a very safe class. The men at present outside the representative pale, who would support the publicans had they the power, are a greatly lower class, who, though in some instances they may pay as high a rent, pay it by the month or the week, and who would almost always lack the qualification of being settled for a twelvemonth in the same dwelling. Universal suffrage might, and we dare say would, strengthen the publican cause; whereas a judiciously limited extension of the suffrage would have the effect of virtually weakening it, by, of course, leaving it just what it is, while it greatly strengthened the opposition to it.

The Roman Catholic party in Edinburgh is another comparatively new party. We remember that in 1824— the year in which we first saw the Scottish capital—the bakers of the place and its Irish Papists were at feud, and that the bakers, being the more numerous and powerful party of the two, had the better. Times have since changed: the Edinburgh Irish are now to be reckoned by thousands, —from fifteen to twenty mayhap,—and their franchise-holders amount to from two to three hundred. And yet an extension of the franchise to the five-pound level would do exceedingly little for them. They consist almost exclusively of two classes,—a broker and petty-dealer class, whose shops, and sometimes their dwelling-houses, are of value enough to bring them within the limits of the ten-pound qualification; and a class of unskilled laborers, who live gregariously in humble hovels, to which a five-pound qualification would not nearly descend, and who, besides, shift their dwellings, on the average, some four or five times every twelvemonth. One has but to scan for a few minutes the congregation of the Roman Catholic chapel in Lothian Street, as the motley crowd defiles on their dismissal towards the Cowgate, in order to see that in this party, as in the other, it is from universal suffrage, and not from a judiciously restricted extension of the franchise, that there is aught to be feared. The qualification would require to descend very low indeed ere it could reach the class, comprising nine tenths of the whole congregation, who wear their weekday clothes on Sunday because they have no other, and are able, at the utmost, to greet the day with a clean shirt; and as for the few respectably-dressed men in black among them,—that, though on but the ordinary level of the attendants at Protestant places of worship, catch the eye here as the magnates and aristocrats of their church,—they are all voters already. As opposed to the Papists, however, the Protestant party would gain in strength, and that very considerably, by such a limited extension of the franchise as the one we specify. At least not more than one third of the men who attend anti-popish lectures, and whose presence at Mr. John Hope's meeting made it respectable in point of appearance and numbers, at present possess the franchise; but they are a decent, well-conditioned class, and the houses they inhabit do not fall, save in exceptional cases, beneath a five-pound rental, but, on the contrary, average very considerably above it. It may be safely presumed that any few accessions of strength which popery might receive from a five-pound extension of the franchise would be balanced at least ten times over by the assistance which Protestantism would draw from the accession to political power of this thoroughly respectable antagonist class.

There are two other parties — the "Edinburgh Review" Whigs and the "Blackwood Magazine" Tories — whom the extension of the franchise would leave exactly as they are, unless, indeed, in the exercise of an ingenuity in which they excel all other political bodies, they should fall upon some new mode of manufacturing fictitious votes. Of these two parties the Parliament House forms the central nucleus, and each in turn, as their friends chance to be in power for the time, possess the legal patronage of Scotland.

Judgeships, sheriffships, clerkships, procurator-fiscalships, all the many offices which Government can bestow, and to which gentlemen of the law are alone eligible, with not a few, besides, for which they are as eligible as any other class, are the good and weighty things which, like a great primary planet in the centre of a system, give cohesion and force to the movements of these parties. We make the remark in, we trust, no invidious spirit. Human nature being what it is, these things must and will have their weight and influence. There have been many instances of wholly disinterested *individuals* among both Whigs and Tories; but there never yet was a wholly disinterested *party*, especially when in power; and that patronage which made Lord Dundas in the last age the great centre round which Scotch politics revolved, renders the Parliament House a great political centre now, especially in Edinburgh. We remember seeing, many years ago, an ingenious caricature of the times of Fox and Pitt, which represented the great political system of the country as formed on the plan of the solar system. The Treasury, with its massive bags of guineas, formed the solar centre, and the various British statesmen of the age, inscribed within circles of bright gambouge, of a size proportionate to their influence, revolved as planets around it. Some of the larger ones had their satellites. Georgium Sidus (George III.) possessed as his moons the class of men known as "the friends of the King." Pitt also had his numerous satellites, and so had Fox. There was a good deal of complexity in the system, —

> "With centric and eccentric scribbled o'er,
> Cycle and epicycle, orb in orb."

But the great centre of all — the vast attractive mass towards which all gravitated, and round which all revolved — was the Treasury, with its bushels of golden guineas. And round this attractive circle, alike bright and solid, the great statesmen of London and the smaller statesmen of

Edinburgh will continue to revolve in these as certainly as in former times; and it would be idle to dream of any other condition of things with respect to the old governing parties, whether Whig or Tory. But not the less is it a duty on the part of men who love their country, sedulously to watch over an influence of this biasing kind, and on proper occasions to strive hard to counteract it. And by no class could it be more effectually counteracted than by a class who for themselves could have nothing to look for or expect. And such a class the five-pound householders would scarce fail to approve themselves. A scarlet coat, associated with a letter-carrier's office, might now and then be found for a compliant working man who voted as he was bid; but there could be no loaves and fishes found for so great a multitude as that of the five-pound householders.

Nor would we deem them an unsafe class in the main. They would be found to comprise the great bulk of the membership of all the evangelical churches, but few indeed of the lapsed classes. Nay, we know not that we could draw a better or more practical line of demarcation between these lapsed classes and our useful citizens of the humbler class than that which a five-pound household qualification would furnish. It has been said by a contemporary, more especially by its correspondents, that our aim in supporting Mr. Brown Douglas in the recent contest was simply to aggrandize the Free Church. We see not, however, how, in the political field, the Free Church *could* be aggrandized. We certainly look for no endowments for herself, and ask neither place nor emolument for her ministers or members. We have assuredly no wish to see *her* revolving round the Treasury as her centre. If we have desired to see some of her abler men returned to Parliament, it was not because they were Free Churchmen, but because we knew that on the most important questions of the day their opinions were sound; and if we now desire to see many of her members possessed of the franchise, it is only because we believe they would exercise it safely and well. We simply throw off, on the present occasion, a few suggestions, not as definite conclusions, but as food for thought, — as contributions, too, towards the solution of what we deem an interesting problem. The show of hands at the hustings of last week was greatly in favor of Mr. Brown Douglas; and when led to inquire how best the "declaration of the poll" could be made to agree with the "show of hands," we could bethink us of no better plan than that of a five-pound qualification.

V.

THE STRIKES.

ARTICLE I.

The last twelvemonth has been peculiarly marked in the manufacturing world as a year of strikes and combinations; nor, though there are adjustments taking place, and bands of operatives returning to their employment after months of voluntary idleness, are they by any means yet at an end. Great fires and disastrous shipwrecks are both very terrible things; but so far as the mere waste of property is involved, a protracted strike is at least as formidable as either, and its permanent effects are often incalculably more mischievous. Wreck or conflagration never yet ruined any branch of industry. Were all the manufactured goods in London to be destroyed in one fell blaze, a few months of accelerated industry would repair the loss. The greatest calamity of the kind which could possibly take place would resemble merely the emptying of a reservoir fed by a perennial stream, that would continue flowing till it had filled it again. But the loss occasioned by a long-protracted strike is often of a deeper kind. It not only empties the reservoir, but in some instances cuts off the spring, and in this way robs of its means of supply the town or district whose only resource the spring had constituted. Nations have in this way, when there were competing nations in the field, been permanently stripped of lucrative branches of industry, and become the mere importers of articles with which they had been accustomed to supply their neighbors. In other instances the effects are disastrous, not to the nation generally, but to merely a class of its workers. A partial strike of one section of workmen, on the product of whose labors certain other sections are dependent for employment, disturbs the social machine, and arrests its progress. By a stoppage in the movements of a single wheel or pinion, the whole engine is brought to a stand. The inventive power is quickened, through the necessity thus created, to originate some mode of supplying the place of the refractory bit or segment; the ingenuity exerted at length proves successful; wood, iron, and leather are made to perform the work of human nerve and muscle; and a province of industry is divested of its living workmen, and occupied by dead machines. We believe one of the last instances of this kind furnished by the history of strikes took place in the flax manufacture. Simple as the work of the heckler may seem, it was long found impossible to supersede him by machinery. In drawing the tangled flax through the bristling hedge of steel employed in disentangling and straightening its fibres, the human hand had a nice adaptability to the ever-varying necessities of the tuft in process of being sorted, which for so long a period could not be communicated to the movements of the unconscious machine, that the mechanist at length fairly gave up the attempt, and sat down in despair. A series of strikes, however, on the part of the hecklers, roused him anew to the work. Necessity at length proved the mother of invention. After repeated failures, he ultimately succeeded in making a most accomplished heckler of wood and metal, who never strikes work so long as he gets a few shovelfuls of coal to consume; and the flesh-and-blood hecklers, driven out of the field, have had to seek in other countries, and in other walks of exertion, the employment which, in consequence of his overmastering competition, they can no longer secure in their own.

Strikes are unquestionably great evils. In the case of the hecklers, what they effected was, not the ruin of the flax trade in Scotland, but simply the ruin of the class of mechanics that lived by the heckle. A series of strikes among the sawyers had a similar result. Circular saws, driven by machinery, entered the field on the side of the masters, and the recusant sawyers of flesh and blood went to the wall in the competition that ensued. But in both cases the trade of the district in which the strikes occurred was not permanently injured. Wood-and-metal hecklers and sawyers, with the strength of giants in their iron arms, and that fed on coke and charcoal, took the place of the greatly feebler human ones; and that was just all. But in certain other cases the result has been, as we have intimated, more disastrous. During the season of strikes and combinations that followed the passing of the Reform Bill, a combination of the ship-carpenters of Dublin, accompanied by more than the ordinary Irish violence and coercion, was completely successful. The terrified masters broke down, and, yielding to the terms imposed, gave their workmen the wages they demanded. But though they escaped in consequence the bludgeon and the brickbat, they could not escape the ordinary laws of trade and manufacture. They of course looked for the proper return from the capital invested in their business; they expected the proper remuneration for the time, anxiety, and trouble which it cost them. Profit was as indispensable to them as wages to their operatives. They found, further, that on the new terms, and with the competition of the western coast of Britain, especially that of the ship-builders of Liverpool and the Clyde, to contend with, profit could no longer be realized, and so they had to shut up their workyards, one after another; and

Dublin has now scarce any trade in ship-building. Its ship-carpenters have become very few, and of consequence very weak; and, no longer able to dictate terms as before, they have to work for wages quite as low as in any other part of the United Kingdom. But though carpenter-work may now be had as cheaply in the Irish capital as in Liverpool or Glasgow, the trade, fairly scared away, failed to return; the dock-masters of the Clyde and the Mersey kept a firm hold of what they had got; and all that was accomplished by the successful strike of the Dublin ship-carpenters was simply the ruin of the ship-carpentry of Dublin. Nor would the result have been different had the combination been more extensive. Had it included all the carpenters of Britain and Ireland, the competition in ship-building would have lain between, not the opposite sides of the Irish Channel, but the opposite sides of the German Ocean; our merchants would have purchased their vessels not from the Clyde and the Mersey, but from the dock-yards of the Baltic and the Zuyder Zee; and our British carpenters, instead of being, as of old, the fabricators of navies, might set out, shovel in hand, for the railways, and become navvies themselves. Unless the originators of the strikes of the country were also the makers of its laws, and could reintroduce the protective system, very successful strikes could have but the effect of striking down the trade of the empire, and prostrating its commerce.

And yet, disastrous as strikes almost always are, it cannot be questioned that the general principle which they involve is a just one, — quite as just as that of the masters who continue to resist them. In the labor market, as in every other, it is as fair to sell dear as to buy cheap; and it is in no degree more unjust for five hundred, five thousand, or fifty thousand men to agree together that they shall demand a high price for their labor, than it is for five or for one. The laws framed to compel working men to labor at whatever rate of remuneration legislators may choose to fix, — and in this country the terms *legislators* and *employers* have in the main been ever synonymous, — are properly regarded as evidences of a barbarous and unscrupulous time. The unquestioned right of the working man, is, however, of all others one of the most liable to abuse. It is greatly more so than the corresponding right of his employers. Both possess the same common nature; and it is quite as much the desire of the one to buy labor cheaply as of the other to sell it dear. But there is an amount of responsibility attached to the position of the masters which has always the effect, in at least a free age and country, of keeping their combinations within comparatively the same bounds. Masters of a morally inferior cast cannot control their fellows. Should they even be a majority, and should they agree to fix a rate of wages disproportionately low compared with their own profits, a few honest employers, instead of incurring loss by entering into competition with them, and raising the hire of their workmen, would soon appropriate to themselves their gains by robbing them at once of their workers and their trade. Competition on the side of the masters forms always the wholesome corrective of combination. Nor dare the combiners take undue means to overawe and control the competitors. Their amount of property, and their general standing in consequence, give them a stake in their country which they dare not forfeit by any scheme of intimidation; a regard, too, to the general interests of their trade, imposes upon them its limits; and thus supposing them to be quite as unscrupulous and selfish as the worst workmen that ever lived, as no doubt some of them are, there are in the nature of things restrictions set upon them which the workman, often to his disadvantage, escapes. On him the lowliness of circumstances virtually confers a power, if he has but the hardihood to assert it, of overawing competition. And we find from the history of all strikes that he always does attempt to overawe it. During the last thirty years he has shot at it, thrown vitriol upon it, rolled it in the kennel, sent it to Coventry, persecuted it with clumsy but very relentless ridicule, and subjected it, where he could, to illegal fines. Masters have no doubt the same nature in them as their men; but from their position they cannot, or dare not, attempt putting down competition in this way. Their position is that of the responsible few, while that occupied by the operative classes is the position of the comparatively irresponsible many; and, from the little stake which the latter possess in the property of the country individually, and from their conscious power in the mass, they are ever under the temptation of overstretching their proper liberties of combining, to carry out their own intentions, into a wild license, which demands that their neighbors and fellows shall not, either singly or in parties, exercise the liberty of carrying out theirs. There have been several glaring instances of this species of tyranny during even the present strikes. But one instance of the kind may serve as a specimen of the class. We quote from the Stockport correspondent of a London paper:—

"At a large mill not three miles from this, where upwards of a thousand hands are engaged, one of the weavers did not choose to subscribe to the weekly delegate's tax towards the unfortunate Preston strike. In consequence, one evening this week, when the mill stopped, he was watched in passing through the large gates into the road, was immediately knocked down and blindfolded, his arms pinioned, and his legs tied fast together, and, thus disabled, was carried through the population of the place, mobbed by hundreds upon hundreds, shouting, yelling, and execrating, not a soul daring to interfere, as any resistance to these proceedings would probably have cost the poor fellow his life. I know the man well, as an honest, sober, hard-working operative, and feel grieved that he should be thus persecuted.

"You may say, Why do not the masters protect such men, and put down such tyranny? Simply because *they dare not;* such interference being sure to be followed by a general turn-out, and, very likely, by destruction of property by fire or otherwise. These are sad realities; and I cannot but conclude that the above outrage has been a natural sequence to the visit of one of the Preston delegates to the heads of that very mill during last week. My own life would not be safe, were it known that I had told this circumstance to one connected with what the delegates call the 'vile hireling press.'"

It is one of the grand disadvantages of these strikes that their management and direction are almost always thrown into the hands of a class of men widely different in character from the country's more solid and respectable mechanics. We had to record in one brief paragraph, a few numbers since, the flight of two delegates of the Preston movement, — the one with twenty-five pounds of the defence fund in his possession, the other with one hundred and sixty. And such are too generally the sort of men that force themselves into prominence in these movements. Inferior often as workmen, low in the moral sense, fluent as talkers, but very unwise as counsellors, they rarely fail to land in ruin the men who, smit by their stump oratory, make choice of them as their directors and guides. Too little wise to see that the most formidable opponent which any party can arouse is the moral sense of a community, violence and coercion form invariably the clumsy expedients of their policy. And so, for the success which a well-timed strike, founded on just principles, would be almost always certain to secure, they succeed in but achieving from their unfortunate constituencies discomfiture, either immediate or ultimate. It is really the least mischievous of these strike-leaders that, like the Preston delegates, run away with the funds. We find in strikes, as they ordinarily occur, the disastrous working of exactly the same principle which has rendered the revolutions of the Continent such unhappy abortions. Who can doubt that the revolutions, like some of the strikes, had their basis of real grievances? But their leaders lacked sense and virtue; their wild license became more intolerable than the torpid despotism which it had supplanted; and in the reaction that ensued, the sober citizen, the quiet mechanic, the industrious tiller of the soil, all the representatives of very influential classes, found it better, on the whole, again to submit themselves to the old tyranny than to prostrate themselves before the new.

ARTICLE II.

Capital and labor are joint values, invested together for the production of a common result, which result is the *selling price* of the article brought into the market. In a commercial point of view, it matters little what the article may be. It may be corn, cattle, or agricultural produce; it may be raw cotton, cotton yarn, or calico; it may be iron-stone, pig-iron, bar-iron, steel, or hardware goods, large or small, a needle or an anchor, an iron spoon, or a magnificent iron ship that transports thousands of men in ease and comfort, or thousands of tons of goods with safety and celerity; it may be a picture, a statue, a piece of music, a poem, or any other work of art; it may be a dwelling, varying from the wretched holes in which modern society stores away so large a portion of the working population, bringing them up in an atmosphere of physical deterioration and moral corruption, to the lordly dwelling or the royal palace, where luxury seems almost to have exhausted her inventions, and left no wish ungratified or unprovided for; it may be a book or a magazine, a quarterly or a newspaper; it may be, in fact, anything whatever produced for the market and exposed for sale. In the articles there may be a thousand varieties; but there is always a permanent object which is common to them all — the selling price. The object of the manufacturer — and every producer may be termed a manufacturer; the farmer, for instance, is as much a manufacturer of grain, of horses, sheep, and cattle, as the cotton-spinner of cotton yarn, or the ship-builder of a ship — is not merely to produce a given article, but to produce an article that will realize a selling price, which selling price ought to repay the cost of production, and leave a profit on the transaction. This is the great commercial principle which pervades the ordinary world of industry. True there are exceptions, because there are labors expended and objects produced which have an end and purpose differing altogether from the purposes of commerce. The end of commerce is gain, — profit to those who are engaged in it. But gain, though absolutely necessary where men live in a world of exchange and competition, may have a higher counterpart, — the gain, not of ourselves, but of others. Hence all works of charity, benevolence, and moral instruction originate in a higher principle than that of commercial gain. So also in the region of literature, which abounds with what the mercantile world would term unprofitable speculations. Books are produced from many various motives, entirely separated from the commercial principle. Some authors produce books from a desire to enlighten their fellow-men; some from the spontaneous desire to give utterance to the native voice of genius — the "Paradise Lost," for instance; some from a love of fame; some from a miscalculated estimate of their powers. In almost every department of art there are artists who regard excellence as higher than profit, and who pursue it sometimes to their own loss; just as there are philosophers who pursue their inquiries after truth without regard to the accident of remuneration; and just as there are inventors who perfect machines and processes, with minds so ardently bent on the realization of their special idea that they sacrifice fame and fortune to an achievement that may have great results, or no results, as the chance may be, yet which bring to themselves no element of worldly prosperity. Thomas Waghorn's overland route to India, and Morgan's paddle-wheel, are notable instances of skill and perseverance which brought no commercial reward; the first saving hundreds of thousands of pounds annually to this country, and leaving Mr. Waghorn's widow on a pension of fifty pounds a year; the other being a most beautiful piece of mechanism, which cost the fortune of the inventor, and left him, we believe, in a commercial sense, a ruined man.

Accidental exploits of this kind, however, are merely the pioneerings of commerce, — the voyages of discovery into new regions, which may prove Arctic with unprofitable snows, or Australian with untold treasures of wool, copper, and gold. In commerce, as well as in geography, there are invasions of a hitherto unknown territory, — new speculations, like new expeditions, opening up new fields of enterprise and industry. Columbus discovers a new world, but reaps small advantage from a deed that is unsurpassed in the annals of adventure. On the other hand, a chemist, experimenting on sugar, finds that certain substances will refine it, and straightway he reaps a princely fortune from the accidental revelation. In commerce, however, as well as in geography, there is an old world as well as a new, — a region of beaten paths and customary ways, as well as a region of emigration, into which the old world pours the enterprising or the unemployed of its population. In commerce there is an every-day old world of buying, selling, and getting gain, — of manufacturing for the ordinary necessities of the race, — of producing multitudes of articles which are the joint productions of capital and labor. In this manufacturing world there are two parties, — the employers and the employed. The first brings his money or his money's worth — his land, his houses, his materials, his credit, and his power of waiting for a return. The latter bring their skill and labor, their knowledge, their practice, — in short, their power of doing the thing that is requisite to produce the article. Capital and labor, then, are joint investments; but they are, in the present constitution of society, antagonistic to each other. Whether a plan might be devised by which this antagonism should be obviated, as a superfluous and unnecessary encumbrance, we cannot as yet say. Such a plan, if such be possible, is the great desideratum of the commercial world; and we have at least the satisfaction of knowing that at no anterior period of history has it been sought for with the same ardor as at present, or with the same probability of a successful issue. Leaving that question alone, however, we need not hesitate to affirm that at present capital and labor are commercially antagonistic when employed together in the production of the same work, — capital perpetually endeavoring to reduce the price of labor, and labor perpetually endeavoring to enhance its own market value. In this case there is nothing unreasonable or improper. No doubt there are evils incidental to the system, and occasional cases where the principle is pushed to an extreme which is morally wicked and fraudulent; as, for instance, where capital takes advantage of the penury of the laborer, and accords him only a starvation price; or where the laborer takes advantage of some distressed situation into which the capitalist has fallen, — say war, shipwreck, fire, or many other calamitous conditions, — and refuses to give his labor except at an exorbitant rate, such as ought not to be required between man and man. These cases are the abuses of the system, and we pass them over. But so long as competition is the regulating principle of the commercial world, capital and labor must be to this extent antagonistic, — that each will endeavor to obtain as large a share as possible of the selling price of the produced article; and the portion which the one obtains cannot be also obtained by the other. The question, then, is to ascertain the proper proportion that should be allotted to capital, and the proper proportion that should be allotted to labor, when they are jointly employed. At first sight this may appear a simple question. If five pounds' worth of labor and five pounds of capital are embarked in a dining-table which sells for fifteen pounds, we might say that the expenditure had been equal, and that the price obtained should be shared equally. This may seem an easy way of solving the difficulty; but unfortunately these easy methods of solution are utterly incapable of application. There are fluctuations in supply and demand which alter the value of every single item; and unless we could make supply and demand absolutely constant, we could never apply a rule which proceeded on the distribution of the selling price. A capitalist may conduct his business for years, paying regularly for the labor he employs, and yet barely clear his expenses, when suddenly some accidental circumstance causes a great demand, a great rise in price, and he is fairly entitled, not only to the profits of the present time, but, to some extent, to the profits of past unfruitful years, and this not because the present should pay for the past, but because the present is actually a portion of the past; that is, the capitalist calculates on an average of profits, which average may extend over a very considerable period. Recently we have had two instances of this kind, namely, in shipping property and in cattle-dealing. About two years since both of these businesses became highly remunerative; but previously they had been carried on almost without profit, and in many cases with loss. Those engaged in them were fairly entitled to a certain amount of extra profit, because they had submitted to a term of years in which their returns were far below the average, and they required to recover the legitimate value of their previously-expended time and capital. Another instance we may cite, as showing the extent of fluctuation. During the Australian gold mania, seamen, hoping to reap some of the advantages, were willing to ship for the voyage *to* Australia, for, say a shilling per month; whereas seamen shipping *from* Australia (Melbourne) obtained ten, twenty, and, we believe, in some cases, thirty pounds per month. In the first case the capitalist made a large profit out of the laborer; in the second the laborer made a large profit out of the capitalist; but in neither case was the absolute return — the selling price — made the criterion. The criterion was — as it always must be so long as free competition is the ruling principle — the relation of the demand to the supply, and of the supply to the demand.

That these fluctuations are evils, and attended with many inconveniences, there can be little doubt; and if any specific proof were required, we might point at once to the fact that they lead to those most unhappy and most unprofitable exhibitions called *strikes*. When prices rise, and the laborer thinks that he receives less than his due proportion of the returns, he strikes; and when prices fall, so that the returns will not cover the cost, the capitalist strikes, — that is, he ceases to employ labor; and in so doing he *strikes* quite as much as the laborer does when he refuses to employ capital. In fact, the whole commercial world is always in a modified state of strike. When prices rise, it is because the seller has struck against the buyer; and when prices fall, it is because the buyer has struck against the seller. A few infatuated individuals will attempt to resist the ordinary law of supply and demand; and hence we hear of some English farmers who have kept their wheat in stack for ten years, to be eaten up by rats and mice, rather than sell at the current market value. But the great majority must always succumb to the market, and take the current rates, whether those rates be the price of labor, or the price of capital, or the price of produce. In general it is the capitalist who receives the sale price, and who had charge of the money; and it is this circumstance, perhaps as much as any other, that creates a jealousy in the mind of the working men that the capitalist appropriates more than his just share. Let us reverse the picture, however, and look at the other side. The banks advance money to parties who want capital. These parties are, in fact, the laborers, and the bank is the capitalist. Now, these persons often engage in very lucrative transactions with the bank money; yet the bank never seeks to share in the unusual profits. The bank for its money asks only the current rate of interest, regulated by the supply of cash in the market and the demand for the same. In this case, it is not the capitalist, but the laborer, who receives the sale price, and who has the whole and sole charge of the returns. The bank is not jealous that its money has been employed to advantage; on the contrary, the bankers know that the better the return, the more does business thrive, and the more demand will there be for the bank money, leading to a higher rate of interest and to the prosperity of the capitalist. So also with the workmen who labor for the capitalist. They ought to rejoice that the capitalist derives large returns, because those large returns afford them a security that their labor will be employed, and that their wages will be as high as the competition in their particular branch of business will prudently afford. Strikes, then, for a rise of wages, are destructive resorts to an extreme and hazardous remedy. Theoretically they are wrong and unnecessary in a free country, where everything is open to free competition; and practically they have, we believe, in almost every case on record, shown themselves to be perfectly useless. They have never done good; and though we no more deny the right of men to strike than we deny the right of the master to discontinue his business when it no longer pays, yet we are thoroughly assured that they never will do good. They will always do more harm to the workmen who strike than to the masters who are struck against. We would therefore calmly but seriously recommend working men to refrain from strikes, more especially when the object is a rise of wages; and these cautions we deem the more necessary at the present time, as we greatly regret to perceive that the Dundee ship-carpenters have shown a disposition to engage in the same course that has been carried on so disastrously at Preston. We hope that they will not be misled by vague anticipations, and, above all, that they will not manifest a spirit of unreasoning obstinacy, which will certainly tend to defeat the very purpose they have in view. We must say a word, however, on another matter. A strike for a rise of wages is not likely to be attended with success. The *time* is altogether another question. If workmen are in a position to strike for a rise of wages, they are also in a position to strike for shorter time; and our own experience enables us to affirm that as much work and as good work will be done in a short week as in a long one. We should rejoice to see the Saturday half-holiday absolutely universal; and we recommend all workmen to strive for this great reform, as incomparably more valuable to themselves, to their wives, and to their children, than any additional shillings that they hope to obtain. If men strike at all, let them strike for the Saturday half-holiday, and the good wishes of the whole community will go with them. To every workman we would say, look to the Saturday half-holiday as one of the most precious things that you can possibly acquire: get it by all means. Work hard, faithfully, honestly, like a man; but by all means get the Saturday half-holiday; and when you get it, be sure to make a good use of it.

VI.

THE COTTAGES OF OUR HINDS.

We presented to the reader on Saturday last, in our report of the late half-yearly meeting of the Highland and Agricultural Society, the remarks of two very estimable noblemen on the cottages of the country, especially the cottages of hinds, and on the best means of improving them. It was stated by the one noble speaker, and reiterated by the other, that in order to render cottages immensely better than they are at present, it is not at all necessary that they should be rebuilt. The rebuilding of them, in the greater number of instances, might be impossible, and in all cases it would be at least very inconvenient. But if proprietors had thus little in their power regarding them, much might be done by the humble inmates in the way of dividing their single rooms when their accommodation chanced to be greater, and in imparting to them an air of general comfort. It was held that on this point, therefore, the premiums of the Society ought specially to be directed. The proprietary of the country could not be expected to help their poor laborers on a large scale, by providing them with suitable dwellings (a single cottage might cost fifty pounds); but then they were ready to encourage them in any feasible way of helping themselves. A room twelve feet by sixteen might be regarded as a very pretty sort of problem; and if a man and his wife, with some eight or ten children, could contrive to solve the difficulty by residing in it with comfort and decency, they should be by all means rewarded for their ingenuity by a premium from the Society. Now, we would be very unwilling to indulge on this subject in aught approaching to severity of remark; nor, were it otherwise, would we single out two of the more benevolent noblemen of our country as objects on which to be satirical. Scarce any Scottish nobleman has done so much for his humbler dependants as the Earl of Rosebery: we have been informed that on his own property every cottage has its two comfortable apartments, and that many of them have three. Nor is his Grace the Duke of Buccleuch other, we believe, than a well-meaning man. As we deem the matter one of considerable importance, however, we shall take the liberty of soliciting the attention of the reader to a piece of a simple narrative, which bears on it very directly.

We passed the summer and autumn of 1823 in one of the wildest and least accessible districts of the northwestern Highlands. The nearest public road at that period was a long day's journey away. Among the humbler people we met with only a single man turned of forty who understood English. It was, in truth, a wild, uncultivated region, brown and sterile, studded with rock, blackened with morasses, and cursed with an ever-weeping climate. The hills of hard quartz rock — of all the primary formations the most unfavorable to vegetation — seemed at least two thirds naked; and their upper peaks, bleached by sun and storm, showed, from the pale hue of the stone, as if ever covered by a sprinkling of fresh-fallen snow. The Atlantic, specked by the northern Hebrides, stretched away from an iron-bound coast; and here and there, though far between, a group of dark-colored cottages, that rather resembled huge molehills than human dwellings, occupied some of the deeper inflections, where, for a short interval, the cliffs gave place to a strip of sand or pebbles, or an outlying group of skerries formed a sort of breakwater to ward off the violence of the sea. Every little village had its few boats and its few green patches of cultivation. Some of the latter, scarcely larger than onion beds, seemed to stand out from amid the brown heath like islands in the ocean; and both the boats and the patches served as indices to show how the poor inhabitants of so barren a region contrived to live. Could we travel back into the past, amid the rich fields of the Lothians, for full ten centuries, we would fail to arrive at so primitive a state of things with regard to the common arts of life as existed only nineteen years ago in this wild district. In the little straggling village in our immediate neighborhood every man was a fisherman, and in some degree an agriculturist; and yet there was neither horse nor plough among all its twenty families. The ground was turned up by the long-handled spade, still known in some parts of the Highlands as the *cas-chrom;* and the manure was carried out in spring, and the produce brought home in autumn, mostly by women in slip-bottomed creels. All the other arts practised in the village reminded one of a remote age. We have seen the poor Highland women bending under their burdens of turf or manure, and employed at the same time in spinning with that most primitive of implements, the distaff and spindle. Some of the boats, caulked with moss, like the ancient Danish vessel disinterred some ten or twelve years ago out of the silt of an English river, were furnished with sails of woollen, anchors constructed of wood and stone, and tackle spun out of the fibres of moss fir. The little patches of cultivation were suited to remind one, from their size, of the fields described by Gulliver; but they had, besides, a peculiarity all their own; — the ground abounded with stones, many of them by much too bulky to be removed. To save as much space as possible, each of the larger masses had its pyramid of smaller stones piled upon it to the height of four or five feet, and there were patches in which these pyramids lay well-nigh as thickly grouped together as tents in an encampment. A man of some little imagination might have supposed that one of the many Scotch witches of the seventeenth century had passed the way in the time of harvest, and transformed all the newly-reaped shocks into accumulations of stone. Such was the agriculture of the district: it was the agriculture of the first ages, — the fruit of the very first lesson which man had derived from experience, on setting himself to force a living from the soil. Nor, it may be well supposed, could the art of the builder in such a country be greatly in advance of the art of the agriculturist. The human dwellings were quite as rude as the fields. But we shall describe one, just as a specimen of the whole.

On the first evening of our arrival in the district, we accompanied an acquaintance, to secure the services of a Highlander whom we were desirous to engage as a laborer, and who lived in the nearer village. Twilight was falling, but there remained light enough to enable us to examine the surrounding forms of things. The cottage we sought was a low, long, dark building, whose roof and walls sloped in nearly the same angle, without any aperture for windows, except along the ridge of the roof, and with a door raised little more than four feet above the threshold. In these northwestern regions, where there falls about twice as much rain as on any part of the eastern coast, and where, at some seasons, the almost incessant showers beat at an angle of inclination varying from thirty to sixty, it is imperatively necessary to render the side-walls of a building as impervious as the roof; and hence the slope of the walls, — a slope given them by filling up a bulwark of solid turf against the comparatively erect line of stone. Our first step into the interior was into a pit fully two feet in depth. In this outer chamber, according to the custom of the district, the ashes produced by the turf and peat burnt during the year had been suffered to accumulate, for the purposes of manure; and as it was now early in summer, the place had been but lately cleared out. It was intensely dark, and filled with smoke; and we had some difficulty in finding the inner door, the threshold of which we found raised to the level of the door without. A step brought us into what proved to be the middle apartment of the cottage. A fire of turf, enlivened by a few pieces of moss fir, blazed on a flat stone in the middle of the floor, with no protecting back to screen any part of the building, so that the flames shone equally all around on the rude walls and the equally rude furniture. On one side the fire sat the master and builder of the mansion, — a strongly-built, red-haired, red-whiskered Highlander, — with two boys, his sons; on the other, the mistress, — a thin, sallow woman, — with her three daughters. The woman was busied in spinning with the primitive implement to which we have already alluded; now twirling the spindle half at arm's length, and now coiling up the thread. Her girls were teasing wool, which they stored up in a large spherical basket, wattled all round, except at a little square opening. A cloud of smoke, thick and flat as a ceiling, rested overhead; and there hung, as if dropping out of it, a dark drapery of herring-nets. The inner walls, as shown by the red glare of the fire, were formed of undressed stone, uncemented by mortar; but the interstices had been carefully caulked with dried moss. The furniture was somewhat of the scantiest. There were a few deal-seats, and a rude bed-frame in a corner, half-filled with heath, — the sleeping-place of the boys; a few wooden cogs occupied a recess behind the woman; and there was a large pot suspended over the fire from the roof. But what we chiefly remarked was, that the place, rude as it was, had what by much the greater number of the dwellings of our south-country hinds have not, — the luxury of an inner apartment: the wicker door opened through a stone wall; the thick turf roof was at least water-tight, except where, beside the gables (not over the fire), there were two openings to admit air and light, and to give egress to the smoke. Our readers would smile were we to associate ideas of comfort with such a dwelling. Certain it was, however, that its inmates could do so; and all can at least associate ideas of decency with it. The construction of Red Murouch's house was quite as primitive as the tillage of his little croft, or the tackle of his boat, or the distaff and spindle employed by his wife. His grandfather removed by twenty generations had lived, in all probability, in just such another; but it served Murouch quite as well as its antitype had served his remote ancestor. Besides, if he wished it better or larger, could he not improve or add to it? There was space enough outside; vast abundance of stone everywhere, and wood in the neighboring hollow; and Murouch, unsophisticated, like all his neighbors, by the scheme of dividing labor, which, while it adds to the skill of the community, lowers mightily that of the individual, was a master of the entire art of building such houses.

Just six months after quitting the Highlands, we were residing in one of the richest districts in the Lowlands of Scotland — one of those centres of cultivation from which the art of the agriculturist has spread itself over all the more accessible portions of the kingdom. The rent of land in the neighborhood averaged somewhat above five pounds per acre; the yearly rental of the parish in which we lived was estimated at about twenty-eight thousand pounds. The Scottish metropolis lay not three hours' walk away. Considerably more than two hundred miles intervened between us and the scene of our last year's labor. We have often thought whether it would not be equally correct to say that we had travelled in advance of it at least a thousand years. The whole seemed, viewed in recollection from amid the fertile fields of the south, as if belonging rather to the remote past than to the present. Even the most unpractised eye could not fail being struck by the superior style of the husbandry in the *modern* district. How very close the plough had contrived to skirt the well-dressed fences! How straight the furrows! — how equal the braird! How thoroughly had the land been cleared of weeds! And then, what an air of snugness seemed to pervade the farm-houses of the district, and how palpably had the experience of ages been concentrated on the means and appliances of their several steadings! The jealous neatness, too, with which the various gentlemen's seats in the neighborhood were kept, their general style, the appearance of the surrounding grounds, their woods, and gardens, and belts of shrubbery — all testified to the elegant tastes and habits of the possessors. Whatever belonged immediately to the upper classes had but one character — comfort gilded by the beautiful. And there was much, doubtless, in the very sight of all this for the poor man to enjoy. We still entertain a vivid recollection, distinct as a picture, of the beautiful vista in a gentleman's woods, — tall, green, finely arched, close over head as the roof of a cathedral, — through which we could see, almost every evening, as the twilight faded into darkness, the Inchkeith light twinkling afar off, like a star rising out of the sea. The noble grove through which it shone was scarce a hundred yards distant from the humble cottage in which we lodged.

But the cottage was an exceedingly humble one. It was one of a line on the wayside, inhabited chiefly by common laborers and farm-servants, — a cold, uncomfortable hovel, consisting of only a single apartment, — by many degrees less a dwelling to our mind, and certainly less warm and snug, than the cottage of the west-coast Highlander. The tenant, our landlord, was an old farm-servant, who had been found guilty of declining health and vigor about a twelvemonth before, and had been discharged in consequence. He was permitted to retain his dwelling, on the express understanding that the proprietor was not to be burdened with repairs; and the thatch, which was giving way in several places, he had painfully labored to patch against the weather by mud and turf gathered from the wayside. But he wanted both the art and the materials of Red Murouch. With every heavy shower the rain found its way through, and the curtains of his two beds, otherwise so neatly kept, were stained by dark-colored blotches. The earthern floor was damp and uneven; the walls, of undressed stone, had never been hard-cast; but, by dint of repeated whitewashings, the interstices had gradually filled up. They were now, however, all variegated by the stains from the roof. Nor had the pride of the apartment, its old-fashioned eight-day clock or its chest of drawers, escaped. From the top of the drawers the veneers were beginning to start, in consequence of the damp; and the clock gave warning, by its frequent stops and irregularities, that it would very soon cease to take further note of time. The old man's wife, still a neat, tidy woman, though turned of sixty, was a martyr to rheumatism; and her one damp and gousty room, with its mere apron-breadth of partition interposed between it and the chinky outer door, was not at all the place for her declining years or her racking complaint. She did her best, however, to keep things in order, and to attend to the comforts of her husband and her two lodgers; but the bad roof and the single apartment were disqualifying circumstances, and they pressed on her very severely. It was well remarked by his Grace the Duke of Buccleuch, that "the keeping of lodgers along with families in cottages where there is scarce room for the family itself, is a great evil." It is even so — a very great evil. But, my Lord Duke, there are still greater evils which press upon the indigent. These poor old people had very slender means of living, and they found it necessary to eke them out in any honest way. Their lodgers, too — humble, hard-working men — could not afford a very sumptuous lodging-place, nor were there any such in the neighborhood, even if they could. There are stern necessities that press upon the poor in matters of this kind, which we sincerely trust your Grace may never experience, but of which all would be the better of knowing just a very little.

And this was all that civilization, in the midst of a well-nigh perfect agriculture, and amid the exercise of every useful and elegant art, had done for the dwelling of the poor hind. The rude husbandry of the western-coast Highlander had been left more than a thousand years behind; manufactures had made marvellous advances since the relinquishment of the distaff and spindle; trade had imported many a luxury since woollen sails and wooden anchors had been abandoned; every umbrageous recess had its scene of elegance and comfort; the homes of the poor had alone remained stationary, and worse than stationary, — they had sunk below the level of semi-civilization. But we are building perhaps on a solitary instance, — attempting to found a grievance on a needle-point. Would that it were so! Our description is far above the average, however exaggerated it may seem. Take, by way of proof, from a very admirable little work on the subject by the Rev. Dr. W. S. Gilly of Norham, a description of the hovels on the border, deemed quite good enough by the proprietary of the country for their own and their tenants' hinds. He selects a single group as a specimen of the whole.

"Now for a more detailed description of that species of hut or hovel — for it is no better — which prevails in this district. I have a group of five such before my mind's eye. They belong to the same property, and have all changed inhabitants within eighteen months. The property, I may add, is tenanted by one of the best and most enterprising farmers in all England. They are built of rubble, loosely cemented; and, from age and the badness of the materials, the walls look as if they would scarcely hold together. The chinks gap open in many places, and so widely that they freely admit every wind that blows. The chimneys have lost half their original height, and lean on the roof with fearful gravitation. The rafters are evidently rotten and displaced; and the thatch, yawning in some parts to admit the wet, and in all parts utterly unfit for its original purpose of giving protection from the weather, looks more like the top of a dunghill than of a cottage. Such is the exterior; and when the hind comes to take possession, he finds it no better than a shed. The wet, if it happens to rain, is making a puddle on the earth floor. This earth floor, by the by, is one of the causes to which Erasmus ascribed the frequent recurrence of epidemic sickness among the cottars of England more than three hundred years ago. It is not only cold and wet, but contains the aggregate filth of years from the time of its being first used. The refuse and dropping of meals, decayed animal and vegetable matter of all kinds, these all mix together, and exude from it. Window-frame there is none. There is neither oven, nor copper, nor shelf, nor fixture of any kind. All these things the hind has to bring with him, besides his ordinary articles of furniture. Imagine the trouble, the inconvenience, and the expense which the poor fellow and his wife have to encounter before they can put this shell of a hut into anything like a habitable form. This year I saw a family of eight, — husband, wife, two sons, and four daughters, — who were in utter discomfort, and in despair of putting themselves into a decent condition, three or four weeks after they had come into one of these hovels. In vain did they try to stop up the crannies, and to fill up the holes in the floor, and to arrange their furniture in tolerably decent order, and to keep out the weather. Alas, what will they not suffer in the winter! There will be no fireside enjoyments for them. They may huddle together for warmth, and heap coals on the fire; but they will have chilly beds and a damp hearthstone; and the cold wind will sweep through their dismal apartment; and the icicles will hang by the wall, and the snow will drift through the roof, and window, and crazy doorplace, in spite of all their endeavors to exclude it."

Great as they may seem, however, these are merely physical evils; and they are light and trivial compared with the horrors which follow. These miserable cabins consist, in by much the greater number of instances, as in the cottage of the poor old hind, of but a single room. We again quote: —

"And into this apartment are crowded eight, ten, and even twelve persons. How they lie down to rest, how they sleep, how unutterable horrors are avoided, is beyond all conception. The case is aggravated when there is a young woman to be lodged in this confined space who is not a member of the family, but is hired to do the field-work, for which every hind is bound to provide a female. It shocks every feeling of propriety to think that in a room within such a space as I have been describing civilized beings should be herding together without a decent separation of age and sex. So long as the agricultural system in this district requires the hind to find room for a fellow-servant of the other sex in his cabin, the least that morality and decency can demand is, that he should have a second apartment, where the unmarried female and those of a tender age should sleep apart from him and his wife."

The following simple story places the degradation to which the poor hind and his family are subjected, in consequence of the wretched accommodation provided for them, in a light painfully strong. We may truly remark with the poet, in this case, without metaphor, that misery makes strange bedfellows.

"Last Whitsuntide, when the annul lettings were taking place, a hind who had lived one year in the hovel he was about to quit called to say farewell, and to thank me for some trifling kindness I had been able to show him. He was a fine, tall man, of about forty-five, — a fair specimen of the frank, sensible, well-spoken, well-informed Northumbrian peasantry, — of that peasantry of which a militia regiment was composed which so amazed the Londoners when it was garrisoned in the capital many years ago, by the size, the noble deportment, the soldier-like bearing, and the good conduct of the men. I thought this a good opportunity of asking some questions, — where he was going, and how he would dispose of his large family (eleven in number). He told me they were to inhabit one of these hinds' cottages, whose narrow dimensions were less than twenty-four feet by fifteen; and that the eleven would have only three beds to sleep in, — that he himself, his wife, a daughter of six, and a boy of four years old would sleep in one bed, — that a daughter of eighteen, a son of twelve, a son of ten, and a daughter of eight, would have a second bed, — and a third would receive his three sons, of the age of twenty, sixteen, and fourteen. 'Pray,' said I, 'do you not think that this is a very improper way of disposing of your family?' 'Yes, certainly,' was the answer; 'it is very improper in a *Christian* point of view; but what can we do until they build us better houses?'"

It were needless to expatiate on this picture; it is quite enough that we hold it up to the reader. There is much to militate against the character of the poor hind all over the country. His very situation is adverse, however comparatively favorable the circumstances with which it may chance to be surrounded. When aggravated by the horrors of the bothy system, deterioration is inevitable; nor can any one honestly or rationally hold that the gross cruelty which consigns him to situations such as the one described — situations wholly subversive of that nice delicacy of feeling which is at once the safeguard and ornament of virtue — does not furnish a necessary item in his degradation. Mark the effects. In an interesting report, on farm-servants, of the very reverend the Synod of Perth and Stirling, published in October last, we find the following astounding passage. It embodies a piece of moral statistics in connection with this hapless class, as furnished by the returns of thirty-nine parishes: —

"Of the public scandals chargeable on farm-servants, the proportion varies considerably in different parishes; but in all of them, with three exceptions, the number chargeable on that class of the parishioners is larger — in some of them much larger — than on all the others put together, although in no one instance does that class constitute anything like a majority of the inhabitants. In those three cases where the scandals among the farm-servants are fewer than those among the other classes, the proportion of the whole number of farm-servants to the other, and especially the working classes, is exceedingly small. It requires to be particularly noticed, that in *one* parish, the scandals which have occurred of late among the farm-servants are reported to be *nine tenths* of the whole."

"Where is thy brother Cain? — the voice of thy brother's blood crieth unto me from the ground." This surely is not one of the matters in which our aristocracy do well to study a niggard economy. With all due respect, therefore, for the excellent and benevolent nobleman who advocated an opposite view of the case in the meeting of last week, we must be permitted to say, that it will not do to speak of forty-pound impossibilities and twenty-pound inconveniences, when the morality of the country is thus at stake. It will not do merely to propose premiums for introducing beds with woollen screens in front into the one miserable apartment of the poor neglected hind, or to incite him to task his ingenuity in partitioning the narrow area in which he is compelled to cram his family. Pecuniary sacrifices *must* be made by the proprietary of the country, even should they have to part, in consequence, with one or two superfluous horses or a few supernumerary dogs. Mere alterations will not do. In the language in which Watts, in one of his less-known lyrics, describs the leprous house, they must —

> "Since deep the fatal spot is grown,
> Break down the timber and dig up the stone."

VII.

THE BOTHY SYSTEM.

MOST of our readers must know what the bothy system is. A very considerable number of the farm-steadings of the country, built on the most approved plan, with roomy courts and sheds for the breeding of cattle, and stables constructed on the best possible principle for the horses,—with all, in short, that the modern system of agriculture demands, — have no adequate accommodation for the laborers by whom the farms attached to them are wrought. The horses and cattle are well provided for, but not the men. A wretched outhouse, — the genuine bothy, — furnished with a few rude stools, a few deal bedsteads, a few bowls of tin or earthen ware, a water-pail, and a pot, serves miserably to accommodate some eight or ten laborers, all of them, of course, single men. Here they kindle their own fire, cook their own victuals, make their own beds. The labors of the farm employ them from nine to ten hours daily; the grooming and feeding of their horses at least an hour more. The rest of their time falls to be passed in their miserable home. They return to it often wet and fatigued, especially in the briefer and stormier months of the year, just as the evening has fallen, and find all dark and chill: the fire has to be lighted,—in some districts even the very fuel to be procured; the water to be brought from the well; the hasty and unsavory meal to be prepared. It is scarce possible to imagine circumstances of greater discomfort. The staple food of the laborer is generally oatmeal, cooked in careless haste, — as might be anticipated in the circumstances, — by mixing a portion in a bowl with hot water and a little salt; and often for weeks and months together there is no change in either the materials of this his necessarily heating and unwholesome meal, or in the mode of preparing it. The farmer, his master, in too many instances takes no further care of him after his labors for the day are over. He represents merely a certain quantum of power purchased at a certain price, and applied to a certain purpose; and as it is, unluckily, power purchased by the half-year, and abundant in the market, there is no necessity that it should be husbanded from motives of economy, like that of the farmer's horses or of his steam-engine; and therefore little heed is taken though it should thus run to waste. The consequences are in most cases deplorable. It used to be a common remark of Burns, — no inadequate judge, surely, — that the more highly cultivated he found an agricultural district, the more ignorant and degraded he almost always found the people. Man was discovered to have deteriorated at least as much as the corn and cattle had improved. Now, in Scotland there has been a very obvious reason for this. The altered circumstances of the country rendered inevitable the introduction of the large-farm system, and broke down our rural population, composed almost exclusively of what we still term the small tenantry, — a moral and religious race, — into two extreme classes, — gentlemen farmers and farm-servants. The farmers composed, of course, but a comparatively small portion of the whole; nor, though furnishing many high examples of intelligence and worth, can we equal them as a body with the class which they supplanted. Hitherto they have lived less in the "eye" of the great "Taskmaster." They took their place, not in the front of the common people, but in the rear of the aristocracy: they passed, to employ the favorite proverb of the poet whose remark we are attempting to illustrate, from the "*head of the commonalty to the tail of the gentry.*" The other and greatly more numerous class proved much more decidedly inferior. The tenant of from fifteen to fifty pounds per annum necessarily occupied a place in which, in accordance with the distinguishing characteristic of the species as rational creatures, he had to look both before and after him. He had to think and act; to enact by turns the agriculturist and the corn-merchant; to manage his household, and to provide for term-day. He was alike placed beyond the temptation of apeing his landlord, or of sinking into a mere ploughing and harrowing machine. But, in many instances, into such a machine the farm-servant sunk. Still, however, there remained in his lot circumstances favorable to the development of the better parts of his nature. There is much in having a home; nor was he placed beyond those ennobling influences of religion which are scarce less necessary for enabling man rightly to perform his part in this world than to prepare him for another. Chiefly, however, from motives of a miserable economy, the unnatural bothy system was introduced, and with the disastrous effects described. It promised to spare some of our landlords the expense of providing cottages; and some of their tenantry expected to have their farms more cheaply wrought by single than by married men. We have seen more than the mere outsides of bothies, and know from experience, that though they may be fit dwellings for hogs and horses, they are not fit dwellings for immortal creatures, who begin in this world their education for eternity.

Nearly twenty years ago, we lived for a short time in an agricultural district in the north of Scotland, on the farm of one of the first introducers of the bothy system into that part of the country. He has been dead for years, nor do we know that any of his relatives survive. He had been a bold speculator in his time, and had risen, with the rise of the large-farm system, into the enjoyment of a very considerable income; but instead of regarding it as mere capital in the forming, — the merchant's true estimate of his gains, — he had dealt by it as the landed gentleman does in most cases with his yearly rental. His style of living had more than kept pace with his means; a change had taken place in his circumstances at that eventful period, so very trying to many of similar character, when England, at the close of her long war with France, ceased to be the workshop and general agency-office of Europe; and he was now an old man, and on the eve of bankruptcy. The appearance of his steadings and fields consorted well at the time with his general circumstances. The stone fences were ruinous; the hedges gapped by the half-tended cattle. Harvest was just over, and on his farm at least it had been a miserably scanty one; but it would have been somewhat better with a little more care. In walking over one of his fields, we counted well-nigh a dozen sheaves scattered about among the stubble, that seemed to have fallen from the carts at leading time, and were now fastened to the earth by the grains having struck their shoots downward and taken root. His steadings, though they wore a neglected look, were of modern, substantial masonry, and well designed, — the stables roomy, the cattle-courts and sheds formed on the most approved plan. Very different, however, was the appearance of the building in which his farm-servants found their sort of half-shelter. Some twenty or thirty years before it had been a barn; for it had formed part of an older steading, of which all the other buildings had been pulled down to make way for the more modern erection. It was a dingy, low, thatched building, bulged in the sidewalls in a dozen different places, and green atop with chickweed and stonecrop. One long apartment, without partition or ceiling, occupied the interior from gable to gable. A row of undressed deal-beds ran along the sides. There was a fire at each gable, or rather a place at which fires might be lighted, for there were no chimneys; the narrow slits in the walls were crammed with turf; the roof leaked in a dozen different places; and along the ridge the sky might be seen from end to end of the apartment. We learned to know what o'clock it was, when we awoke in the night-time, by the stars which we saw glimmering through the opening.

It was, in truth, a comfortless habitation for human creatures in a wet and gusty November, and the inmates were as rugged as their dwelling-place was rude. We need hardly say that none of them could regard it as a home. It was the gloomy season of the year, when the night falls fast, abridging the labors of the day; and ere they returned to their miserable hovel in the evening, all was deep twilight without, and all darkness within. The fuel had to be procured, the fire to be kindled, water to be brought from the well, and the unsavory meal to be prepared; and all this by men stiff with fatigue, and not unfrequently soaked with wet. It was no easy matter at times to light the fire: the fuel often got damp, and, when at length lighted, burnt dead and cheerless. There was a singular want, too, of the ordinary providence among the inmates, and it could be shown in a matter slight as this. No provision was made in the morning for the fire of the night. If the rain fell, their fuel and their tempers were just so much the worse in consequence; and that was all. Does the reader remember Crabbe's admirable stroke of nature in his "Phœbe Dawson?" He describes the poor thing as almost heartbroken in her misery, and yet struggling with it in patient silence; but a single drop serves to make the full cup run over. When dragging herself painfully along the green, with her broken pitcher in her one hand and sustaining her child with the other, she sinks ankle-deep in a quagmire. The mischance, slight as it may seem, is the single drop which more than fills the cup, and she bursts out into a hysteric fit of weeping. We have seen matters quite as slight rouse into fierceness, in the bothy, tempers already soured by bitter discontent. The inmates, if careless of their master's interests, were scarce less careless of their own comforts. A little hot water poured on a handful of oatmeal, with a sprinkling of salt, furnished the thrice-a-day meal. Had the materials at their command been more luxuriant, we question much whether they would have taken the trouble to prepare them. It seems natural for men in such circumstances to be careless of themselves, and equally natural for them to avenge on the cause of their general discomfort the irritating effects of their own indifferency and lack of care. There was a large amount of rude sarcasm in the bothy, and, strange as it may seem, a great deal of laughter. It has been remarked by, we think, a French writer, that the people of despotic governments laugh more than those of free states. We never heard the name of the farmer mentioned among his servants without some accompanying expression of dislike; we never saw one of them manifest the slightest regard for his interest. They ill-treated his horses, neglected his cattle, left his corn to rot in the fields. Some of them could speak of his approaching ruin with positive glee. What we would fain have said to him then may not be without its use to others now. "You, in your utter selfishness, have spoiled the men whom you employ; and they, in turn, are spoiling your horses and cattle and corn, and glorying in the ruin which is just on the eve of overtaking you. All right. There is no getting above the natural laws. Alkalies neutralize acids; dense bodies invariably descend when placed in fluids lighter than themselves; and men, when they are spoiled, spoil all other things."

Scarce any one except Crabbe could have done full justice to the interior of the bothy. We remember there was a poacher, — a desperate, thick-set, black-visaged fellow, — who used to steal in about midnight with his gun, when all was dark and quiet, and draw himself up into one of the beds. He was of the stuff that felons are made of, — beyond comparison more a criminal than any of the inmates of the bothy; and his occasional presence served to show, by the force of contrast, that the others were nothing worse than just useful members of society, of the average character, lucklessly spoiled. It was the bothy system that had made them what they were. The fact, however, seems not unworthy of being noted, that the poacher should have come to harbor in such a place. He was a man living in a state of warfare with the upper classes, — a black-fisher and a breaker into game preserves; but no inmate of the bothy thought a whit the worse of him for his trade. He annoyed only people of the same class with their master, and could there be harm in that? Immediately after dinner, especially when the fuel was bad, most of the bothy-men disappeared. There was a small village about a mile away, to which they generally resorted. It had its smithy and its public house; and in the latter there were rustic dances got up at least once a fortnight, at which all the men of the bothy were sure to attend. A young jemmy lad — the beau of the party, who used at times to wear his Sunday coat of red tartan at the plough, and who, had he been born to a more fitting sphere, would haply have smoked cigars and sported moustaches on Prince's Street — had quite a knack at getting up these entertainments, and in providing his companions with partners from all the farmhouses round. It was generally late in the morning, on such occasions, ere they got home; and the unsteady tread as they groped along the floor for their beds, or the previous fumbling at the latch, gave evidence in most cases that the protracted merry-making had terminated in drunkenness. But we find we must abridge our description. We may sum up the whole by remarking that the evils of the bothy system are of a threefold character, — economic, intellectual, religious. Our agriculturists are, fortunately, becoming convinced of the first, — a conviction which may lead, in time, through the abolition of the system, to the removal of the others. It is scarce possible for the inmate of a bothy to cultivate his mind. The bothy is a place in which the cogitative faculties fall asleep; the higher sentiments of our nature fare no better. As for religion, it may be enough to remark, that we have not yet seen a bothy in which the Sabbath *could* be properly kept: the ploughman who entertained a due reverence for the Sabbath would walk out into the fields. Cobbett, during his short stay in this country, acquainted himself with the system, and was by much too quick-sighted not to detect its evils. "Better," he said, in his own extreme style,— "better the fire-raisings of Kent than the bothy system of Scotland." We are far from reiterating the remark. We would deem, on the contrary, fire-raisings, such as those of Kent, one of the worst consequences that could result from it, though perhaps not one of the most improbable. We may be permitted to ask, however, whether the Scottish Church is much to be blamed for having endeavored to lay on such a system what Wordsworth well terms "the strong hand of her purity?"

VIII.

THE HIGHLANDS.[1]

"It is very sad that the people of this fine wild country have not got enough to eat; but, depend on't, we will collect no more money for them in England. We have already done our best to help them, and they must now help themselves." Such was the remark of a comfortable-looking Englishman whom we encountered, a few weeks ago, among the wilds of the northern Highlands; and, judging from the indifferent success which has attended the recent efforts to form a second fund in behalf of the suffering Highlander, it seems to represent pretty fairly the average feeling and general determination of the country on the subject. Charity on a large scale, and directed on distant objects, soon exhausts itself. It is competent, if thoroughly roused, to grapple with the necessities of one famine, and to do a very little for a second; but a third wearies it; and, should famine become chronic, it leaves

[1] At this date, 1862, the depopulation of the Highlands is still rapidly going on. Not half a mile from the spot where we write, in the northwest Highlands, many families were ejected from their holdings but a few months ago. *The factor*—that dreaded middleman of the people—came with the underlings of the law, with spade and pickaxe, and left literally not one stone upon another of their poor cottages standing. I can see a miserable hovel into which several families have crowded, who had before separate holdings of their own. I have no hesitation in saying that the proprietor ought to be held legally bound, in such cases, either to provide other home accommodation or the means of emigration. Such scenes ought not to be allowed to disgrace a Christian country. But even where the inhabitants are allowed to remain on their miserable and insufficient crofts, the able-bodied—that is, the choicest of the population—are rapidly emigrating. "There is not a lad *worth anything*," said a person, the other day, who had just left a very large strath at some twenty miles distance,—"there is not a lad worth anything that is not going away to New Zealand, or some other place."

The people are, indeed, oppressed with a sense of utter poverty, and a total

it to devastate unheeded, and ends where it is said to begin, by exerting itself at home. Nor do we see, man being the impulsive creature that he is, how his charity, if voluntary and at large, is to be made to act other than paroxysmally and at wide intervals. We must be content, we are afraid, to accept it as a fact, that, even should our poor Highlanders not have enough to eat for several years to come, there will be very little more money collected for them in England or elsewhere; and that, however great the difficulty which attaches to the "state of the Highlands" problem, it is a difficulty with which our own country, and in especial the Highlands themselves, must be prepared to grapple, undiverted by any vain hopes of eleemosynary aid from without.

The difficulty is certainly very great, and it has been vastly enhanced by the late years of famine. We are old enough to remember the northern Highlands, rather more than thirty years ago, when there were whole districts of the interior, untouched by the clearing system, in possession of the aboriginal inhabitants. And if asked to sum up in one word the main difference between the circum-

inability to rise above it. In many places their circumstances are made as wretched as possible, on purpose to starve them out. There are a few proprietors — such as Sir Kenneth M'Kenzie of Gairloch — who respect the feelings of those who have been for generations located on their properties; but these are *very* few. It is but justice, too, to the present and late noble proprietors of Sutherland to say, that, notwithstanding the melancholy clearings, — for which, of course, they individually are not responsible, — such of their small tenantry as remain are not rack-rented. They are, in fact, very leniently dealt with in this respect. But nothing can ever make the Highlander what he was, but that interest in the soil which he has lost. Every Highlander formerly was possessed of all those feelings which constitute much that is valuable in the birthright of true gentlemen, — a long-descended lineage, a sense of status, and property, and an intense attachment to home and country. We fear that we have seen nearly the last of this noble race on the battlefield of the Crimea; and that soon, unless a marvellous revolution takes place, the so-called Highland regiments may be Irish, or what they please, but not *Highlanders*. But if the mountains and moors only were let for deer-shootings, and the soil proper were restored to its children in farms capable of supporting families, this calamity might yet be averted; nor would the proprietors, in the long run, be the losers, in a pecuniary point of view. We are disposed to think the contrary would be the case.

L. M.

stances of the Highlander in these and in later times, our one word would be, that most important of all vocables to the political economist, — *capital.* The Highlander was never wealthy; the inhabitant of a wild mountainous district, formed of the primary rocks, never are. But he possessed on the average his six, or eight, or ten head of cattle, and his small flock of sheep, and when — as sometimes happened in the high-lying districts — the corn crop turned out a failure, the sale of a few cattle or sheep more than served to clear scores with the landlord, and enabled him to purchase his winter and spring supply of meal in the Lowlands. He was thus a capitalist, and possessed the capitalist's peculiar advantage of not living "from hand to mouth," but on an accumulated fund, which always stood between him and absolute want, though not between him and positive hardship, and enabled him to rest during a year of scarcity on his own resources, instead of throwing himself on the charity of his Lowland neighbors. And in these times he never *did* throw himself on the charity of his Lowland neighbors. Nay, in what were emphatically termed the "dear years" of the beginning of the present and the latter half of the past century, the humbler people of the Lowlands, especially our Lowland mechanics and laborers, suffered more than the crofters of the Highlands, and this mainly from the circumstance that, as the failure of the crops which induced the scarcity was a corn failure, not a failure of grass and pasture, the humbler Highlanders had what the humbler Lowlanders wanted, — sheep and cattle, — which continued to supply them with food and raiment; while the others, depending on corn almost exclusively, and accustomed to deal with the draper for their articles of clothing, were reduced by the high price of provisions to great straits. In truth, the golden age of the Highlands was comprised in that period which extended from shortly after the suppression of the rebellion of 1745, and the abolition of the hereditary jurisdictions, down till the commencement of the clearance system. It is to this period that Mrs. Grant's description of Celtic habits and of Celtic character belong, and which give one the idea of so contented, and, in the main, so comfortable a people, that, save for our own early recollections when we lived among the Highlanders, we would be disposed to suspect that the good lady had drawn on her imagination for the coloring of her pictures. Previous to the long wars of the first French revolution, the people of our country generally did not work so hard as they do now. One set of mechanics, such as our weavers, had not to contend with machinery, and earned good wages in comparatively "short hours;" another class, such as masons and carpenters, had not to work, as now, under the competition of the estimate system, but wrought easily on day's pay. The Highlander, whose labors were more prevailingly pastoral than agricultural, wrought still less than either class; but having less to compete with, the little which he did work served his turn. And as his mode of life was favorable to the development of the military spirit, — a spirit which the traditions of the country served mightily to foster, — great numbers of the young men of the country, of a very different class from those that usually enlist in England and the Lowlands, entered the army, and our Highland regiments were composed of at once the best men and the best soldiers in the service. It was early in this period that the eloquent Chatham could boast, in his place in Parliament, that, indifferent whether a man's cradle had been rocked to the south or north of the Tweed, he had seen high military merit among the Scottish mountains; and that, calling forth from amid their recesses, to the service of the country, a "hardy and dauntless race of men, they had conquered for it in every quarter of the globe."

With the wars of the first French revolution there was a great change introduced into the country. The wheels of its industry were quickened by the pressure of taxation, and by the introduction of a system of competition with machinery, on the one hand, that lengthened the term of labor by reducing its remuneration, and with the "estimate system" on the other. Nor was it in the nature of things that the Highlands should long remain unaffected by this change. The price of provisions rose in England and the low country, and with the price of provisions the rent of land. The Highland proprietor naturally enough bethought him how his rental was also to be increased; and, as a consequence of the conclusion at which he arrived, the sheep-farm and clearing system began. Many thousand Highlanders, ejected from their snug holdings, employed their little capital in emigrating to Canada or the States; and there, in most cases, save in very inhospitable localities, as in the Cape Breton district, the little capital increased, and a rude plenty continues to be enjoyed by their descendants. Many thousands more, however, fell down upon the coasts of the country, and, on moss-covered moors or bare, exposed promontories, little suited for the labors of the agriculturist, commenced a sort of amphibious life as crofters and fishermen; and there, located on an ungenial soil, and prosecuting with but indifferent skill a precarious trade, their little capital dribbled out of their hands, and they became the poorest of men. Meanwhile, in some parts of the Highlands and Islands a busy commerce sprung up, which employed, much to the profit of the landlord, several thousands of the inhabitants. The manufacture of kelp rendered tracts of barren shore and inhospitable islets of more value than the richest land in Scotland; and, under the impetus given by full employment, and, if not ample, at least remunerative pay, population increased. Suddenly, however, free trade, in its first approaches, destroyed the trade in kelp; and then the reduction of the salt-duties, and the discovery of a cheap mode of manufacturing soda out of common salt, secured its ruin beyond the power of legislation to retrieve. Both people and landlords experienced in these, the kelp districts, the evils which a ruined commerce always leaves behind it. Old Highland families have disappeared, in consequence, from among the aristocracy and landowners of the country; and the population of extensive islands and seaboards of the country, from being no more than adequate to the employment furnished, suddenly became oppressively redundant. It required, however, another drop to make the full cup run over. The potato is of comparatively modern introduction into the Highlands. We were intimate in early life with several individuals who had seen potatoes first transferred from the gardens of Sutherland and Ross to the fields. But during the present century potatoes had become the staple food of the Highlander. In little more than forty years their culture had increased fivefold; for every twenty bolls reared in 1801, there were a hundred reared in 1846; and when in the latter year the potato blight came on, the poor people, previously stripped of their little capitals, and divested of their employment, were deprived of their food, and ruined at a blow. The same stroke which did little more than slightly infringe on the comforts of the people of the Lowlands, utterly prostrated those of the Highlands; and ever since, the sufferings of famine have become chronic along the bleak shores and rugged islands of at least the northwestern portion of our country. Nor is it perhaps the worst part of the evil that takes the form of clamorous want. Wordsworth, in describing a time of famine in which the fields for two years together "were left with half a harvest," tersely says, that

> "Many rich
> Sank down, as in a dream, among the poor,
> And of the poor many did cease to be."

We fear that during the famines of the last five years not a few of our Highland poor *have* ceased to be, if not in consequence of absolute starvation, in consequence at least of the severe course of privation to which they have been exposed. But their wants are now all provided for; and it is a more disastrous though less obtrusive fact, that so heavily has the famine borne on a class that were not absolutely the poor when it came on, that they are the absolutely poor now. It has dissipated the last remains of capital possessed by the *people* of the Highlands, and placed them in circumstances of prostration too extreme to leave them any very great chance of recovering themselves, or rather in circumstances from which, in the present state of the country, recovery for them as a *people* is an impossibility.

Such seems to be the present state of the Highlands. Where are we to look for the proper remedies? Alas! in the body politic, as in the natural body, injuries may be easily dealt, for which it may be scarce possible to suggest a cure. In travelling over an extensive Highland tract last autumn, we had a good deal of conversation with the people themselves. Passing through wild districts of the western coast, where the rounded hills and scratched and polished rocks gave evidence that the country had been once wrapped up in a winding-sheet of ice, we saw the soil for many miles together — where the bare rock had any covering at all — composed of two almost equally hopeless ingredients. The subsoil was formed of glacial debris, — the mere scrapings of the barren primary rocks; and over it there lay a stratum, varying generally from six inches to six feet, of cold, wet, inert moss, over which there grew scarce even a useful grass, except perhaps the "deer's hair" of the sheep-farmer. And yet, on this ungenial soil, representative of but vegetable and mineral death, — the dead ice-rubbish and the dead peat, — we saw numerous cultivated patches in which the thin green corn or sickly-looking potatoes struggled with aquatic plants, — the common reed and the dwarf water-flag. No agriculturist, with all the appliances of modern science at command, would once think of investing capital in such a soil; and yet here were the poor Highlanders investing at least labor in it, and their modicum of seed-corn. And we are not to wonder if the tillers of such fields be miserably poor, and fail to achieve independence. There was a locality pointed out to us, in a barren quartz-rock district, in which the indestructible stone, that never resolves into soil, was covered by a stratum of dark peat, where the proprietor had experimented on the capabilities of the native Highlanders, by measuring out to them amid the moor, at a low rent, several small farms, of ten or twelve acres apiece. But in a moor composed of peat and quartz-rock no rent can be low. No farmer thrives on a barren soil, let his rent be what it may; and so the speculation here had turned out a bad one. The quartz-rock and the peat proved pauper-making deposits; and while the tenants paid their rents irregularly and ill, the demands made on the poor-rates by the hangers-on of the colony came to be demanded very regularly indeed, and were beginning to overtop the nominal rent in their amount. "How," we have frequently inquired of the poor people, "are you spending your strength on patches so miserably unproductive as these? You are said to be lazy. For our own part, what we chiefly wonder at is your great industry. Were we at least in your circumstances, we would improve upon your indolence by striking work, and not laboring at all." The usual reply used to be: "Ah, there is good land in the country, but *they* will not give it to us." And certainly we did see in the Highlands many tracts of kindly-looking soil. Green margins, along the sides of long-withdrawing valleys, which still bore the marks of the plough, but now under natural grass, seemed much better fitted to be, as of old, scenes of human industry, than the cold, ungenial mosses or the barren moors. But in at least nineteen cases out of every twenty we found the green patches bound by lease to some extensive sheep-farmer, and as unavailable for the purposes of the present emergency, even to the proprietor, as if they lay in the United States or the Canadas.

So far as we could see, the effects of recent emigration had not been favorable. The poor-rates were heaviest in the districts from which the greatest numbers had emigrated. Unless emigration be so enforced as to become a sort of indiscriminate banishment, — and in these days of poor-laws it would not be easy to enforce it, even in the Highlands, — it will be the more vigorous and energetic portion of the community that will seek for a home in other countries, and the feeble in mind and body that will be left behind. We were much struck by the casual statement in a newspaper paragraph, that, of several hundred emigrants from Lewis who arrived in Canada this season, there was scarce one who was not under thirty. It was the *elite* of the island that went, while its pauperism staid behind. The pauperism of the Highlands will not willingly cross the Atlantic; it would be going from home much more emphatically than the vigorous emigrant. There are poor-laws in Scotland, but none in the backwoods. But on a subject at once so extensive and so difficult we can do little more than touch. We regretted to find, during our late visit, that the military spirit is at present so dead in the Highlands that the recruiting party of one of the most respectable Highland regiments under the Crown succeeded in enlisting, during a stay of several months, only some ten or twelve young men, in a county charged with an unemployed and suffering population. In popish Ireland as many hundreds would have enrolled in the time; and this disposition on the part of the Irish has crowded the British army with a preponderating proportion of Roman Catholics, who, in the event of such a religious war as may one day break out to convulse Europe, could be but little depended upon on the side of Protestantism and the Queen. We fain would see a revival of the old military spirit of the Highlands, both on their own account and on that of the country. The condition of the British army is at the present time one of comfort and plenty compared with that of the general population of the northwestern parts of Scotland; the prospect of retirement, with a snug pension, some one-and-twenty years hence, is a better prospect than any poor Highland crofter or cottar can rationally entertain; and we would much prefer seeing some twenty thousand of our brave countrymen enrolled in the army, as at once its best soldiers and best Protestants, than lost forever to the country in a colony that in a few years hence may exist as one of the States of the great North American republic.

IX.

THE SCOTCH POOR-LAW.

We have never yet been able to see any foundation for the assertion of Paley, that "the poor have the same *right* to that portion of a man's property which the *laws* assign to them, that the man himself has to the remainder." *Right* cannot be created by law where right did not exist before; and in the poor-laws, as now administered in England, we have a striking illustration of the fact. No law can give to one man a *right* to take another man, guilty of no crime save poverty, and in debt to no one, and shut him up in prison. Poverty is surely not so grave an offence as to merit a punishment so severe. And yet certain it is, that a legal right of this character exists in England at the present day. It exists as surely as the other legal right asserted by Paley; nor does it in the least alter the state

1 A poor-law edict indeed "become inevitable for Scotland!" But alas for its consequences! One who was session-clerk for fourteen years in a parish as large as three or four of the smaller English counties, tells me that in all those years the proprietors, four in number, gave just one five-pounds in all to assist the poor. Now they give about five hundred a year, while the people are taxed to the amount of another five hundred. This would be little matter, if the condition of the poor were improved; but it is unmistakably and undeniably a hundred times worse. Nothing like the thousand pounds named finds its way into their pockets: collectors, inspectors, law expenses, etc., swallow up a great part of it. But, worse than all, the kindly charities of the poor towards the poor are quite frozen up. Formerly paupers were assisted with a little milk, potatoes, and fish; now the industrious poor, irritated by the poor-law tax, will contribute nothing towards the support of their poorer neighbors. The cry is, Go to the Poor's

of the case that the prison is called a workhouse. If the poor, simply in their character as poor, had any such right to a portion of the property of their more fortunate countryfolks as that which their more fortunate countryfolks themselves have to the remainder, no legislator, Scotch or English, would dare clog that right with so degrading a condition. The laboring man has a right to be paid for his labor. Where is the despot who would venture to affirm, that, in order to make that right good, the laboring man would require to go into prison? His right was made good when he completed the stipulated work; and it is the lack of all such solid right on the part of the pauper, in his character as a pauper, that enables British legislators to attach conditions to the fulfilment of his ill-based claims, which even Turkish or Persian despots would not dare to attach to the claims of the creditor who demanded some debt legally due to him. The conditions by which the legal right described by Paley may be clogged at pleasure, demonstrate that it is not a reality, but a fiction. The deserving poor have, indeed, a claim upon their wealthier brethren; but it is a claim which human laws cannot enforce without entirely altering its character; it is a claim which bears reference to the divine law alone, and to man's responsibility to his Maker.

Let us analyze this matter: we deem it one of considerable importance. It seems to be mainly from this want

Board. Even the sympathies of children towards their parents are dried up. This is universally spoken of as a new and shocking phase of things. It is not uncommon for young people to get married on the very day their parents go to the poor-house. In towns the state of matters is, if possible, worse. The assistance rendered by the Poor's Board becomes an absolute premium on vice. No hand is stretched out towards the *struggling* poor, because character is made of no account; but vice and improvidence urge their claims unblushingly, and they dare not be disregarded. This is very disheartening to well-disposed individuals of the better classes who take an interest in the condition of the poor: the more so that the poor themselves are so well aware of it. "Ah! *we* will get no help," say those who strive to maintain a little outward decency; "but let us first get drunk, and then sell everything that is left to us, and then we shall be sure of it!" It is easier to create evils by unwise legislation than to cure them Nevertheless, some checks upon such an unwholesome state of things ought to be devised.

of a solid claim on the one side, and this consequent right of enforcing disagreeable conditions on the other, that compulsory assessments have so invariably the effect of setting at variance the classes on whom they are levied, with the class for whose support they are made.

We resided and labored in this part of the country for a summer and autumn about eighteen years ago, at a time when wages were high and employment abundant. There was much dissipation among the working classes of the period; and one of our brother workmen, Jock Laidlie, was an extreme specimen of the more dissipated class. Pay day came round once a fortnight, and then we were sure to lose sight of Jock for about three days. When he came back to resume his labor, he had always a miserable, parboiled sort of look, as if he had been simmering for half an hour in a caldron over a slow fire. He was invariably, too, in that wretched state of spirits which in those days workmen used to term "*the horrors;*" and as men can't get parboiled and into "*the horrors*" for nothing, it was found in every instance that Jock's whole wages had been dissipated in the process. And such, fortnight after fortnight, was the course pursued by Jock. Now, employment, though easily enough procured in summer and autumn in Jock's profession, was always uncertain in winter, even when the winter proved fine and open; and when frosts were keen and prolonged, and the snow lay long on the ground, there was no employment for even the more fortunate. It was essentially necessary, therefore, in the busier seasons, to make provisions for the season in which business failed. For our own part, we were desirous, we remember, to have the *winter* all to ourselves; and when Hallow-day came round, and employment failed, we found ourselves in the possession of twelve pounds, which we had laid by just as its price, if we may so speak. Twelve pounds released us from the necessity of laboring for twice twelve weeks. Twelve pounds were sufficient to purchase for us leisure and independence — two very excellent things — from the end of October to the beginning of May; and we were desirous to employ the time thus fairly earned in cultivating a little inheritance which, in lesser or larger measure, descends to all, and of which no law of appropriation can rob even working-men, but which, unless resolutely broken in, and sedulously improved, must lie fallow and unproductive, — of no benefit to the possessor, and useless to the community. Jock Laidlie had not laid by a single farthing; we, on a very small scale, were a capitalist determined on making an investment. Jock was a pauper; and here, in a state of great simplicity, in comes the question at issue, — Had Jock Laidlie any right to our twelve pounds?

To not one copper farthing of it, say we. It was all our own, — all honestly earned by the sweat of our brow. We had never claimed any right to share with Jock in a single gill; we had never tasted his whiskey; we had never enjoyed one whiff of his tobacco; we had never meddled with *his* earnings; he had no right to intermeddle with ours. But Jock Laidlie had an aged mother, who, without any fault on her part, was miserably poor, just because Jock had failed in his duty to her. Had Jock Laidlie's mother any right to our twelve pounds? No — no right. It might doubtless be a duty to help the poor suffering woman; but her claim upon *us* was merely a claim on our compassion. She had no *right;* nor had any third party a right to thrust his hand into our pocket, and, out of our hard-earned twelve pounds, to assist Jock Laidlie's mother.

But if this was the true state of things with regard to the earnings of a single summer and autumn, accumulated with an eye to the coming winter, could there be any new element introduced simply by multiplying the summers and autumns some thirty or forty times, and by making their accumulated earnings bear reference, not to the winter of the year, but to the winter of life? Assuredly not, say we. The principle would remain intact and unchanged, however largely the seasons or the earnings might be multiplied. But suppose, further, that these earnings of forty years were to be invested in a house or a piece of land, would not Jock Laidlie or his mother have some right to share in them then? Would not their conversion into earth and stone, or into stone and lime, derive a right to Jock or Jock's mother? Paley has a very elaborate argument on the subject, from which he seems to arrive at the conclusion that it would. "All things," says this writer, "were originally common. No one being able to produce a charter from heaven, had any better title to a particular possession than his next neighbor. There were reasons for mankind agreeing upon a separation of this common fund; and God, for these reasons, is presumed to have ratified it; and as no fixed laws for the regulation of property can be so contrived as to provide for the relief of every case and distress which may arise, these cases and distresses, when their right and share in the common stock were given up or taken from them, were supposed to be left to the bounty of those who might be acquainted with the exigencies of their situation, and in the way of affording assistance. And therefore, when the partition of property is rigidly maintained against the claims of indigence and distress, it is maintained in opposition to the intention of those who made it, and to *His* who is the supreme proprietor of everything, and who has filled the world with plenteousness for the sustenance and comfort of all whom He sends into it." Does not Jock Laidlie or his mother acquire a claim to intromit with earnings transmuted into land, in virtue of all this fine philosophy, and this original compact on which it professes to be founded? No, not the shadow of a claim. We insist, in the first instance, upon Jock Laidlie's producing proof of this compact. We never heard of it before. Paley tells us that "when the partition of property is rigidly maintained against the claims of indigence, it is maintained in opposition to *the intention of those who made it.*" It is imperative, say we, that Paley prove that intention. To what records does he refer?

To what histories? Wherever man exists one degree above the savage state, there land is appropriated. It is appropriated in China in the far east, and in America in the far west; it is appropriated on towards the Antarctic in New South Wales, and far to the north, on the coasts of Iceland and the White Sea. In some of these countries the appropriation took place no later than yesterday; in some of the others it took place full thirty centuries ago. But from which of them, we marvel, could Paley or Jock Laidlie prove the existence of the compact? Do the settlers in the backwoods take axe in hand, to impart value to their newly appropriated acres by a long course of severe labor, with the intention affirmed by the philosopher? Do they recognize a right in the Jock Laidlies of the country to intromit with their buckwheat or their Indian corn now? Or do they yield to future Jock Laidlies a prospective right to intromit with the buckwheat or Indian corn of their descendants, when all the country shall have been appropriated and cleared? Most assuredly not. Or are the evidences of any such intention on the part of our ancestors embodied in the older records of those pieces of land in our own country which we occasionally see in the market at an upset price of from thirty to forty years' purchase? No. There can be but one answer to questions such as these. Paley's compact is altogether a fiction; and a citizen of York or of Bagdad might as well lay claim to some island of the Indian Ocean or the Pacific, on the ground that his townsman Robinson Crusoe or his townsman Sinbad the Sailor had once taken possession of it in behalf of the people of York or of Bagdad, as the poor lay claim to free support on the ground that such was the intention of the first appropriators of the soil. The first appropriators of the soil had no such intention.

The poor have indeed claims on the compassion of men; and these claims, when their poverty is the result of misfortune, are very strong. God speaks in their behalf in his word. He speaks in their behalf in the human heart, which His finger has made. When He gave laws to His chosen people of old, He forbade them to reap the corners of their fields, or to gather again the loose ears which fell from the hands of their reapers, that the fatherless and the stranger might pluck and eat, and that the poor gleaner might not ply in vain her tedious labors. But He gave to the poor no right in the property of his neighbor which the poor could assert before the civil magistrate. No third party was permitted to step in and determine what amount of assistance the pauper was entitled to receive or the rich necessitated to give. To himself alone did God reserve the right of being legislator and judge in the case; and under his wise management a genial charity, that softened and improved the heart, and clarified the whole atmosphere of society, did not degenerate into an odious tax, redolent of bitter discontent and ill-will; the bowels of compassion were not sealed up among those whom he had blessed with substance; nor did the children of poverty degenerate into mean and ungrateful paupers. In mercy to the poor, He gave them no such rights as those contended for by Paley.

Now, it is this felt want of right to support on the part of the poor that communicates, as we have said, to those who are compelled to support them, a right of enforcing disagreeable conditions. No man has a right in this country to put another man in prison simply because he is poverty-stricken and grows old. But any man has a right to say to any other man who is destitute of support, and yet has no legitimate claim to be supported, Go into prison, and I will support you there. From the invariable tendency of a poor-law not only to perpetuate itself, but also to increase mightily in weight, by adding to the improvidence and destitution of every country in which it is established, checks are found necessary: from its tendency to harden men's hearts, these checks are almost always of a barbarous character; and hence the workhouse check. The law, as it stands in England at present, empowers one man to take another man, guilty, it may be, of no other crime save poverty, from the wife with whom he has perchance lived in happiness for many years, and the circle of mayhap an attached family, and to shut him up in a prison under the rule of a despotic jailer, and among the very refuse of mankind. And what does it give the poor man in return, as the price of his liberty, and all that he enjoyed from the sympathy and society of a circle in the round of which his attachments lie? It gives him Paley's *right* of the poor — food, shelter, and clothing; for the two rights — the *right* of putting in prison and the *right* of being supported there — have come to be balanced against each other. It gives him miserable rations of the coarsest food, scanty in quantity, mayhap unwholesome in quality; and the share of a truck-bed, with, it may be, some poor diseased wretch, as loathsome in mind as in person, for his bedfellow. Such is the character of the English check. Nor can we doubt that in Scotland, naturally a much poorer country, — a country, too, in the possession of at least as hard-hearted an aristocracy as that of the sister kingdom, and in which, if once thoroughly contaminated by the influence of a poor-law, pauperism must increase enormously, — some check at least equally severe will come to be devised. The atmosphere of the English poor-houses is tainting all England with unwholesome disaffection and discontent; it is making bitter everywhere the heart of the poor man against the middle classes and the aristocracy; and, truly, no wonder. The poor-law bastilles at the last election furnished the grand topics of Chartist vituperation in England against the Whigs; and we are of opinion that the man requires to be a sanguine speculator indeed who ventures to surmise that their introduction into Scotland will have the effect of "sweetening the breath of society" there. The effect will be directly the reverse. The enactment of a Scottish poor-law must of necessity widen that gulf, so perilously broad already, which separates the upper from the lower classes.

There is one misguided and very numerous class on whom it must be brought peculiarly to bear, and whom we deeply pity. We are, we trust, friendly to Chartists, though determinedly hostile to Chartism. The principle is ruining thousands and tens of thousands of our working-men. It is an *ignis fatuus*, leading them astray in quest of an imaginary and unrealizable good, when, in many cases at least, some real good lies full within their reach, but of the very existence of which, blinded by the Chartist hallucination, they have no perception. Scotland was always a poor country, narrow in its resources, and at times grievously oppressed. It never yet succeeded in employing all its people. But in times when religion was prized, and education not neglected, the effects of the pressure were rather favorable than otherwise. It thrust out on every side an intelligent, energetic, trustworthy people, who made room for themselves everywhere. Continental Europe knew them in all its cities, — England, Ireland, the colonies, the whole world. Ere taking leave of their country, they stood on the elevation of the parish school and the parish church; and, discerning advantage at a great distance over the face of the globe, they bent their steps direct upon it. And in virtue of the same process, those who remained behind were fitted for improving to the utmost the resources within their reach at home. There are thousands of Scotchmen in the present day, men with the same blood in their veins, who are wasting their energies on the five points of the Charter, engaged in dreaming a disturbed and unhappy dream about unrealizable political privileges, which, even if attainable, would be useless; and precipitating themselves, meanwhile, on the poor-house. Let the reader just try to imagine a poor-law bastille existing under the more stringent and repulsive checks of the system, and filled with superannuated Chartists. Of all writers, Crabbe alone was fitted to do justice to the miseries of such a prison so filled. It would be truly "a hell upon earth." The transition from a state in which aspirations after universal suffrage are deemed of but a lower and comparatively commonplace kind, and in which all existing institutions are denounced as far beneath the ideal of true liberty or the standard of free-born men, to a state compared with which the despotism of Turkey or Morocco would be liberal, and the degradation of ordinary slavery not at all subversive of the dignity of man's nature, could be compared to only those transitions described by Milton: —

> "When all the damned
> Are brought to feel by turns the bitter change
> Of fierce extremes, — extremes by change more fierce, —
> From beds of raging fire to starve in ice."

But we cannot shut our eyes to the fact that a poor-law has become inevitable in Scotland. Though unable to recognize that right on the part of the poor for which Paley contends, there exists a right to legislate in their behalf on the part of the state which we cannot avoid recognizing. The state has as decidedly a right to impose a tax in order that a portion of its subjects may not be destroyed by starvation, as it has a right to impose a tax in order that a portion of its subjects may not be destroyed by an invading enemy; and there are cases in which the enactment of a poor-law may be imperatively a duty. In such a state of things as that drawn by Goldsmith in his allegory of Asem the Man-hater, for instance, where, in a country whose inhabitants were devoid of all pity, the diseased and the aged were suffered to perish by the wayside, a poor-law would have been the state's only alternative. It would have been as much its duty to interpose a tax between its perishing people and destruction in a case of this kind, as it would be its duty to levy a tax for carrying on a war of defence against a merciless enemy, who was ravaging its territories with fire and sword. There is another principle on which a state may well interfere. During the great plague of Marseilles, the living, sunk in the indifference of despair, would no longer bury their dead, and fifteen hundred bodies lay rotting in the sun outside the city gate, adding, by their poisonous effluvium, at once to the horrors and the intensity of the contagion. The magistracy interfered, and compelled their interment; and who can doubt that the magistracy did right in the case? Now it seems unquestionable that, among our neglected poor, diseases originate which, like the effluvium of the dead at Marseilles, spread infection and death through all classes of the community; and the circumstance derives to the state both a right and a duty in behalf of all its people to remove, through a provision for the poor, the distress and squalor in which the evil originates. Now, in the present state of Scotland we recognize an urgent necessity, on both these principles, for state interference in behalf of the poor. They are perishing for lack of bread; they are spreading deadly contagion through our lanes and alleys; the system of compulsory support is a coarse, inadequate system; it will have by and by to be connected with some repulsive check, in order that the capital and industry of the country may not be swallowed up by its lean and blighted poverty. But, however coarse, however inadequate, however productive it may prove of fierce discontent or miserable degradation, it is the only system in the field at present. A poor-law, we repeat, has become inevitable in Scotland. The controversy between contending systems exists among us no longer. Dr. Alison still occupies his ground; Dr. Chalmers has withdrawn.

Truly, it is enough to make one's heart swell to think how the gigantic exertions of this great and good man in behalf of his country have been met in this cause. Were we to say that the poor of Scotland are on the eve of perishing in utter degradation, from a lack of faith in the efficacy of the gospel of Christ on the part of our influential classes, the remark would no doubt be deemed over extreme and severe; and it would be a remark open, doubtless, to objection, — not, however, from its severity, but from its tame and inexpressive inadequacy. It is the condemnation of the class most influential in directing the destinies of our country, not that, in the indifferency of unbelief, they have stood aloof and done nothing, but that they have risen in maniac hostility, and overpowered those who were straining all their energies in their behalf. Not since the days of Knox did any venerable father of the Church of Scotland so exert himself in bringing Christianity to the people by the erection of congregations and the planting of churches, as Dr. Chalmers has done. Never has merchant so travailed to fill his coffers, or statesman so labored to consolidate his power, as this man has travailed and labored, in season and out of season, to bring the blessings of the gospel to the poor, the degraded, and the forgotten. In ten years the Church of Scotland saw two hundred places of worship added to her communion. And how have these his weapons — forged to bear down the crime and ignorance, and, with these, the poverty of the country — been dealt with? Let our law-courts tell, in the first instance; let our aristocrats who stand by applauding their decisions, declare in the second. Who was it that, when the state and the aristocracy of the country refused to endow his churches, and when the industrious and religious poor came forward for the purpose with their coppers, widows with their mites, and toil-worn laborers and mechanics with pittances subtracted from their scanty wages, — who was it that made prize of their humble offerings, and confiscated them, on behalf of the pauperism of the country, forsooth? There was an irony in the pretext which those who employed it could not have fully understood at the time, but which they will come to appreciate by and by. And who, through the Stewarton and Auchterarder decisions, have fully completed what the Brechin decision began? Truly, the parties who had most at stake in the exertions of the champion who took the field in their behalf have been wonderfully successful in disarming and forcing him aside; and all that is necessary for them now is, just to be equally successful in grappling with the o'ermastering and enormous evils which he set himself so determinedly to oppose. We trust, however, that they will no longer attempt deceiving the country, by speaking of a moral force as a thing still in the field, in opposition to the merely pecuniary force recommended by Dr. Alison. The moral force is in the field no longer; Dr. Alison stands alone.

For the present, however, we must conclude. Very important questions of morals are on the eve of becoming questions of arithmetic in Scotland; and the wealth of the country, though it may find the exercise a reducing one, will be quite able to sum them up in their new character. Let us just touch one two of them by way of specimen. We have adverted oftener than once to the evils of the bothy system. They are going to take the form of a weighty assessment; and our proprietary may be induced to inquire into them in consequence. There is another great evil to which we have not referred so directly. All our readers must have heard of vast improvements which have taken place during the present century in the northern Highlands. The old small farm, semi-pastoral, semi-agricultural system was broken up, the large sheep-farm system introduced in its place, and the inland population of the country shaken down, not without violence, to the skirts of the land, there to commence a new mode of life as laborers and fishermen. And all this was called improvement. It was called great improvement not many years since, in most respectable English, in the pages of the "Quarterly Review." And we heard a voice raised in reply. It was the scrannel voice of meagre famine from the shores of the northern Highlands, prolonged into a yell of suffering and despair. But, write as you may, apologists of the system, you have ruined the country, and the fact is on the eve of being stated in figures. The poor-law assessment will find you out.

X.

PAUPERISM.

The utterly miserable are always unsafe neighbors. In former days, when a barbarous jurisprudence, with its savage disregard of human life, extended to our prisons, and every place of confinement in the kingdom was a stagnant den of filth and wretchedness, the contagious disease originated in these receptacles of horror and suffering, and which from this circumstance bore the name of the jail-distemper, frequently burst out on the inhabitants of the surrounding town or village, and carried them off by hundreds at a time. It is recorded, that, after a criminal court had been held on one occasion, in the reign of James VI., at which the celebrated Lord Bacon took some official part, a malignant fever broke out among the persons who had attended, which terminated fatally in the case of several of the jury and of some of the gentlemen of the bar, and that the philosophic Chancellor expressed his conviction that the contagion had been carried into the court-room by a posse of wretched felons from the tainted atmosphere of their dungeon. Self-preservation in these cases enforced the dictates of humanity: the same all-powerful principle enforces them still. It is more than probable that the misery of the neglected classes occasionally breaks out upon that portion of our population which occupies the upper walks in society, in the form of contagious disease, — in the form of typhus fever, for instance: there can be no doubt whatever that it often breaks out upon them in the form of crime.

But where is the true remedy to be found? It was comparatively an easy matter to ventilate our prisons, and to introduce into them the various improvements recommended alike by the dictates of humanity and prudence. But how are the suffering masses to be ventilated, and their condition permanently improved? It does not do to grope in the dark in such matters. It is well, surely to meet with the evil in *its effects* when it has become utter misery and destitution, and to employ every possible means for relieving its victims. It is infinitely better, however, to meet with it in *its causes*, — to meet with it in the forming, and to check it there. It was not by baling back the waters of the river that Cyrus laid bare the bed of the Euphrates; it was by cutting off the supply. Where are the sources of this fearfully accumulated and still accumulating misery to be found? At what particular point, or in what particular manner, should the enlightened benefactor of the suffering classes interfere to cut off the supply? The reader anticipates a truism, — one of those important and unquestioned truths which, according to Goethe, seem divested of their proper effect, as *important* just from the circumstance of their being *unquestioned*, and which, gliding inefficiently along the stream of universal assentation, are allowed to weigh less with the public mind than the short-lived and unfruitful paradoxes of the passing time. Instead, however, of laying down a principle, we shall simply state a few facts of a kind which many of our humbler readers — the "men of handicraft and hard labor" — will be able fully to verify from their own experience, and that embody the principle which seems to bear most directly on the subject.

We passed part of two years in the neighborhood of Edinburgh immediately before the great crisis of 1825, and knew perhaps more about the working classes of the place than can well be known by men who do not live on their own level. The speculations of the time had given an impulse to the trading world. Employment was abundant, and wages high; and we had a full opportunity of seeing in what degree the mere commercial and trading prosperity of a country — the mere money-welfare which men such as Joseph Hume can appreciate — is truly beneficial to the laboring portion of the community. We shall pick out, by way of specimen, the case of a single party of about twenty workmen, engaged at from twenty-four to twenty-seven shillings per week, most of them young, unmarried men, in the vigor of early manhood. Remember we are drawing no fancy sketch. Fully two thirds of that number were irreligious, and in a greater or less degree dissipated. They were paid by their employer regularly once a fortnight, on the evening of Saturday; and immediately as they had pocketed their wages, a certain number of them disappeared. On the morning of the following Wednesday, but rarely sooner, they returned again to their labors, worn out and haggard with the excesses of three days grossly spent, and without a single shilling of the money which they had earned during the previous fortnight. And such was the regular round with these unfortunate men, until the crisis arrived, and they were thrown out of employment in a state of as utter poverty as if they had never been employed at all.

There was a poor laborer attached, with a few others, to the party we describe, whose wages amounted to about half the hire of one of the mechanics. His earnings at most did not exceed fourteen shillings per week. This laborer supported his aged mother. On Sundays he was invariably dressed in a neat, clean suit; he occasionally indulged, too, in the purchase of a good book; and we have sometimes seen him slip, unnoticed as he thought, a few coppers into the hands of a poor beggar. And yet this man saved a little money. We lived nine months under the same roof with him; and as we were honored with his confidence and his friendship, we had opportunities of seeing the character in its undress. Never have we met with a man more thoroughly a Christian, or a man who felt more continually that he was living in the presence of Deity. Now, in the ordinary course of events, and debarring the agency of accident, it is well-nigh as impossible that men such as this laborer can sink into pauperism, as that men of the opposite stamp can avoid sinking into it. The dissipated mechanics, with youth and strength on their side, and with their earnings of twenty-four and twenty-seven shillings per week, were yet paupers in embryo. It is according to the inevitable constitution of society, too, that vigorous working-men should have relatives dependent upon them for sustenance, — aged parents or unmarried sisters, or, when they have entered into the marriage relation, wives and families. And hence the mighty accumulation of pauperism when the natural prop fails in yielding its proper support.

We have another fact to state regarding our old acquaintances, which is not without its importance, and in which, we are convinced, the experience of all our humbler readers will bear us out. Some of the most skilful mechanics of the party, and some, too, of the most intelligent, were among the most dissipated. One of the number, a powerful-minded man, full of information, was a great reader; there was another, possessed of an intellect more than commonly acute, who had a turn for composition. The first, when thrown out of employment, and on the extreme verge of starvation, enlisted into a regiment destined for some of the colonies, whence he never returned; the other broke down in constitution, and died, before his fortieth year, of old age. What is the proper inference here? Mere intellectual education is not enough to enable men to live well, either in the upper or lower walks of society, and especially in the latter. The moral nature must also be educated. Was Robert Burns an ignorant or unintelligent man? or yet Robert Ferguson?

Facts such as these — and their amount is altogether incalculable — indicate the point at which the sources of pauperism can alone be cut off. The disease must be anticipated; for when it has passed to its last stage, and actually *become pauperism*, there is no remedy. Every effort which an active but blind humanity can suggest in such desperate circumstances is but a baling back of the river when the floods are rising. If there be a course of moral and religious culture to which God himself sets his seal, and through which even the dissipated can be reclaimed, and the uncontaminated preserved from contamination, — a course through which, by the promised influences of a divine agent, characters such as that of our friend the poor laborer can be formed, — that course of moral and religious culture is the only remedy. The pauperism of Scotland, in its present deplorable extent, is comparatively new to the country; and certain it is, that in the last age the spirit of anti-pauperism and of anti-patronage were inseparable among the Presbyterian people. There is a close connection between the non-intrusion principle and the formation of characters such as that of our friend the laborer. What were the religious sentiments of the class, happily not yet forgotten in our country, who bore up in their honest and independent poverty, relying for support on the promise of their Heavenly Father, but who asked not the help of man, and who, in so many instances, would not receive it even when it was extended to them? To what party in the church did the poor widows belong who refused the proffered aid of the parish, — if they had children, lest it should be "cast up" to them in after-life, — if they had none, "because they had come of honest people?" Much of what was excellent in the Scottish character in the highest degree arose directly out of the Scottish Church in its evangelical integrity; much, too, of what was excellent in the main, though perhaps somewhat dashed with eccentricity, arose out of what we may term the church's reflex influences.

XI.

PAUPER LABOR.

We hold that the only righteous and practical check on adult pauperism, the only check at once just and efficient, is the compulsory imposition of labor on every pauper to whom God has given, in even the slightest degree, the laboring ability. One grand cause of the inefficiency of workhouses arises mainly from the circumstance that their names do not indicate their character. The term workhouse has become a misnomer, seeing that it designates buildings in which, for any one useful purpose, no work is done. We say for any useful purpose; for in some cases there is work done in them which is of a most mischievous, pauper-producing kind. They enter, in the character of competitors, into that field of unskilled, or at least very partially skilled labor, which is chiefly occupied by the self-sustaining classes that stand most directly on the verge of pauperism; and their hapless rivals, backed by no such bounty as that upon which *they* trade, sink in the ill-omened contest, and take refuge within their walls, to assist in carrying on that war against honest industry in which they themselves have gone down. Folly of this extreme character in the management of the pauperism of the country admits of no apology, from the circumstance that it is as palpable as it is mischievous. The legitimate employment of the inmates of a workhouse we find unmistakably indicated by the nature of their wants. What is it that constitutes their pauperism? Nature has given them certain wants, which, from some defect either in character or person, they themselves fail to supply; they lack food and they lack raiment; and these two wants comprise the wants of a poorhouse. Then let the direct supply of these wants be the work of a poorhouse, — its direct, not its circuitous work, — not its work in the competition market, to the inevitable creation of more paupers, but its work in immediate connection with the soil, out of which all food and all raiment are produced, and with the wants of its own inmates. The organization of labor in society at large we regard as an inexecutable vision. In even the most despotic nations of Europe that compulsory power is wanting which must constitute — man being what he is — the moving force of organized labor; but within the precincts of a workhouse the compulsory power *does* exist; and there, in consequence, the organization of labor is no inexecutable vision, but a sober possibility. It would impart to our workhouses their proper character, by not only furnishing them with an efficient labor check, and converting them into institutions of discipline, in which the useless member of society, that could but would not work, would be compelled to exert himself in his own behalf; but it would also convert them into institutions in which a numerous pauper class, of rather better character, — too inefficient, either from lack of energy or of skill, to provide for themselves, amid that pressure and bustle of competition which obtains in society at large, — might, by being shielded from competition, and brought into immediate contact with the staple of their wants, become self-supporting. All that would be necessary in any poorhouse would be simply this, — that its class of raiment-producers should produce clothes enough for both themselves and its sustenance-producers; and that its sustenance-producers should, in turn, produce food enough for both themselves and its raiment-producers. And, brought fairly into contact with the soil and its productions in the raw state, — with their wants reduced to the simple natural level, the profits of the trader superseded, the pressure of taxation removed, the enormous expenses of the dram-shop cut off by that law of compulsory temperance which the lack of a command of money imposes, — we have little fear but that many of those institutions would become self-supporting, or at least very nearly so. The country would still have to bear some of the expense of what has been well termed its heaven-ordained poor, — the halt, the maimed, and the fatuous; but be it remembered that these always bear a definite proportion to the population; and that the present alarming increase in the country's pauperism is not a consequence of any disproportionate increase in that modicum of its amount which the heaven-ordained poor composes.

So much for the country's adult pauperism. With regard to its juvenile pauperism, the labor scheme is more important still. The country has many poor children living at its expense in workhouses, or boarded in humble cottages in the country; and there are many more that either want parents, or worse than want them, that are prowling about its larger towns, and scraping up a miserable livelihood by begging or theft. Unless in the season of youth — ere the mind becomes rigid under the influence of habit, and takes the set which it is to bear through life — these juvenile paupers and vagabonds be converted into self-sustaining, honest members of society, they will inevitably become the adult paupers or criminals of the future, and the country will have to support them either in poorhouses or penal settlements, or, worse still, to pay executioners for hanging them. Of all non-theological things, labor is the most sacred; of all non-ethical things, labor is the most moral. The working habit — the mere homely ability of laboring fairly and honestly for one's bread — is of more value to a country, when diffused among its people, than all the other gifts — be they hills of gold or rocks of diamonds — that can possibly fall to its share. And if its people, or any very considerable part of them, possess not that habit and ability, it matters not what else it may possess: there is an element of weakness in its constitution for which no amount of even right principle among them will ever form an adequate compensation. There is, we believe, no part of her Majesty's dominions in which there is more right principle than in the Highlands of Scotland; but, from causes which it might be a mournful, but certainly no uninstructive task to trace, their people possess the working habit and ability in a comparatively small degree; and so they can do exceedingly little for the propagation of the principles which they hold, and, when disease touches the root of the potato, they find themselves in circumstances in which, save for the charity of their neighbors, they would perish. Principle, even when held truly and in sincerity, as among many of our poor Highlanders, is not enough of itself; and the mere *teaching* of principle in early life, in lessons which may or may not be received efficiently and in truth, must of itself be still less sufficient. Even if the best churches in the country had the country's vagabond and pauper children subject to their instruction, — supposing the thing possible, though, of course, if the churches did not feed them, it is not; and supposing, further, that they turned them out on society, the course completed, destitute of industrial habits or skill, — what would be the infallible result? The few converted to God by a vital change of heart, — and in all ages of the church the numbers of such have been proportionably few, — would no doubt either struggle on blamelessly through life, or, sinking in the hard contest, would resign life rather than sustain it by the fruits of a course of crime; but the great bulk of the others would live as paupers or criminals: they would be simply better instructed vagabonds than if they had been worse taught. The welfare of a country has two foundations. Right principle is the one; and the other, and scarce less important foundation, is industrial habit combined with useful skill. And in order to obviate the great danger of permitting juvenile paupers to grow up into adult paupers and criminals, it is essentially necessary that the skill should be communicated to them, and the habits formed in them. And hence the importance of the scheme that, by finding regular employment for the youthful paupers of the country, would rear them up in honest, industrial habits, and thus qualify them for being useful members of society.

It has been alleged against Presbyterianism by excellent men of the English Church, — among the rest by Thomas Scott the commentator, — that in its history in the past it has been by much too political, and has busied itself too engrossingly with national affairs. There can be little doubt that its history during the seventeenth and the latter half of the sixteenth century is very much that of Scotland. Presbyterianism *was* political in those days, and fought the battles of civil as certainly as those of religious liberty. During a considerable part of the eighteenth century it was *not* political. From the suppression of the Rebellion of 1745 to the breaking out of the great revolutionary war, the life led by the Scottish people was an exceedingly quiet one, and there were no exigencies in their circumstances important enough to make large demands on the exertions of the patriot or the ingenuity of the political economist. The people of the empire rather fell short than exceeded its resources, and were somewhat less than sufficient to carry on its operations of agriculture and trade; and hence the comfortable doctrine of Goldsmith and Smollett regarding population, — a comfortable doctrine, for it never can obtain save when a nation is in comfortable circumstances. The best proof of the welfare of a country, they said, was the greatness of its population. It was unnecessary in such an age that Presbyterianism should be political. The pauperism which had deluged Scotland immediately after the Revolution had been all absorbed; the people, in at least the Lowlands, were a people of good working habits; and in the Highlands little work served; and all that had to be done by such of the ministers of religion in the country as were worthy of the name was to exert themselves in adding right principle and belief in relation to the realities of the unseen world, to the right habits in relation to the present one that had already been formed among the people of their charges. But with the revolutionary war and the present century the state of matters greatly altered. Pauperism began mightily to increase ; the recesses of our large towns, that some forty or fifty years before used to pour out to the churches, at the sound of the Sabbath bells, a moral and religious population, became the foul dens in which a worse than heathen *canaille* festered in poverty and ignorance ; habits of intemperance had increased twenty-fold among the masses; the young were growing up by thousands in habits of idleness and crime to contaminate the future ; even the better people, placed with their children in perilous juxtaposition with the thoroughly vitiated, were in the circumstances of men in health located per force in the fever-ward of a hospital. The Scottish Highlanders, too, ruined by the clearing system, had come to be in circumstances greatly different from those of their fathers ; and it had grown once more necessary that the Presbyterian minister should, like his predecessors of the sixteenth century, interest himself in a class of secular questions that are shown by experience to be as clearly allied to spiritual ones as the body is to the soul. The one great name specially connected with this altered state of things, and the course of action which it demands, is that of Chalmers, — Chalmers, the true type and exemplar of the Presbyterian minister as specially suited to the exigencies of the time. But there are other names. The late Dr. Duncan with his savings banks, Guthrie with his ragged schools, Begg and Mackenzie with their dwellings for the working classes, Tasker in his West Port laboring in the footsteps of his friend the great deceased, must be regarded as true successors of those Presbyterian ministers of the seventeenth century who identified themselves with their people in all their interests, and were as certainly good patriots as sound divines. And there are signs in the horizon that their example is to become general. We have scarce met a single Highland minister for the last three or four years,

— especially those of the northwestern Highlands, — who did not ask, however hopeless of an answer, "What is to be done with our poor people?" The question indicates an awakening to the inevitable necessity of inquiry and exertion in other fields than the purely theological one; and one of these, in both Lowlands and Highlands, is that in which Chalmers so long labored. The case of the poor must be *wisely* considered, or there will rest no blessing on the exertions of the churches.

But we must bring our remarks to a close; and we would do so by citing an instance, only too lamentably obvious at the present time, of how very much, in our mixed state of existence as creatures composed of soul and body, a purely physical event may affect the religious interests of a great empire. The potato disease was a thing purely physical. It seemed to have nothing of the nature of a missionary society about it; it did not engage missionaries, nor appoint committees, nor hire committee-rooms, nor hold meetings; and it seemed to have as little favor for popish priests as for Episcopalian curates or Presbyterian ministers. And yet, by pressing out the popish population of Ireland on every side, and surcharging with them the large towns of England, Scotland, and the United States, it has done more in some three or four years for the spread of popery in Britain and America than all the missionary societies of all the evangelistic churches of the world have done for the spread of Protestantism during the last half-century. He must be an obtuse man who fails to see, with such an example before him, how intimately associated with the ecclesiastical the secular may be.

XII

THE CRIME-MAKING LAWS.

If there was a special law enacted against all red-haired men and all men six feet high, red-haired men and men six feet high would in a short time become exceedingly dangerous characters. In order to render them greatly worse than their neighbors, there would be nothing more necessary than simply to set them beyond the pale of the constitution, by providing by statute that whoever lodged informations against red-haired men or men six feet high should be handsomely rewarded, and that the culprits themselves should be lodged in prison, and kept at hard labor, on every conviction, from a fortnight to sixty days. The country would at length come to groan under the intolerable burden of its red-haired men and its men six feet high. There would be frequent paragraphs in our columns and elsewhere to the effect that some three or four respectable white-haired gentlemen, varying in height from five feet nothing to five feet five, had been grievously maltreated in laudably attempting to apprehend some formidable felon, habit and repute six feet high; or to the effect that Constable D. of the third division had been barbarously murdered by a red-haired ruffian. Philosophers would come to discover, that so deeply implanted was the bias to outrage and wrong in red-haired nature, that it held by the scoundrels even after their heads had become bald and their whiskers gray; and that so inherent was ruffianism to six-feet-highism, that though four six-feet fellows had, for the sake of example, been cut short at the knees, they had remained, notwithstanding the mutilation, as incorrigible ruffians as ever. From time to time there would be some terrible tragedy enacted by some tremendous incarnation of illegality and evil, who was both red-haired and six feet high to boot. Of course, to secure the protection of the lieges, large additions would be made to the original statute; and thus the mischief would go on from bad to worse, unmitigated by the teachings of the pulpit or the press, and unrestrained by the terrors of the magistracy, until some bold reformer, rather peculiar in his notions, would suggest, as a last resource, the repeal of what ere now would come to be very generally lauded as the sole safeguards of the public peace and the glory of the Constitution, — the anti-red-hair, anti-six-feet-high enactments. And after the agitation of some fifteen or twenty years — after articles innumerable had been written on both sides, and speeches without number had been spoken — the enactments would come to be fairly rescinded, and the tall and the red-haired, in the lapse of a generation or two, would improve, in consequence, into good subjects and quiet neighbors.

Is the conception too wild and extravagant? Let the reader pause for a moment ere he condemns. England little more than a century ago was infamous for the number of its murders committed on the highway. Hawksworth's story, in the "Adventurer," of the highwayman who murdered a beloved son, just restored, after a long absence, to his country and his friends, before the eyes of his father, and then threw the old man a shilling, lest, said the ruffian, he should be stopped at the tolls, was not deemed out of nature at the time. It was, on the contrary, quite a probable occurrence in the days of Jack Sheppard, Turpin, and Captain Macheath. About an age earlier, as shown by the "London Gazette," one of the oldest of English newspapers, there were from six to eight murders perpetrated yearly by foot-pads on the public roads; and paragraphs such as the following, which we extract from this ancient journal, were comparatively common: "On the 23d of this month [March, 1682], three highwaymen, two on horseback and one on foot, set upon two persons on Hindhead Heath, in Surrey, one of whom they mortally wounded, and took from them a black crop gelding near fifteen hands high;" or such notices as the following, inserted as a general citation of witnesses, by the keeper of the Newgate: — "Whereas many robberies are daily committed on the highways, to the great prejudice of his Majesty's subjects, — these are to give notice that there has lately been taken and are now in the custody of Captain Richardson, Master of his Majesty's jail at Newgate, several supposed highway robbers, of whom here followeth the names and descriptions," etc. Such was the state of things in times when the earlier British novelists, desirous of making the incidents lie thick in their fictions, gave them the form of a journey, and sent their heroes travelling over England. The evil, however, was at length put down, partly through the marked improvement which took place in the police of the country, but still more through the great increase of its provincial newspapers, and the vast acceleration in the rate of its travelling, — circumstances which have united to render the escape or concealment of the highwayman impossible. And so highway murder has become one of almost the rarest offences in the criminal register of the country. Very different is the case, however, with murders of another kind. Our newspapers no longer contain in their English corner paragraphs at all resembling those we have just quoted, by way of specimen, from the "London Gazette," and which so strike in the perusal, as characteristic of an age only half escaped from barbarism; but they exhibit, instead, their paragraphs, to the barbarity of which the accommodating influence of custom can alone reconcile the reader, and which will be held, we trust, in less than half an age hence, to bear as decidedly the stamp of savageism. Within the last few years there have been no fewer than twenty-five gamekeepers murdered in England. The cases were all ascertained cases; coroners' juries sat upon the bodies, and verdicts of wilful murder were returned against certain parties, known or unknown; and these were, of course, but the murders on the one side. We occasionally hear of the death of a poacher; and all our readers must remember a late horrible instance, in which an unfortunate man of this class, captured after a desperate resistance, was found to be so dreadfully injured in the fray that his bowels protruded through his wounds. But in by far the greater number of cases the poor wounded wretch has strength enough left to bear him to his miserable home, and the parish hears little more of the matter than that there has been a brief illness and a sudden death. It is quite bad enough that Hawksworth's story of the highwayman should be a not improbable one in the times of the first two Georges; it is still worse that Crabbe's story of the rival brothers who killed each other in a midnight fray, in which the one engaged in the character of a poacher the other in that of a gamekeeper, should be as little improbable in the times of William and Victoria.

Be it remembered, too, that the peculiar barbarism of the modern period is greatly more a national reproach than that of the ancient. The older enormities were enormities in spite of a good law; the newer enormities are enormities that arise directly out of a bad one. There is sound sense as well as good feeling in the remark of Mrs. Saddletree on the law, in Effie Dean's case, as laid down by her learned husband the saddler. "The crime," remarked the wiseacre to his better half, "is rather a favorite of the law, this species of murder being one of its own creating." "Then, if the law makes murders," replied the matron, "the law should be hanged for them; or if they would hang up a lawyer instead, the country would find nae faut." All the twenty-five ascertained murders to which we have referred, and the at least equally great number of concealed ones, were crimes of the law's making, — murders which as certainly originated in the law, and which, if the law did not exist, would as certainly not have been, as the supposed crimes of our illustration under the anti-red-hair, anti-six-feet-high statutes. No murders arise out of the killing of seals and sea-gulls; why should there arise any murders out of the killing of hares and pheasants? Simply because there is a pabulum of law in the one case, out of which the transgression springs, and no producing pabulum of law in the other. There can be nothing more perilous to the morals of the people than stringent laws, that, instead of attaching their penalties to actual crime, and having, in consequence, like the laws against the housebreaker and the highwayman, the whole weight of the popular conscience on their side, create the crime which they punish, and have thus the moral sense of the country certainly not for, mayhap against them. They become invariably, in all such cases, a sort of machinery for converting useful subjects and honest men into rogues and public pests. Lacking the moral sanction, their penalties are neither more nor less than a certain amount of peril, which bold spirits do not hesitate to encounter, just as a keen sportsman does not hesitate to encounter the modicum of risk which he runs from the gun that he carries. It may burst and kill him; or in drawing it through a hedge a sprig may catch the trigger, and lodge its contents in his body; or it may hang fire, and send its charge through his head half a minute after he has withdrawn it from his shoulder. Accidents of the kind happen in sporting countries almost every month, — for such is the natural law of accident in the case; but there is no moral stigma attached, and so men brave the penalty every day. And such is the principle, when the law, equally dissociated from the promptings of the moral sense, is not a law of accident, but of the statute-book. Men brave the danger of the penalty, as they do the peril of the fowling-piece. But there is this ultimate difference: without being in any degree a felon tried by his own conscience, the traverser of the statutory enactment becomes legally a felon; he may be dealt with, like the red-haired or six-feet-high felon of our illustration, as decidedly criminal. He is ferociously attacked with lethal weapons as a felon; and, defending himself in hot blood with the resembling weapons, without which his amusements cannot be carried on, he becomes a murderer; or he is apprehended, manacled, tried, condemned, imprisoned, transported, as a felon, and, in passing through so degrading a process, becomes at length the actual criminal which he had been in the eye of the law all along. Few of our readers can have any adequate conception of the immense mass of criminalty created yearly in the empire by this singularly deteriorating process. In the year 1843 there were in England and Wales alone no fewer than four thousand five hundred and twenty-nine convictions under the game-laws. Forty of that number were deemed cases of so serious a nature that the culprits were transported. In all the other cases they were either fined or imprisoned, — the fines taken in the aggregate averaging two pounds sterling, the imprisonments seven weeks. And it is out of this system of formidable penalties that the numerous murders have arisen, and that the game-laws of the country have, like those of Draco, come to be written in blood.

The character of the ordinary Scotch poacher must be familiar to all our readers. "E'en in our ashes," says the poet, "live our wonted fires." There are few things more truly natural to man than a love of field-sports. Voyagers have remarked of the wild dogs of Juan Fernandez, that they hunt in packs. It needs, it would seem, no previous training to make them hunting animals: they are such by nature; and, placed in the proper circumstances, the nature at once develops itself. Now, it would appear as if man were also a hunting animal: the peculiar occupation which the first circumstances of society in almost every country render imperative upon the species, and for which, in an early age of the world, ere the human family was yet dispersed, Nimrod became so famous, is perhaps of all others the most natural to us. What the passion which leads to it is in the aristocracy, the game-laws serve of themselves sufficiently to testify; and the humbler classes feel the impulse as strongly. It is truly wonderful how soon men brought up in a state of civilization accommodate themselves, when thrown by circumstances among a barbarous people, or into a state of seclusion from their fellows, to the life of the hunter, and learn to love it. And the inherent feeling is, of course, as little blamable in the humble as in the wealthy or titled man. We have seen it greatly indulged in by dwellers along the seashore, — farmers, cottars, mechanics, — and almost every more spirited young man in the locality becoming in a lesser or greater degree a marksman. For a certain period, a young fellow of fair character has been shooting east, over the beach, towards the sea, and picking down the scart and the gray goose, the coot and duck, and now and then sending a bullet through the head of an otter or seal. A tempting opportunity occurs, however; and, instead of shooting east, he shoots west, over the beach, towards the land, and lodges his shot, not in a scart or seal, but in a woodcock or hare. Formerly he was in danger from his gun, or in scrambling among the rocks: he is now in danger of being fined, and, should he frequently repeat the offence, of being imprisoned; but in his own estimate and that of his neighbors the one kind of danger is no more connected with any moral stigma than the other. Had he fired west, and wilfully shot a sheep or goat, the case would, of course, be altogether different; but he is merely an occasional poacher, not a scoundrel. And if the game-laws be not strictly enforced in the district, he remains, as at first, a good and useful member of society, in no degree either the better or the worse for now and then shooting a coot or wild goose that has no standing in the game-list, and now and then picking down a partridge or heath hen that has.

But in those parts of England where game are rigidly preserved, and the game-laws strictly enforced, the process is different. The commencement of the poacher's course is nearly the same in both cases. There is the same instinctive love of sport, and the same general conviction that game is not real property, — a conviction which every view of the subject serves but to strengthen and confirm. The Englishman sees that if his neighbor the shopkeeper or banker detects a rascal robbing his till or breaking his strong box, he never once thinks of engaging him as his shopman or cashier; and that, on the same principle, the sheep-feeder or farmer avoids hiring as his shepherd a man notorious for stealing sheep, or declines employing as his farm-servant a man who has been tried and cast for stealing horses. He finds, too, that the fair trader never bargains with habit-and-repute thieves for their stolen goods. But he sees that an entirely different principle obtains among game-preservers. Not a few of them, bent on stocking their preserves, deal freely with poachers for live game; and still more of them, in choosing their game-keepers, prefer poachers — clever, active fellows, extensively acquainted among their own class — to any other sort of persons whatever. Nor, if the poachers be nothing worse than poachers, can there be a single objection to the arrangement, save on the unrecognizable, untenable ground that game is property. It is, however, the tendency of the poacher, in a country where the game-laws are strictly enforced, to *become* something worse. He goes to the woods, shoots or traps game, and finds himself, in consequence, in the circumstances of the red-haired or six-feet-high men of our illustration. He is apprehended and fined; and as his wages as a laborer are small, he has just to go to the woods again, in order — we quote a remark grown into a proverb among the class — that he may seek his money in the place where he lost it. He is again apprehended, and imprisoned for some six or eight weeks, during which time he is occasionally visited by the chaplain of the prison, who tells him he has done wrong, but always, somehow, forgets to quote the text which proves it, and is besides not particularly clear in his argument. He receives, too, visits of a different character, — those of hardened felons; and their lessons impress him much more deeply than the teachings of the chaplain. He is again discharged; but he has now become rather an unsettled sort of person, and fails not unfrequently to procure employment. But the neighboring preserves prove an unfailing resource: he is time after time surprised and apprehended; but he at length becomes weary of passive submission; the hour is late, the thicket dark and lonely, the gamekeeper alone; they are simply man to man; and in the scuffle which ensues the keeper is baffled and beaten off. Better a brief fray than a heavy fine or a long imprisonment. The poacher's associates, ere he has reached this stage, are chiefly desperate men. "There are notorious poachers," says Mr. Bright, in his speech on the game-laws with which he prefaced his motion for a parliamentary committee on the subject, "who have by a long succession of offences and imprisonments been driven out almost from the pale of society, — a kind of savages, living in hovels, or wherever they can find shelter. One of this outcast class was recently tried at the assizes for an act of incendiarism." Such company can have, of course, no tendency to improve a man's morals, or to increase his tenderness of human life. He engages in the forest in one fray more; and he who commenced his career as a law-made criminal, and free of moral stain in the abstract, terminates it in the character of an atrocious felon in the sight both of God and man, — a red-handed murderer, through whom two human lives have been lost to society, — that of his victim and his own.

It must be miserable policy that balances against the lives of human creatures and the morals of thousands of our humbler people, the mere idle amusements of a privileged class, comparatively few in number, and who have a great many other amusements full within their reach. Even were their claims to the game of the country clear, — and all know that a right of property in wild animals *can* be constituted by taking and keeping them, as Cowper did his hares, — still, did these claims interfere with the public good, they ought of necessity to give way. Justice, as certainly as humanity, demands the sacrifice. We are much pleased, in this point of view, with an anecdote related by Mr. Jesse in his "Gleanings in Natural History," an exceedingly interesting volume, from which the reader may learn that there are many other ways of deriving amusement from animals besides killing them. "One of the keepers in Richmond Park informs me," says the naturalist, "that he has often heard his father, who was also a keeper, mention that, in the reign of George II., a large flock of turkeys, consisting of not less than three thousand, was regularly kept up as part of the stock of the park. In the autumn and winter they fed on acorns, of which they must have had an abundant supply, since the park was then almost entirely wooded with oak, with a thick cover of furze; and although at present eleven miles in circumference, it was formerly much larger, and connected with extensive possessions of the Crown, some of which are now alienated. Stacks of barley were also put up in different places of the park for their support; and some of the old turkey-cocks are said to have weighed from twenty-five to thirty pounds. They were hunted with dogs, and made to take refuge in a tree, where they were frequently shot by George II. I have not been able to learn how long they had been preserved in the park before his reign; but they were totally destroyed towards the latter end of it, in consequence of the danger to which the keepers were exposed in protecting them from poachers, with whom they had many bloody fights, being frequently overpowered by them." Here we have a pleasing instance of even the monarch of the country yielding up his amusements in order that the lives of his servants might not be endangered. David would not drink of the water which was, he said, "the blood of the men that went for it in jeopardy of their lives," and so he "poured it out unto the Lord."

XIII.

IS GAME PROPERTY?

When we last walked out through several of our busier Edinburgh streets into the country, we did not see a single article in the shop-windows or elsewhere which we did not at once recognize as property, and of whose general lineage, as such, we could not give some satisfactory account. Human skill and labor had been employed upon them all, from the nicely-fashioned implement or machine in which the baser metals had become more valuable than silver, or the elaborate strip of gossamer-like tissue in which the original vegetable fibre had been made to outprice its weight in gold, to the wild intertropical nut or date gathered from their several palms under the burning sun of the African or Asiatic desert, or the costly furs of the Arctic hunter, purchased by the adven urous merchant of a civilized country amid the wild wastes of Lapland, or on the icy confines of Baffin's Bay or Mackenzie River. All was property on which the eye rested, — that of individuals or the community; — houses, churches, public halls, the paved streets, the lamps, the railings, the shrubs and flowers in the squares and gardens, the very stones on the macadamized road, — all was property.

As we cleared the suburbs, with their reticulations of cross walls, their scattered trees, and their straggling houses, there opened upon us a wide extent of country, with its woods and fields, its proprietors' seats, and its farm-steadings. And here was property of another kind, — property in land, emphatically termed by our laws — in contradistinction to the portable valuables which we had just seen in passing outwards, in the shops, and on the persons of the passengers, — *real* property. And real property the land of the country unquestionably is, — more obscure in its lineage, mayhap, than the furs furnished in barter by the American Indian, or the flowered piece of netting elaborated to order by the incessant toil, prolonged for months, of the poor lace-maker, but obscure merely on the principle through which the early history of an ancient people or long-derived family is obscure, — obscure simply because its beginnings reach far beyond the era of the annalist and the chronicler. It has been property so long that the metaphysician can but surmise how it became such; nor can the historian decide which of the philosopher's many guesses on the subject is the best one. We incline to the solution of Locke, though in some respects inadequate, in preference to that of Paley, who holds, most unphilosophically we think, that the real foundation of right in the case is the law of the land. Law of the land! We could as soon believe that a son was the producing cause through which his father came into being, or that a daughter was the producing cause of her mother's existence. Property in the land existed long ere there were laws in the land. Cain must have been as certainly the proprietor of the field which he rendered valuable by incorporating his labor with its soil, as Abel of the flock which his labor had tamed or reared. Both the land and the animals were general gifts to the species from the Beneficent Giver of all; and the *individual* right was fairly constituted in the one case by the man who broke in the animals from their state of original wildness, and in the other by the man who cleared and tilled and sowed the hitherto uncultivated waste, and converted it into a patrimony worthy of being bequeathed to his children. There must have been at least as much labor expended in the case of the agriculturist as in that of the shepherd; and, if the poets are to be regarded as authorities, — and there are instances in which they wonderfully approximate to the truth, —

considerably more. Paley tells us that the first partition of an estate which we read of was that which took place between Abram and Lot, — "If thou wilt take the left hand, then I will go to the right; or if thou depart to the right hand, then I will go to the left." Had he examined his Bible just a little more carefully, he would have found that the transaction was not a partition of land, — for Abram had none at the time, — but a mere temporary arrangement regarding the occupation for a certain term of a certain extent of common; that the portions of land in that country with which, according to Locke, human labor had been mixed up, had already, in consequence of the incorporation, become property; and that when Abram desired the field of Machpelah, with the sepulchral "cave that was in the end thereof," he had to purchase it of the proprietor for "four hundred shekels of silver." If the sole foundation of men's rights to their landed properties was, as Paley holds, the law of the land, — if there had been no previous foundation of right on which the law itself rested, — we would have to regard as miserably inadequate and precarious indeed the tenures of our lairdocracy, and to recognize the aspirations of the levelling Chartist and the agrarian ten-acre man as at once rational and fair. The right which the law had created at one time it might without blame disannul at another; for if the law did not rest on a heaven-derived justice, but was itself a primary foundation, and rendered just whatever rested on *it*, justice would of course be as variable in its nature as opinion among the law-making majorities of the country; and so it would not be more than equally just for the Conservative majorities of to-day to secure their estates to the existing proprietors, than for the Chartist majorities of to-morrow to break up these estates into single fields, and give a field apiece to the working-men of the country. The law of the land cannot create property: it can merely extend its sanction and protection to those previously existing rights of property on which all legislation on the subject must rest, or be mere enacted violence and outrage, abhorrent to that ancient underived justice which existed ere man was, and which shall long survive every merely human law.

Nay, even in cases where man's labor has not yet been incorporated with the soil, — on wide moors and among rugged hills, where he has neither ploughed nor planted, — it is for the benefit of the species that individual rights of proprietorship should exist and be recognized. The proprietor virtually holds, in many such cases, not merely in his own behalf, but in that of the country also. We were never more forcibly struck by the fact than when travelling several months ago in the mainland of Orkney, in a locality where the properties are small, and there exists a vast breadth of undivided common. Wherever the rights of individual proprietors extended, we found land of some value; we at least found vegetation and a vegetable soil. On the common, on the contrary, there was almost no vegetable soil, and scarce any vegetation. The upper layer of mould, scanty at first, had been stripped off by repeated parings, and carried away for fuel; and for hundreds of acres together the boulder clay lay exposed on the surface, here and there mottled by a tuft of stunted heath, but covered by no continuous carpeting of even moss or lichen. Were such the state of the entire island, it would be wholly uninhabitable: it is the rights of individual property alone that have preserved Pomona to its people. Even a wood of any value is never suffered to grow on a common, unless, perchance, in the uninhabited recesses of a country: no peasant ever dreams of sparing a sapling in order that it may expand into a tree for the benefit of his neighbor's children. The winter is severe, and, standing in need of fuel, he cuts the promising plant down by a stroke of his bill, and, fagoting it up with several hundred others, he carries it home to his fire. Property in land is, we repeat, *real* property, — property held not merely for the benefit of individual proprietors, but also for the best interests of the community; for, did all the land belong to all, it would be of no value to any.

Such were some of our reflections as we walked on from field to field into the open country. In approaching a small stream that divided the lands of two proprietors, we startled a hare that had been couching amid a plot of turnips. It ran downwards for a few score yards along a furrow, stopped short, looked round, resumed progress, cleared the little stream at a bound, and was then lost to our view amid a brake of furze that skirted one of the fields of the neighboring proprietor. As we walked on, and, after crossing the streamlet, were rising on the hillside, beside a field laid down with wheat, we raised a covey of partridges. They went whirring above our head, and, reversing the course of the hare, flew over the stream, and settled in a second field of wheat, just beside the turnip one. That hare and these partridges were, it seems, property; and we had witnessed on this occasion a curious transferrence of valuables that had taken place without bargain or agreement on the part of any one. Up to a certain moment the hare had belonged to one proprietor; when we had first started it, and when it was running along the furrow, and when it had turned round to reconnoitre, it had belonged to the proprietor of the turnip-plot; but no sooner had it cleared the stream, than it straightway belonged to the proprietor of the wheatfield and the furze-brake. And, as if to make the first amends for the loss which he had just sustained, the partridges we had raised, from being the property of him of the field and the brake, had, on flying over the runnel, become the property of him of the turnip-plot. Certainly a strange mode of conveyancing! It seemed equally strange, too, that the turnips on which the hare had just been feeding, and the wheat which expanded the crops of the partridges, did not belong to either of the proprietors, but were the property of certain third parties called tenants. We saw within view at the time a considerable number of the tame animals. Enclosed within a fold of stakes and network, in a corner of the turnip-plot, there was a flock of sheep bearing on their necks a certain red mark to distinguish them from those of any other sheep-owner; and a half-dozen cattle were picking up their sustenance for the day amid the furze of the brake. The cattle belonged to the farmer who rented the brake, and the sheep to the owner of the turnips. The one could recognize his cattle, the other his sheep. If the cattle crossed the stream into the turnip-plot, or the sheep broke loose, and, o'erleaping the runnel from the opposite side, did damage to the sprouting wheat, or picked the brake bare, either tenant would have a legitimate claim for damages done his property, but there would be no actual transfer of property in the case. The sheep would have an owner equally on both sides of the streamlet, in the tenant whose red mark they bore; and the cattle, whether in the furze-brake or the turnip-field, would be equally the property of the tenant who farmed the brake. Certainly, if the game of the country be property, it must be property of a very anomalous kind. Is it personal, or real? We find it conveyanced from one nominal owner to another, without these owners knowing aught of the matter; we find that they have no marks by which to distinguish it; we find that, unlike all other live stock, it is fed on food not theirs; we find that they can give no account of its origin or lineage in relation to themselves,—it was neither gifted to them nor bought by them; it runs away from them, and beyond a certain point they dare not follow it; it is brought to them when dead, and, unable to recognize it as theirs, they purchase it on the ordinary terms. It is not personal property; it is not real property; it belongs to an entirely different category: it is simply *imaginary* property.

We are acquainted with an extensive district in the north of Scotland in which some thirty years ago there was not a single wild rabbit. Rabbits there had once been in the locality, though at a very early period. The laborer, in running his ditches through a sandy soil, or casting up the foundation of some farmhouse or stone fence, laid open, not unfrequently, underground excavations greatly larger than those of the mole, with here and there a blackened nest-like bunch of decayed grass and leaves, huddled up far from the light, and here and there a few minute bones strewed along the passages; and he would point out the remains to his employer, and say that the site had been once that of a rabbit-warren. But the rabbits themselves had become as thoroughly extinct in the locality as the wolf or bear. About a quarter of a century since, however, one of the minor proprietors of the district, a gentleman possessed of some two or three hundred acres, let loose a few pairs of rabbits; and so enormous has been the increase, that, over a space of some two or three hundred square miles, rabbits abound; and of that large area, scarcely one thirtieth part is in the hands of the proprietary; it is farmed by tenants who pay large rents. To whom belong the millions of rabbits by which it is infested, and who gobble up yearly many hundred pounds' worth of the produce? To the proprietor who originally turned them loose? Alas! no: the two or three pair, — the progenitors of the whole, — that, so long as they were in his possession, were assuredly his, would have scarce brought him half a crown in the market; besides, he has long since sold his little property, and left that part of the kingdom. His claim would be exactly that of the Italian boy, who, having turned loose his two tame mice in a granary, came back some twenty years after, and found their descendants twenty millions strong. Do they belong, then, to the proprietors of the district in general? On what plea? They were not theirs originally; they have been supported, not on their produce, but on that of their tenants. The non-farming, non-resident proprietors have not a particle of property in them; they are simply a certain amount of the grass, corn, and turnips of the farmers and farming proprietors, converted into animal food, and running about on all fours. They are mischievous vermin when alive, which no one ought to be prevented from destroying, and which the farmer has a positive right to destroy; and, when dead, they ought surely, just like the fur-bearing animals of Siberia or Hudson's Bay, to be the property of the man who has taken the trouble of killing them. All quite right, says the game-preserver. You are, however, rather unfortunate in your illustration; rabbits are not game. We are quite aware of that fact, we reply, and might have chosen what you would have deemed a better illustration. In Pomona, twenty years ago, there were no hares. A young man, the son of a proprietor, procured a very few from the mainland of Scotland; and hares have in consequence become comparatively common in Orkney, just as rabbits have become common in the Black Isle; and, in proportion to their number, they do as much mischief. It is the part of the game-preserver to show how or why the hares, in such circumstances, should have become property, and the rabbits not. Wherein lies the difference between two tribes of animals that so nearly resemble each other? There can be but one reply; the law has made the hare property, which means simply, say we, that the game-laws exist, — a fact which it requires no profound process of argumentation to demonstrate. We would never once have thought of writing our present article if the game-laws did not exist. But the unreal and imaginary property, which has no other foundation than human enactment, — which the law makes to-day and unmakes to-morrow, — which a few years ago comprised the wild rabbit, and which a few years hence will not comprise the wild hare, — is property of an eminently precarious nature. It resembles property in ice in a warm summer. Laws which are themselves not founded in moral right and the nature of things form but unsolid foundations for aught else. There was a law in Russia, enacted in the days of the capricious Paul, which rendered it imperative on the male portion of Paul's subjects to wear small-clothes, and empowered the police to cut short at the knees the trowsers of the refractory. There was a law in Great Britain in the days of George II., that made it treasonable for a Scotch Highlander to wear tartan. Put neither the one law nor the other was based on the principles of ever-enduring justice. Independently of conventional enactment, it is no more a moral offence to wear trowsers than to knock down a partridge, or to sport tartans than to shoot a hare; and so trowsers are now worn in Russia, and tartans in the Highlands.

Our views on this subject are in no respect novel: they do not belong to the times of the Chartist and the leveller. They have, on the contrary, been long embodied in our literature. The conventional game-laws had never the effect of creating in Britain a conventional morality, that learned to respect these laws as its code and standard. On this point our masters of fiction — the men whose special work it was to draw character as they found it, draperied in the manners of their age, and modified by its opinions — are high authorities. When Goldsmith requires for the purposes of his story to get a thoroughly honest fellow into Newgate, he makes him knock down a hare. When Fielding — an honorable magistrate at least, however lax in other matters, and a determined enemy of thieving — wishes to bring his hero into trouble without rendering him culpable, he sends him, with all the eagerness of the young sportsman, after a covey he had started on his benefactor's grounds, into the grounds of a neighboring proprietor, and makes him kill them there. "The Edwardses of Southhill," says Mackenzie, — "and a worthy family they were!" — how came these same worthy Edwardses to be ruined? Young Edwards, "who was a remarkably good shooter, and kept a pointer," knocked down a partridge one day in the field of his neighbor, a country justice, and so the ruin was quite a matter of course. But there is no end of such instances; and the report on the game-laws shows on how broad a basis of reality these adepts in fictitious narrative (the prose-*makers*) founded their inventions. Unfortunately, in not a few cases a poacher *becomes* a bad character, and a source of loss and annoyance to the community; but it is not in the beginning of his career, when he is simply a poacher, that he is in any degree a bad character. He is in most cases either an adventurous young fellow, a "good shooter," like young Edwards, and fond of sport, like the game-preserving proprietors whom he annoys, or else some poor man out of employment, with a wife and family dependent on him, and much in terror of the neighboring workhouse. The evidence of Mr. M. Gibson, Inspector of Prisons in England, is peculiarly valuable on this head: "There are certainly many," he says, "who poach and are sent to prison, who would not commit a robbery." "There are poachers," he adds, "from the love of adventure and of sport, who are the most irreclaimable of any; there are poachers from poverty; and there is the young man, always in the fields, who from early life has set his bird-trap, and cannot resist the impulse of subjugating the wild animals." Such is Mr. Gibson's opinion of a numerous class of poachers; and their opinion of themselves seems, as might be expected, not greatly worse than his. "Have you had any opportunity," he is asked by the committee, "of ascertaining the opinions of chaplains and officers of prisons at all generally as to the operation of the present game-laws?" The reply is eminently worthy of being carefully noted and pondered. "Yes," he says; "with regard to the effect on the prisoners, the opinion of the chaplains generally is, that they can produce no moral effect whatever upon them under the game-laws; that they leave the prison only to return; frequently replying to the proffered advice by saying that *the game was made for the poor as well as the rich, and that God made the birds of the air and the fishes of the sea for all.*" It so happens, curiously enough, that Judge Blackstone, and most of the philosophic thinkers which the country has yet produced, were of the same opinion; but, more curious still, not a few of even the more zealous game-preserving proprietors seem also to entertain it, though of course in a greatly more covert style. They are indisputably gentlemen, and would neither employ as their servants habit-and-repute thieves, nor yet act the part of the Jonathan Wilds of the last age by being receivers of stolen goods. And yet there are two facts which come fully out in the evidence. They have no hesitation whatever in employing as gamekeepers and gamewatchers active habit-and-repute poachers; and hundreds of them, when stocking their preserves, drive a trade with the poachers that are still actually such, in live leverets and pheasants' eggs. Now these live leverets and pheasants' eggs cannot be property, or else these same game-preserving proprietors would to a certainty be not gentlemen, but scoundrels. By their doings at least they virtually decide the question against themselves.

XIV.

THE FELONS OF THE COUNTRY.

It is very generally felt that life and property are less secure in this country at the present time than they were some eight or ten years ago. In the course of nearly a century Britain had greatly changed its character for the better, in the degree of security which the civil magistrate afforded to the peaceable subject. So late as the year 1750, it was unsafe to walk at night the streets of our larger towns; and the man who sauntered unprotected after sunset into their quieter suburbs, or traversed even their more frequented approaches, might be almost certain of being struck down and robbed, if not murdered. Fielding, who was not only a great novelist but also one of the most efficient magistrates that ever lived, relates in his narrative of the earlier stages of that illness which ultimately carried him off, that the symptoms were much aggravated by the fatigue which he incurred in long examinations regarding the street robberies and murders of London, in especial by the examinations respecting "*five* different murders, all committed within the space of a week by different gangs of street robbers." The materials of his comparatively little-known volume, "The Life of Jonathan Wild," were collected during this period of crime and outrage; nor does the work, as a whole, exaggerate the actual state of things at the time. Another of his works he entitled an "Inquiry into the Increase of Thieves and Robbers,"—"a work which contains several hints," says Sir Walter Scott, "which have been adopted by succeeding statesmen, and some of which are worthy of still more attention than they have received." If an "increase" of the robber class actually took place at the time, as the title indicates, matters must have been bad indeed; for, about an age earlier, so sadly were the roads that approach the metropolis infested by highwaymen, as to be scarce at all passable by the solitary traveller. "Whatever might be the way in which a journey was performed," says Macaulay, "the travellers, unless they were numerous and well armed, ran considerable risk of being stopped and plundered. The mounted highwayman, a marauder known to our generation only by books, was to be found on every main road. The waste tracts which lay on the main routes near London were especially haunted by plunderers of this class. Hounslow Heath, on the great western road, and Finchley Common, on the great northern road, were perhaps the most celebrated of these spots. The Cambridge scholars trembled when they approached Epping Forest, even in broad daylight; and seamen who had just been paid off at Chatham were often compelled to deliver their purses on Gadshill." Long after the times that Macaulay describes, long after the times of Fielding too, even in country districts, the law served but imperfectly to protect the peaceable subject from the housebreaker and the highwayman. Cowper's graphic description, written in the year 1783, must be familiar to all our readers: —

"Now, ere you sleep,
See that your polished arms be primed with care,
And draw the night-bolt: ruffians are abroad,
And the first 'larum of the cock's shrill throat
May prove a trumpet summoning your ear
To horrid sounds of hostile feet within.
Even daylight has its dangers; and the walk
Through pathless wastes and woods, unconscious once
Of other tenants than melodious birds
Or harmless flocks, is hazardous and bold."

But a gradual improvement took place, especially in the larger towns. The great increase of newspapers, which recorded every act of violence and outrage as it occurred, and set the whole country on its guard, — that quickening of the postal arrangements which soon overtook and distanced the culprit in his escape, — the admirable organization of the police, effected by the act of Sir Robert Peel, — above all, the outlet furnished through the discovery of Botany Bay, and its appropriation as a penal colony for the country, — had all their effect in producing a favorable change; and, while a great increase took place in the list of minor offences, — a consequence of the growth of what are known as the lapsed classes, — crimes of blacker dye, perpetrated by professional felons, became considerably more rare and less atrocious than in an earlier time. During the first two decades of the present century a few terrible cases occurred. The Williams murders of 1812, and the general panic they occasioned, must be remembered by some of our older readers; and such as belong to a later generation may find their startling effects reproduced in some degree by the vigorous pen of De Quincey, in his grim but singularly powerful essay, "Murder considered as one of the Fine Arts." The murder by the M'Keans, also permanently recorded by the same graphic writer, belongs to a somewhat later period, and is marked by similar circumstances of atrocity. We do not refer to the Burke and Bishop murders, which may be considered as wholly *sui generis;* nor yet to those of the Thurtle or Tawell class, which occurred in private society, and lay outside what may be regarded as the professional pale. Within that pale great improvement took place; robbery accompanied by violence became rare, and robbery accompanied by murder rarer still. The streets and lanes of our larger cities might be traversed in comparative safety at all hours; the great bulk of offences committed against the person were offences committed under the influence of drink, — quite a bad enough symptom of the condition and morals of a great portion of the humbler classes, but in several material respects greatly preferable to that class of offences against the person which obtained in the days of Fielding, and respecting which he had to conduct, as has been said, five examinations in a single week. The means, too, by which the darker class of crimes has been suppressed in our own days are equally in advance of those to which the novelist — unrivalled, as his writings show, in his knowledge of the worse traits and specimens of human nature — had been compelled to have recourse a century ago. In the introduction of the "Voyage to Lisbon," he relates that, when consulted by the Premier of the day, the Duke of Newcastle, respecting the best mode of putting down the robbers and murderers of the metropolis, he could advise nothing better than the employment of money in corrupting their associates. "I had the most eager desire," we find him saying, "of demolishing these gangs of villains and cut-throats, which I was sure of accomplishing the moment I was enabled to pay a fellow who had undertaken for a small sum to betray them into the hands of a set of thief-takers whom I had enlisted into the service, all men of known and approved fidelity and intrepidity. After some weeks," he adds, "the money was paid at the treasury; and within a few days after two hundred pounds of it had come into my hands, the whole gang of cut-throats were entirely dispersed, seven of the thieves were in actual custody, and the rest driven, some out of the town and others out of the kingdom."

For the last six or eight years, however, there has certainly been *no* improvement of the nature which took place in the criminal records of the country during the previous quarter of a century; on the contrary, the course has been retrograde; and at the present time we seem as if passing to the state of matters which obtained during the days of Justice Fielding and Jonathan Wild. Murders have been committed during the last month of the old mercenary class, that, in circumstances of merciless barbarity, do not yield to any in the "Newgate Calendar;" assaults on the person for the same object have rendered the new term *garrotting* a completely naturalized one of familiar use;

and housebreakings on a large scale have become such common events that almost every succeeding newspaper records their occurrence. In some cases the respectable trader goes to his bed *square with the world*, and rises in the morning a ruined man. And yet never was there a time when certain of the causes which formed so powerful a check on crime in the past were so influentially in operation as now. Never were there so many newspapers to spread over the country the intelligence of every offence in all its details, and to direct public attention on the offenders; never was there a time when such intelligence could be transmitted with even a tithe of the present speed,— the act of Sir Robert Peel has certainly not been suffered to fall into desuetude; and never had the country a more active or intelligent magistracy. What, then, can be the more than neutralizing causes of such various circumstances of advantage, under which crime of what we have termed the professional class is so obviously on the increase? The question is easily answered. The causes are two. In the first place, that change through which Britain no longer possesses penal colonies has led to a great accumulation of criminals in the country; and it has got, in consequence, into the unhealthy condition of living subjects when the natural evacuations are stopped; and in the second place, the ticket-of-leave system — a system essentially false in principle in the circumstances — has greatly exaggerated the evil. We cannot, however, agree with those who give a paramount place to the latter cause. Were it to be abolished to-morrow, and criminals imprisoned for the shorter periods, — whether five, seven, or fourteen years, — in no case released until the close of the legitimate terms recorded in their sentences, — the master evil would still remain. The felon, now let loose upon the public at the end of some two or three years, would in the other case not be let loose upon it until the end of five years, or of seven, or fourteen; but ultimately he *would* be let loose upon it; and, even if inclined to live honestly, he would have quite as little chance of procuring the necessary employment at the end of the longer as of the shorter term. There is only one way in which the master evil in the case is to be remedied. The old means of evacuation must, at whatever cost, be procured. Britain, whatever difficulties may lie in the way, must again have recourse to the scheme of penal colonies, or both life and property must continue to remain insecure. And, though difficulties do lie in the way, we do not see that they are by any means insurmountable. Half the trouble which our ancestors had in extirpating the native wolves would suffice to rid us of a greatly more formidable class of wild beasts, — the incorrigible criminals. It is surely not at all necessary that a penal colony should be a paradise. It was no advantage, but, on the contrary, much the reverse, that during even the healthiest state of the country the incipient felon looked with longing eyes on the representations of New South Wales given in the print-shop windows, and then went off to qualify himself by some bold act for a free passage. A penal colony should be simply a country in which the discharged felon could earn his bread by the sweat of his brow, just as our humbler people do at home, and in which the circumstances of the community would be such as to render the life of the marauder not only a more dangerous, but also a more toilsome and difficult one than that of the honest worker who labored fairly for his bread. And a colony of this character ought not to be difficult to find. The country once heard a great deal about the Falkland Islands. Rather more than eighty years ago (1771), it was on the eve of entering, mainly on their account, into a war with France; and on that occasion Johnson wrote his famous tract to dissuade Britain from the contest, by showing that the islands were of really little value, and would be dearly purchased at such a price. But now that all dispute regarding them has ceased, — for the last quarter of a century they have been in the uninterrupted possession of this country, — they might be found very valuable as a penal colony. They have an area of about thirteen thousand square miles; their mean temperature during the year is exactly that of Edinburgh, with summers, however, a little warmer, and winters a little colder, than our Scotch ones; their surface is green; the grass-lands are peculiarly luxuriant, and form such a paradise for cattle that the tame breeds are becoming wild in the interior, and promise to be very numerous; and the bays and sounds which indent the coast abound in fish. Further, so imperfectly are they colonized, that though the expense of maintaining them costs the country about six thousand pounds per annum, their entire exports fall short of four thousand. In fine, at a very slight sacrifice these islands could be converted into a hopeful penal colony, that would fully absorb the more dangerous criminals of the country for a quarter of a century to come.

But while recognizing the lack of penal colonies, and the consequent accumulation of our criminals within the country, as the main causes of that increase of serious crime against both the person and property which has taken place during the last eight or ten years, we must not undervalue the influence of the other cause, — that ticket-of-leave system which has let loose so many dangerous felons on society ere half their terms of punishment had expired. The principle of the system is utterly false and unsolid in all its circumstances and details. A fond mother was once heard addressing her son as follows: "Be a good, religious boy, my little Johnnie; fear God, and honor your parents; and I will give you two pretty red-cheeked apples." Nor is it difficult to say what sort of a religion would be the effect of such a promise. Little Johnnie's two apples'-worth of the fear of God and the honor of parents would be a very hypocritical fear, and a very fictitious honor. And the ticket-of-leave system proceeds wholly on the same principle. Be religious and moral, it virtually says to the convict, for a given time, and you will get, when it has expired, the two red-cheeked apples. It has a grand disadvantage, too, over the scheme of the fond mother. She might no doubt succeed in making little Johnnie a little hypocrite; but the two apples, when made over to him, if really good ones, might be productive of further hurt to neither himself nor the family. Not so the premium for behavior held out to the convict. The proffered reward bears simply to the effect that he is to be let loose on society, to prey upon it anew. There is in reality no scheme in existence by which convicts in the mass can be dealt with as our paper-makers deal with their filth-begrimmed rags. We cannot put them in at the one end of a penitentiary in the soiled state, and take them out white and pure at the other. True, we must not limit the grace of God. It is just possible, however improbable, that little Johnnie, notwithstanding the sad stumbling-block of the two apples, or that a convict, notwithstanding the greatly sadder stumbling-block of the ticket-of-leave system, *might* be in reality converted; but neither on the apple scheme nor any other will there be any wholesale conversions of either the little Johnnies or the greater felons of the country. Regarded as a whole, the latter will enter the penitentiaries as felons, and as felons they will leave them; but if, by seeming to be religious, and by exercising a degree of self-constraint in a place in which there is exceedingly little to tempt, they will have the prospect held out to them of quitting their place of confinement at an early day, the men of strong wills and of self-control among them — always the more dangerous class — will not fail to conform to the conditions. And thus the picked felons will be ever and anon let loose long ere their time, to rob in order that they may live, and to murder in order that their robberies may be concealed. In the brief passage which we have quoted from Sir Walter's "Life of Fielding," we find him remarking, that one of the less known publications of the old magistrate and novelist contained hints, some of which had been adopted, and "some of which are worthy of more attention than they have received." And we would reckon among the latter the hints contained in the chapter entitled, "Of the Encouragement given to Robbers by frequent Pardons." Pardons at the time — a consequence of the extreme severity of the English criminal code — were very numerous and very capricious, though neither so numerous nor so capricious as the ticket-of-leave system has rendered them now. And what were the effects which they produced? Simply this, as determined by a singularly shrewd and sagacious man, who knew more of the matter than any one else, that from the hope of impunity which they created, they hanged ten times more felons than they saved from the gallows, and greatly increased the amount of crime.

XV.

THE LEGISLATIVE COURT.

"On the 22d of Aprile" (1532), says Calderwood, in his "Ecclesiastical History," so recently published, for the first time, by the Wodrow Society, "the Collegde of the Judges was established in Edinburgh," "for judgment of pecuniall and civil causes." "In the beginning," continues the historian, "many things were profitablie devised by them, and justice ministered with equitie. But the event answered not the expectatioun of men; for, seeing in Scotland there be almost no lawes except the acts of Parliament, whereof manie are not perpetuall but temporarie, and the judges hinder what they may the making of such lawes, the goods of all men are committed to the arbitriement and decisioun of fyfteen men that have perpetuall power, which, in truth, is but tyranicall impyre, seeing their own arbitriements stand for lawe."

Such was the objection raised by Calderwood two hundred years ago to the constitution and practice of the Court of Session, at a time when no case of harassing and irritating collision with the ecclesiastical courts had arisen to disturb the equanimity or cloud the judgment of the shrewd old churchman. Such, too, was the decision pronounced regarding it nearly a century earlier by Buchanan, whom, in this significant and very pregnant passage, the ecclesiastical chronicler has been content closely to follow, — so closely, indeed, that the passage may be deemed rather a translation than a piece of original writing. The court was comparatively in its infancy — an institution of about fifty years' standing — when it was characterized by the older historian as an arbitrary erection, opposed in its constitution to the very genius of freedom. And why? It is according to the genius of freedom that a people be governed by laws which they themselves have made. The principle is at once so obvious and fundamental that there is scarce a writer on civil liberty who has not laid it down as his very basis. And it would certainly be no easy matter to conceive of aught in more direct and hostile antagonism to such a proposition, than the proposition that a people should be governed, not by laws of their own making, but by the *legislative* decisions of some fifteen irresponsible judges, chosen by the monarch to "have perpetuall power," and "whose arbitriements should stand for lawe."

Such were some of the grounds of Buchanan's judgment on the "Colledge of Judges;" and they serve to demonstrate the peculiar sagacity of the man, — a sagacity altogether wonderful when we take into account the early period in which he flourished. His reflections on the barbarous torments to which the assassins of James I. were subjected has been instanced by Dugald Stewart, in his "Dissertation on the Rise of Metaphysical Science," as fraught with philosophy of a deeper reach than can be found in the works of any other writer of so early a period. We would place over against it — as scarce less vivaciously instinct with the philosophic spirit, and as even a still better example of that discriminating ability in the political field which enabled him to take his place as an asserter of the just principles of civil liberty so mightily in advance of his age — his remark on the constitution of the Court of Session. It serves at once to remind us of the eulogium of Sir James Macintosh and to justify it. "The science which teaches the rights of man," says this elegant and powerful writer, "the eloquence which kindles the spirit of freedom, had for ages been buried with the other monuments of the wisdom and relics of the genius of antiquity. But the revival of letters first unlocked only to a few the sacred fountain. The necessary labors of criticism and lexicography occupied the earlier scholars, and some time elapsed before the spirit of antiquity was transferred into its admirers. The first man of that period who united elegant learning to original and masculine thought was Buchanan ; and he, too, seems to have been the first scholar who caught from the ancients the noble flame of republican enthusiasm. This praise is merited by his neglected though incomparable tract, 'De Jure Regni,' in which the principles of popular politics and the maxims of a free government are delivered with a precision, and enforced with an energy, which no former age had equalled and no succeeding has surpassed."

A history of the many decisions of the Court of Session that, according to Buchanan and Calderwood, are *legislative*, not *judicial*, — that, instead of explaining existing law, are in reality creations of laws which have no existence save in the decisions themselves, — would form a very curious and a very useful work. It would be well, surely, to know how much of the national code is the production of the "fyfteen men that have perpetuall power, and whose arbitriements stand for lawe," and how much of it has been made by the people themselves, through the people's representatives. It would be at least particularly well to know how much of what is practically the national code is not merely law created by the "fyfteen men" where no law existed before, but law created by them in direct opposition to existing laws, — law directly subversive of the law made by the people. Nor can there be any doubt that the time is coming when such a work will be imperatively called for by the public. Scotland, through the decisions of this court, is on the eve of being placed in circumstances exactly similar to those in which the disastrous wars of five hundred years have placed Ireland. The religion of the country is on the eve of being disestablished, — disestablished, too, at a time when in a state of greater vigor, and more truly popular, than at any other period during the last hundred years ; and as revolutions never occur without at least awakening a spirit of inquiry regarding the causes which have produced them, the period must be inevitably at hand when the legislative decisions of the Court of Session shall be examined, and that with no ordinary degree of attention, in the light of Calderwood and Buchanan.

We have specified on several occasions decisions which, in their character as precedents, have actually become law, — that traverse, and practically abrogate, the statutory law of the kingdom. We adduced one very striking instance when setting against each other the existing mode of provision for the building and repairing of parish churches as settled by decision, and the diametrically opposite mode as arranged and provided by enactment. According to statute, "the parishioners of parish kirks" are charged and empowered to "elect and chuse certain of the most honest qualified men within their parochins," to tax the parish for the expenses of the necessary erection or repair; and in the event of the parishioners "failing or delaying to elect or chuse, through sloth or unwillingness, the power of making such choice or election of such honest qualified men falls to the ecclesiastical authorities." Such is the enacted statutory law on this head, — the people's law. But what is the actual law of precedent in the case, — the law of "the fyfteen?" That any such election "of honest men" would be altogether illegal; that so far are the parishioners or ecclesiastical authorities from possessing any such right of election, that, even were they to make a voluntary contribution among themselves for the repair or improvement of the parish church, they could be legally prevented from lifting a tool upon the building; that, in short, the whole matter of erecting, repairing, improving, is not in the hands of the parishioners or the ecclesiastical authorities, where statute has placed it, but exclusively in hands in which statute never placed it, — in the hands of the heritors. How very striking an illustration of the sagacity of Buchanan!

We need scarce refer to the still more striking illustration which our present ecclesiastical struggle furnishes, — an illustration which, we have said, will scarce fail of being appreciated over the whole empire by and by. We shall venture, however, on one remark. It is not according to the nature of things that the decisions of the Court of Session should traverse statutory enactments, which have originated amid the ebullitions of strong popular feeling, and are in reality embodiments of the popular will, so long as these enactments are recent, and the impulse to which they owed their existence is still predominant in the country as a moving power. Nothing less probable, for instance, than that the court should have reversed any of the more broad and obvious provisions of the Reform Bill when Earl Grey's ministry were still in office, or any of the more thoroughly understood clauses of the Roman Catholic Emancipation Act ere it had attained to a twelvemonth's standing. The state of these measures as *recent* — as measures which had agitated the whole country — whose meanings all the people understood, not so much in their character as statutes as in their character as embodiments of either their own will or the will of the Roman Catholics of Ireland — would have prevented most effectually any judicial reversal of the main principles which they involved. The Court of Session might as safely declare that Ernest of Hanover, not Victoria, is the monarch of these realms, as that ten-pound freeholders have no legal right to vote in the election of members of Parliament, or that at least ten-pound freeholders have no legal right to vote in the election of members of Parliament who are Roman Catholics. The character of such acts, as recent, restricts our judges to the exercise of their purely judicial functions. They cannot, they dare not, reverse them. Taking this obvious principle into account, — and it is certainly not easy to say how any principle *could* be more obvious, — it is of vast importance to ascertain the opinions which our judges held regarding the powers and jurisdiction of the church at a time when both the Revolution and the Union were events as fresh in men's memories as the Reform Bill and the Emancipation Act are now. Hence, in part, the great value of those views and sentiments of our older lawyers on the point, to which we have so often referred. Lord Cullen, with whose admirable tract on patronage most of our readers must be acquainted, was a grown man at the time of the Revolution. His son, Lord Prestongrange, must have remembered the Union as the great event of Scotland in that age. The Lord President Dundas and the Lord President Forbes were lawyers of much the same standing as the latter. Kames, Monboddo, Dreghorn, were all reared at the feet of these men; and though all of them could, no doubt, occasionally unite to their judicial functions those legislative powers which so excited, at an earlier period, the jealousy of Buchanan, all of them must have felt that, regarding the more palpable conditions of those two great events, the Revolution and the Union, they were at liberty to exercise their judicial functions only. The fundamental conditions of these events were present to the national mind as great living principles; they still engaged the feelings of the country; they still exercised its reasoning faculties; they were something other than dead statutory enactments for legislative judges to dissect at will, and on which spruce half-fledged lawyers might try their hand at an amputation, without the necessity of using the tourniquet. Their true meaning was as thoroughly exhibited in the living intellect of the country as in the statute-book itself. And hence, of necessity, the rectitude of judicial opinion regarding them.

Is this view of the matter in any degree a rational one? If so, what estimate must we form of the view taken by Lord Cuninghame in his last note? The church has never yet disputed that the *judicial* sentence of the civil court may legitimately effect a separation between HER *spiritualities* and the *temporalities* of the state; but this, she contends, is the utmost extent to which any such *legiti-*

mate decision can effect her; and in proof of the doctrine she appeals not only to the statutory enactments in which it is embodied, but also to the opinions on the subject of all the Scotch lawyers and more eminent judges of the last century, — men who lived under the direct influence of the immensely important events by which the Constitution of the country had been ultimately fixed at the Revolution and the Union. "There appears to be little doubt," says his lordship, in reply, "that at a certain period in the last century, when ecclesiastical questions first were the subject of discussion in our courts, an opinion *was* entertained by lawyers of learning and reputation, such as Lord Prestongrange, Mr. Crosbie, and others, that such a separation was in certain cases legitimate and competent, and admitted of no remedy in this court. But, able as the persons were, *they had not the benefit of the anxious and elaborate arguments which the questions have undergone in modern times*, and which have thrown a light on cases of this nature that writers at no former period enjoyed." Surely we may be permitted to exclaim, "O unhappy lawyers of the last century! — hapless Henry Home, unlucky Duncan Forbes, unfortunate Monboddo, ill-fated Dreghorn! — O ye Dundases, Cullens, Crosbies, and Prestongranges! — why were ye all born a hundred years too soon? Poor blind gropers in quest of truth, men of deficient law and slender intellect, why were you not fated to imbibe wisdom from the philosophic notes of my Lord Cuninghame, and to inhale at once wit and knowledge from the lucid and sparkling speeches of my Lord Justice-Clerk Hope? Thou, O Kames! hadst thou but lived to see these luminaries, mightest have remained unenlightened thyself notwithstanding, like those very obstinate gentlemen of our own times, Lords Jeffrey and Moncreiff; but in taking measure of the vast intellectual stature of our Hopes and Cuninghames, thou wouldest have at least found it necessary to introduce into thy 'Sketches' one Adam more, and he a giant. And thou, O Monboddo! hadst thou but seen the sort of persons who follow in their train, thou wouldest surely have rejoiced, whatever else thou mightest have done, in the return of the men with tails. But ah! unhappy lawyers, ye lived an age too soon, and so must content yourselves now with just the pity of the Lord Ordinary."

There is assuredly a time coming when our ecclesiastical question, viewed in the clear light of history, shall be judged one of the best possible for illustrating the character of the court in both its judicial and legislative aspects. It will exhibit the Janus-like head of this institution, with its one countenance bent tranquilly upon the past century, and its other countenance breathing war and horror on the present. It will be seen that in the last century, the court, with regard to the church, presented only its judicial aspect: we have shown why. It will be found that it is the legislative aspect which it presents with respect to the church now. And there will doubtless be some interest in marking the exact point at which the one character has been taken up and the other character laid down, with all the various causes which have led to the change. But the prejudices and prepossessions of men interfere, and prevent the question from being one of the best possible illustrations of this in the present time. We have a case before us which at least our antagonists will recognize as happier in its application. It is a case in which the decision arrived at by the court traverses not quite so palpably the laws of the country as the fixed laws of nature. We submitted to our readers, rather more than a week since, the report of a trial which had taken place a short time previous, before the court in Edinburgh, regarding a right to the fishing of salmon in the Frith of Dornoch, and which had gone against the defendant. We stated further that a similar case, involving a similar right to the fishing of salmon in the Frith of Cromarty, had been tried with a similar result a few years before. The principles of both cases may be stated in a few words. Salmon, according to the statutory laws of Scotland, may be fished for in the sea with wears, yairs, and other such fixed machinery; but it is illegal to fish for them after this fashion in rivers. The statutes, however, which refer to the case are ancient and brief, and contain no definition of what is river or what sea. They leave the matter altogether to the natural sense of men. But not such the mode pursued by the Court of Session. In its judicial capacity it can but decide that salmon are not to be fished for in rivers after a certain manner in which they may be fished for in the sea. In its legislative capacity it sets itself to say what is sea and what river, and proves so eminently happy in its definition, that we are now able to enumerate among the rivers of Scotland the Frith of Dornoch and the Frith of Cromarty. Yes, gentle reader, it has been legally declared by that "infallible civil court" to which there lies an appeal from all the decisions of our poor "fallible church," that Scotland possesses two rivers of considerably greater volume and breadth than either the St. Lawrence or the Mississippi! Genius of Buchanan! It is well that thou, who didst so philosophically describe the Court of Session, didst describe also, like a fine old poet as thou wert, the glorious bay of Cromarty!

Some of our readers must be acquainted with the powerful writing of Tacitus in his "Life of Agricola," in which he describes the Roman galleys as struggling for the first time with the tides and winds of our northern seas. The wave rose sluggish and heavy to the oars of the rowers, and they saw all around them, in the indented shores scooped into far withdrawing arms of the sea, evidences of its ponderous and irresistible force. Buchanan must have had the passage in his mind when he drew the bay of Cromarty. He tells us how "the waters of the German Ocean, opening to themselves a way through the stupendous cliffs of the most lofty precipices, expand within into a spacious basin, affording certain refuge against every tempest, and in which the greatest navies may rest secure from winds and waves." The Court of Session, in the wise exercise of its legislative functions, reverses the very basis of this description. The rowers of Agricola must have been miserably in error: the old shrewd historian must have fallen into a gross mistake. The Frith of Cromarty is not the inlet of a mighty sea; it is merely the outlet of an inconsiderable river. It is not an arm of the German Ocean; it is simply a prolongation of the Conon. Prolongation of the Conon! Why, we know a little of both. We have waded a hundred times mid-leg deep across the one, and picked up the large brown pearl mussels from the bottom without wetting our sleeve; we have guided our little shallop a thousand times along the green depths of the other, and have seen the long sea-line burying patch after patch, as it hurried downwards, and downwards, and downwards, till, far below, the lead rested in the darkness, amid shells, and weeds, and zoophytes, rare indeed so near the shore, and whose proper habitat is the profound depths of the ocean. We have seen the river coming down, red in flood, with its dark whirling eddies and its patches of yellow foam, and then seen it driven back by the tidal wave, within even its own banks, like a braggart overmastered and struck down in his own dwelling. We have seen, too, the frith agitated by storm, the giant waves dashing against its stately portals, to the height of a hundred feet; and where on earth was the power that could curb or stay them? The Frith of Cromarty a prolongation of the Conon! Were the Court of Session to put the Conon in its pocket, the Frith of Cromarty would be in every respect exactly what it is, — the noble *Portus Salutis* of Buchanan, — the wide ocean bay, in which the whole British navy could ride at anchor. Is it not a curious enough circumstance, that much about the same time in which the Court of Session, in the due exercise of its legislative functions, stirred up the church to rebellion, it so laid down the law with respect to the Frith of Cromarty, in the exercise of exactly the same functions, that it stirred it up to rebellion also?

Yes, it is a melancholy fact, but it cannot be denied, that this splended sheet of water has been in a state of open rebellion for the last four years. In obedience to its own ocean laws, it has been going on producing its own ocean products, — its prickly sea-urchins, its sea-anemones, its dulce, its tangle, "its roarin' buckies," and its "dead men's fingers;" when, like a good subject, it should have been river-mouth to all intents and purposes, nor have ventured on growing anything less decidedly fluviatile than a lymnea or a cyclas, or a freshwater polypus. It has been so utterly outrageous in some of its doings, that, albeit inclined to mercy, we are disposed to advise the court to deal with it somewhat closely. There might be trouble, perhaps, in bringing it to the bar, — more by a great deal than sufficed to bring the Presbytery of Dunkeld there; but with the precedent of Canute on record, we do not think the court would lower its dignity much below the present level by just stepping northwards to rebuke it. It would be perhaps well, too, to select as the proper time the height of a stiff nor'easter. For our own part, we would be extremely happy to furnish the information necessary to convict, whether geological or of any other kind. We can satisfactorily prove, that no further back than last year, this frith gave admission, in utter contempt of court, to so vast a body of herrings, that all its multitudinous waves seemed as if actually heaving with life; nay, that it permitted them, by millions and thousands of millions, to remain and spawn within its precincts. We can prove, further, that it suffered a plump of whales — vast of back and huge of fin — to pursue after the shoal, rolling, and blowing, and splashing the white spray against the sun; and that it furnished them with ample depth and ample verge for their gambols, though the very smallest of them was larger considerably — strange as the fact may seem — than the present Dean of Faculty. Is all this to be suffered? The Lords of Session must assuredly either bring the rebel to its senses, or be content to leave their own legislative wisdom sadly in question. For ourselves, we humbly propose that, until they make good their authority, they be provided daily with a pail of its clear *fresh* water, drawn from depths not more than thirty fathoms from the surface, and be left, one and all, to make their toddy out of the best of it and to keep the rest for their tea. Nothing like river water for such purposes, and the waters of the Conon are peculiarly light and excellent.

XVI.

THE PEACE MEETINGS.

It is indisputable that peace societies are becoming of importance enough to constitute one of the peculiar features of the time. We learn from Sir Charles Lyell's recent work of travels in the United States, that they appear to be telling on the American mind, albeit naturally a war-breathing mind, combative in its propensities and fiery in its elements. The late peace meetings at Paris, London, Birmingham, and Manchester, seem to have been at once very largely attended and animated by the enthusiasm of a young and growing cause; and newspapers such as the "Times," the "Chronicle," the "Herald," and the "Post," and periodicals such as the "Quarterly Review," evidently deem the movement, of which they are a result, formidable enough to justify the attempt to write it down. It is certain, too, that the substratum of right feeling in which the movement has originated, and which it represents in a rather exaggerated form, is vastly broader and more extensive than the movement itself. There are many thousands both in Britain and America, and not a few in France and Germany, whose judgments may be not at all satisfied by the expedients through which the peace societies propose putting an end to national wars, that yet share deeply in that general dislike of war itself which is happily so marked a characteristic of the age.

There is nothing positively new in what may be termed the main or central idea of the existing peace associations, namely, adjustment of national differences by arbitration, not arms. The true novelty presented lies in the fact that an idea restricted in the past to but single minds should now be operative in the minds of thousands. The reader may find in the works of Rousseau a treatise, originated by the Abbe de St. Pierre, but edited and remodelled by the philosopher of Geneva, entitled a "Project for a Perpetual Peace," in which the expedient of a great European Court of Arbitration for national differences is elaborately developed. We question, indeed, whether any member of the peace societies of the present day has presented to his fellows, or the public generally, the master idea of these institutions in so artistic and plausible a form as that in which it was submitted to the world by Rousseau considerably more than eighty years ago. But though it attracted some degree of notice among the rulers of nations, it failed to attract anywhere the notice of the ruled, — that class of which the great bulk of nations are composed; nor, perhaps, are all the members of peace societies aware how nearly it was realized at one time, and how it yet failed entirely, notwithstanding its plausibility, to work for any good purpose.

Nations can, of course, only act through their governments; and of the European governments in the days of Rousseau, the greater number were arbitrary in their constitution. And in forming his Court of Arbitration, he had of course to admit as its members, governments represented by monarchs possessed of irresponsible power, such as the Kings of France, Spain, Portugal, Prussia, Naples, and Sardinia, and the Emperors of Austria and Russia. He had no other materials of which to form his General

Arbitration Court. Of the nineteeen European states in his list of arbiters, twelve were despotic, and the larger half of the remainder nearly so; and yet, in order to secure the desiderated blessing of peace, he had to lay it down as a fundamental rule, that each state should be maintained by all the others in its internal rights and powers, and that its territories, at the time of the union, should be guaranteed to it entire. On other principles no union of governments could have taken place. To put down war was the object of his proposed confederation, — internal as certainly as foreign war; for of what use would a peace association be under which there could arise such a war as that which raged between Great Britain and its American colonies, or between Austria and Hungary, or as that which deluged the streets of Paris with blood? Nay, under a peace association composed of despotic and semi-despotic governments, no such invasion of one country by the troops of another could have taken place as that of England by William III., which produced the Revolution of 1688. Rousseau's project, if practicable, would have secured peace, but it would have also, of necessity, arrested progress. It would have cursed the world with a torpid, unwholesome quiet, a thousand times less friendly to the best interests of humanity than that mingled state of alternate peace and war under which, with all its disadvantages, the human species have been slowly rising in the scale of intelligence, and securing for themselves constitutional rights and equal laws. Nor were there wanting men among the rulers of the world shrewd enough to see that such was the real character of the scheme; and it was with rulers, not subjects, that that attempt originated to which we have referred, to convert it from an idea into a fact.

A fierce and long protracted European war had just come to a close, — a war productive of greater waste of blood and treasure than any other of modern times, — when three great monarchs met at Paris to originate a peace society on nearly the principles of Rousseau.

These were Alexander of Russia, Francis of Austria, and Frederick William of Prussia. Lord Castlereagh, as the representative of his country, was cognizant of the principles of the association, and warmly approved of them; but it was found that the forms of the British Constitution were such as to prevent the King of England from becoming a member. The document which formed the basis of the confederation was published; and it was found, as might, indeed, be expected from most Christian princes, to be of a greatly higher tone than that which marked the project of Rousseau. It commenced with an announcement of the intentions of the subscribing parties to act for the future on the principles of the gospel, defined to be those of justice, Christian charity, and peace. Then followed three articles, introduced by the scriptural command to all men to consider one another as brethren, which were to the effect, first, that the three contracting princes should remain united to each other by the bonds of a true and indissoluble fraternity; second, that they should conduct themselves to their subjects and armies as the fathers of families; and, third, that all other powers should be invited to join with them in the confederacy. The scheme was hailed throughout Europe as the precursor of a better state of things than the world had yet seen; and liberal politicians everywhere, and more especially in Germany, were filled with the most sanguine expectations of happy results. Most of the European princes became members of this magnificent peace society; and England, though precluded from formally joining itself to it officially, intimated to its members that no other power could be more inclined to act upon the principles which its fundamental articles seemed necessarily to involve. It had its series of congresses; for, curiously enough, its meetings had the same name given them as those of our present peace associations; and at the first of these, held at Aix-la-Chapelle in 1818, there was prepared, and subsequently published by its members, a declaration to the effect that peace was its paramount object. What, asks the reader, was the name borne by this eminently good and truly Christian peace society? Its name was the Holy Alliance, — a name that now stinks in the nostril; and it was in effect a foul and detestable conspiracy against the progress of nations and the best interests of the human species. But such, of necessity, must be the nature and character of every peace association of which the members are governments, if a majority of these be despotic. And if the members of a peace association *be not* governments, they can of course possess no powers of arbitration. In vain may Joseph Sturge and his friends propose themselves as arbiters in any such quarrel as that which recently took place between Austria and the Hungarians, or between France and Rome. The reply made to the pacific Quaker, were there to be reply at all, would be exactly that made by Captain Sword to Captain Pen, —

> "Let Captain Pen
> Bring at his back a million men,
> And I'll talk to his wisdom, and not till then."

And, on the other hand, if governments, we repeat, take up the work of arbitration in such cases, — governments such as those of Russia, Sweden, Prussia, Austria, France, Sardinia, Naples, Spain, and Portugal, — and such are the existing elements for an Arbitration Court, — it is easy to divine how the peace of the world would be preserved: it would be preserved by the putting down of what would be termed *rebellion* in Hungary, and *revolution* in Rome.

Often did Chalmers quote the emphatic words, "first *pure*, then *peaceable*." And very emphatic words they are, and singularly pregnant with meaning. They reveal why it is that peace societies, in the present state of the world, can produce no *direct* results. The nations and the governments must realize the purity ere they can rationally expect the peace. Peace under certain limitations is no doubt a duty. "If it be possible, as much as lieth in you," says the apostle, "live peaceably with all men." But the qualifications of the text are very important ones, — "*if it be possible*," and "*as much as lieth in you*,"— so important that they make a state of peace to be not so much a duty to be accomplished as a gift to be received. "When a man's ways please the Lord," said the wise king, "He maketh even his enemies to be at peace with him." Nor can peace associations alter this state of matters. They cannot by any scheme of arbitration convert the gift simply into a duty, seeing that if they take the existing governments as the elements of their arbitration courts, their plan involves of necessity merely the creation of a new Holy Alliance; and if, on the contrary, they propose first remodelling and reforming the nations, so as to qualify their governments for arbitrating justly, they change their nature, and become revolution societies, — of course, another name for *war* societies.

But, though we can thus promise ourselves no *direct* results from the peace societies of the times, their *indirect* results may be very important. That dislike of war which good men have entertained in all ages, is, we are happy to believe, a fast-spreading dislike. It was formerly entertained by units and tens; it is now cherished by thousands and tens of thousands. And of course the more the feeling grows in any country, which, like France, Britain, and America, possesses a representative government, the less chance will there be of these nations entering rashly into war. France and the United States have always had their senseless war parties. It is of importance, therefore, that they should possess also their balancing peace parties, even should these be well-nigh as senseless as the others. Again, in our own country war is always the interest of a class largely represented in both Houses of Parliament. It is of great importance that they also should be kept in check, and their influence neutralized, by a party as hostile to war on principle as they are favorable to it from interest. We repose very considerable confidence in the common sense of the British people, and so have no fear that an irrational peace party should so increase in the country as to put in peril the national independence; and, not fearing this, we must hail as good and advantageous any revolution in that opinion in which all power is founded, which bids fair to render more rare than formerly those profitless exhibitions of national warfare which the poet of the "Seasons" so graphically describes:—

"What most showed the vanity of life
Was to behold the nations all on fire,
In cruel broils engaged and deadly strife:
Most Christian kings, inflamed by black desire,
With honorable ruffians in their hire,
Cause war to wage, and blood around to pour.
Of this sad work when each begins to tire
They sit them down just where they were before,
Till for new scenes of woe peace shall their force restore."

XVII.

LITERATURE OF THE PEOPLE.

"The Crock of Gold," "Toil and Trial," and a "Story of the West End," are all little works which have been sent us for review during the last few months. "The Crock of Gold" is a story about a poor English laborer, who lived in a damp, unwholesome, exceedingly picturesque hovel, on eight shillings per week; "Toil and Trial" is a story about a poor shopman and his wife, who had to toil together in much unhappiness on the long-hour, late-shutting-up system; and a "Story of the West End" is a story about two poor needle-girls, of whom one sank into the grave under her protracted labor, and the other narrowly escaped degradation and ruin. They are all interesting, well-written little works; but what we would at present remark in incidental connection with them is that very decided change of direction which our higher literature has taken during the last twenty years, and more especially during the last ten. The great-grandfathers and great-grandmothers of the present reading public could sympathize in the joys and sorrows of only kings and queens; and the critics of the day gave reasons why it should be so. Humble life was introduced upon the stage, or into works of fiction, only to be laughed at; or so bedizzened with the unnatural frippery of Pastoral, that the picture represented, not the realities of actual life, but merely one of the idlest conventualities of literature. But we have lived to see a great revolution in these matters reach almost its culminating point. It is kings and queens, albeit subjected to greater and more sudden revolutions than at any former period of the world's history, that have now no place in the literature of fiction. We have our humbler people exhibited instead; and the reading public are invited to sympathize in the sorrows and trials of aged laborers of an independent spirit, settling down, not without many an unavailing struggle, into dreaded pauperism; overwrought artisans avenging their sufferings upon their wealthy masters; and poor, friendless needle women bearing long up against the evils of incessant toil and extreme privation, but at length sinking into degradation or the grave. We are made acquainted in tales and novels with the machinery and principles of strike-associations and trades' unions, and introduced to the firesides of carriers, publicans, and porters.

There is a fashion in all such matters, that lasts but for a time; and what we chiefly fear is, that the present disposition on the part of the reading public to look more closely than formerly into the state of the laboring classes, and to take an interest in their humble stories, may be suffered to pass away unimproved. Wherever there exists a large demand for any species of manufacture, spurious imitations are sure to abound; and when the supply becomes at once greatly deteriorated and greatly too ample, there commences a period of reaction and depression. An overcharged satiety takes the place of the previously existing interest. It is of importance, therefore, — for there are already many spurious articles in the field, — that the still unblunted appetite should be ministered to, not by the spurious, but by the real, and that only the true condition and character of those classes which must always comprise the great bulk of mankind should be exhibited to the classes on a higher level than themselves, on whose exertions in their behalf so very much must depend. Nor would the advantage be all on one side; both the high and the low would be greatly the better for knowing each other. It would tend to contract and narrow the perilous gulf which yawns, in this and in all the other countries of Europe, between the poor and the wealthy, were it mutually *felt*, not merely coldly acknowledged, that God has made them of one blood, and given to them the same sympathies and faculties, and that the things in which they differ are mere superficial circumstances, the effect of accident of position. "I have long had a notion," said the late William Thom, the Inverury poet, "that many of the heart-burnings that run through the SOCIAL WHOLE spring not so much from the distinctiveness of classes, as from their mutual ignorance of each other. The miserably rich look on the miserably poor with distrust and dread, scarcely giving them credit for sensibility sufficient to feel their own sorrow. That is ignorance with its gilded side. The poor, in turn, foster a hatred of the wealthy as a sole inheritance, look on grandeur as their natural enemy, and bend to the rich man's rule in gall and bleeding scorn. Puppies on the one side and demagogues on the other are the portions that come oftenest in contact. These are the luckless things that skirt the great divisions, exchanging all that is offensive therein. 'Man, know thyself,' should be written on the right hand; and on the left, 'Men, know each other.'" These are quaintly expressed sentences, but they are pregnant with meaning.

It is no uninteresting matter to trace, in the various styles of English literature, the part assigned to the people. They cut but a poor figure in Shakspeare. The wonderful wool-comber of Stratford-on-Avon rose from among them; but it would scarce have served the interests of the Globe Theatre in those days to have ennobled, by any of the higher qualities of head or heart, the humble peers and associates of wool-combers; and so, wherever the people, as such, are introduced in his dramas, whether they be citizens of Rome, as in "Coriolanus," or English country folk, as in "Henry VI.," we find them represented as fickle, unthinking, and ludicrously absurd. In the works of his contemporary Spencer we do not find the people at all; but discover, instead, what for nearly two hundred years after his time occupied their place in our literature: we are introduced to shepherds in abundance, Hobbinols,

Diggon Davies, and Colin Clouts, and find much reference made to "huts where poor men lie;" but it is the poor men of the classical Pastoral, who were in reality neither poor nor men, but mere fictions of the poets, — the inhabitants of a Utopia filled with crooks, and pipes, and garlands of flowers.

There have been many criticisms on the Pastoral, some of these by the first names in our literature, — Pope, Addison, and Johnson; but the true secret of the origin of this, the least natural and interesting of all the departments of poetry, we have not yet seen indicated. Like the silver mask of the veiled prophet that gleamed far amid the darkness of the night, and yet covered a countenance too horrible to be bared to the eye, it formed in the ancient literature the mask that at once concealed and represented the face of the people, — a face scarred and deformed by a cruel system of domestic slavery, and so unfit to be uncovered. In every truly national literature the people *must* be exhibited; and if they cannot be exhibited as they are, they must be exhibited as they are not. Hence the pastoral poetry of Rome and Greece: it was the silver mask of a veiled people; and that of England and the other nations of Europe was simply a tame imitation of it. About the middle of the last century the Pastoral proper died out of insanity, and the people began to be exhibited, first in Scotland by Allan Ramsay, who, though he retained in his exquisite drama the old pastoral outlines, looked intelligently around him, and, drawing his materials fresh from among the humble class, out of which he had arisen, gave life, and truth, and nature to the dead blank form. It was perhaps in Scotland that the people *could* be first represented as they really were. The vitalities of the national religion had already placed them on a high moral platform, and the national scheme of education — a result of the national religion — had developed their faculties as thinking men. As the Pastoral gradually disappeared in England, the people began to be exhibited, at first very inadequately and partially, but with certain lineaments of truth. Fielding and Richardson were contemporary. The first, a debauchee and a Bow Street magistrate, had an eye for but what was bad and ridiculous in the popular character. If we except Joseph Andrews, — a sort of male Pamela, drawn rather to caricature Richardson than from any sympathy with good morals and right feeling in a humble hero, — there is not one of the people whom in his character as an artist he exhibits in his works, whom in his character as a magistrate he would not punish as a scoundrel. The staple of his humbler characters is vulgar rascality. Richardson did better as a man, but not greatly better as an artist. His Pamela is rather a picture drawn in his back-parlor from his own imagination than an exhibition of a real character, representative of any section of the people. There is more truth in the humbler characters of Smollett; and, though enveloped in the ridiculous, not a few of them possess what the humbler characters of Fielding want, — right feeling and a moral sense. But even of his own countrymen of the humbler order Smollett could do little more than portray the externals: he was ignorant of the inner life of Scotland, and of those high principles which can impart dignity to even the poorest. A Bunyan or a Robert Burns would have constituted a phenomenon beyond his conception.

It was the part of this last-named genius to assert for the people their true place in British literature, — directly, no doubt, by many of his writings, but not less efficiently by his life, and by the light which his biography has thrown on his humble compeers. It is interesting to observe in the lives of our eminent men how each brings out into full view a group of individuals of whom we would otherwise never have heard. Each, like the sun of a system, possessed in himself the effulgence which renders him visible across the lapse of ages; but that effulgence confers visibility on not only himself, but on many an attendant planet besides, that, save for the reflected light, would miss being seen altogether. We see a Cowper surrounded by the Hesketbs and Austins, the Unwins and the Johnstones; and a Henry Kirke White, by brothers Neville and James, the Maddocks, the Charlesworths, and the Swanns. The light which Burns cast revealed the Scottish peasantry to the literati of Britain as men of no inferior grade or stunted proportions; and the revelation has told upon our literature. Had there been no Burns, it is not very probable that the philosophic hero of the "Excursion" would have been represented as a peddler; nay, we doubt if a man so tinged with Toryism as Sir Walter Scott would have dared to give, under the previous state of things, a heroine so humble as Jeanie Deans to one of his greater productions, or a hero of such lowly extraction as Halbert Glendinning to another. The surprise elicited in the mind of every intelligent man by the introduction to the Scottish people in their true character, which the life and writings of Burns secured, we find well expressed by Lord Jeffrey, in a critique on "Cromek's Reliques," written more than forty years ago. "It is impossible to read the productions of Burns," says this accomplished writer, "without forming a higher idea of the intelligence, taste, and accomplishments of the peasantry than most of those in the higher ranks are disposed to entertain. Without meaning to deny that he himself was endowed with rare and extraordinary gifts of genius and fancy, it is evident, from the whole details of his history, as well as from the letters of his brother, and the testimony of Mr. Murdoch and others to the character of his father, that the whole family, and many of their associates who have never emerged from the native obscurity of their condition, possessed talents, and taste, and intelligence which are little suspected to lurk in these humble retreats. His epistles to brother poets in the rank of farmers and shop-keepers in the adjoining villages, the existence of a book-society and debating-club among persons of that description, and many other incidental traits in his sketches of his youthful companions, all contribute to show that not only good sense and enlightened morality, but literature and talents for speculation, are far more generally diffused in society than is generally imagined, and that the delights and the benefits of these generous and humanizing pursuits are by no means confined to those whom leisure and affluence have courted to their enjoyment. That much of this is peculiar to Scotland, and may be properly referred to our excellent institutions for parochial education, and to the natural sobriety and prudence of our nation, may certainly be allowed; but we have no doubt that there is a good deal of the same principle in England, and that the actual intelligence of the lower orders will be found there also very far to exceed the ordinary estimates of their superiors."

This striking passage suggests to us what we deem the main defect of much of the modern literature in which the working classes are represented. There is no lack of a hearty sympathy on the part of the writers with the feelings of our humbler people; but we are sensible of a feebleness of conception when they profess to grapple with their intellect. They can appreciate the hearts, but fail to estimate at the right value the heads, of those with whom they have to do. And hence pictures true but in part.

The two most remarkable men who rose from among the people during the last century were Robert Burns and Benjamin Franklin; and both have left us autobiographical sketches, in which they refer to the associates of their early days. In what terms do they speak of their capacity? Certainly in terms very different from what the modern novelist or tale-writer would employ. Many of the humble men with whom the great poet and great philosopher came in contact were men from whom they were content to learn. A young lady of literary taste and acquirements would draw a female in the sphere of the authoress of the "Pearl of Days," as perhaps a person of just views and correct feeling; but in describing her intellect, she would of course feel necessitated to let herself down. But we discover, when the authoress of the "Pearl of Days" takes up the pen in her own behalf, and tells her own story, that the young literary lady might not let herself down. She might exercise all her own intellect in portraying that of her heroine, and not find the stock over-great. In like manner, were a modern tale-writer to describe a good weaver, forced by lack of employment to quit his comfortless home, and cast himself with his wife and children upon the cold charity of the world, he might bestow upon him keen sensibilities, a depressing sense of degradation, and a feeling of shame; but his thoughts on the occasion would scarce fail to partake of the poverty of his circumstances. When, however, the weaver Tom tells exactly such a story of himself, not as a piece of fiction, but as a sad truth burnt into his memory, we find the keen sensibility and the sense of shame united to thinking of great power, heightened in effect by no stinted measure of the poetic faculty.

Now, from our knowledge of such cases, and from a felt want, in our modern fictitious narratives, of what we shall term the inner life of the working classes, what we would fain recommend is, that the working classes should themselves tell their own stories. A series of autobiographies of working-men, produced, like the Sabbath Essays, on the competition principle, and rendered, by judicious selection, representative of the various manual trades of the country and its several districts, would form one of perhaps the most valuable, and certainly not least interesting, "Miscellanies," which the enterprise of the "Trade" has yet given to the country. It would constitute, too, a contribution to the domestic history of the period, the importance of which could not be very easily over-estimated. It were well, surely, that the appetite which exists for information regarding the true state and feelings of the working classes should be satisfied with other than mere pictures of the imagination. A series of cheap volumes, such as we desiderate, would furnish many an interesting glimpse into the lives of the laboring poor, and deepen the interest in their welfare already so generally felt. And we are sure the scheme, if attempted by some judicious bookseller, would scarce fail to remunerate.

LITERARY AND SCIENTIFIC.

I.

IMPRESSIONS OF THE GREAT EXHIBITION.

FIRST ARTICLE.

The Exhibition closed upon Saturday last; and one of the most marvellous and instructive sights which the world ever saw now survives only as a great recollection, — as a lesson unique in the history of the species, which has been fairly given, but which, upon the same scale at least, we need scarce hope to see repeated. I spent the greater part of last week amid its long withdrawing aisles and galleries, and, without specially concentrating myself on any one set of objects, artistic or mechanical, set my thoughts loose among the whole, to see whether they could not glean up for future use a few general impressions, better suited to remain with me than any mere recollections of the particular and the minute. The memory lays fast hold of the sum total in an important calculation, and retains it; but of all the intermediate sums employed in the work of reduction or summation it takes no hold whatever; and so, in most minds, on a somewhat similar principle, general results are remembered, while the multitudinous items from which they are derived fade into dimness and are forgotten.

Like every other visitor, I was first impressed by the great building which spanned over the whole, having ample room in its vast areas for at once the productions of a world and the population of a great city. I was one of a hundred and eight thousand persons who at once stood under its roof; nor, save at a few points, was the pressure inconveniently great. If equally spread over the building, all the present population of Edinburgh could, without the displacement of a single article, have found ample standing room within the walls. And yet this greatest of buildings did not impress me as great. In one point at least, where the airy transept raises its transparent arch seventy feet over the floor, and the sunlight from above sported freely amid the foliage of the imprisoned trees and on the play of crystal fountains, it struck me as eminently beautiful; but the idea which it conveyed everywhere else was simply one of *largeness*, not of greatness. There are but two great ideas in the architecture of the world, — the Grecian idea and the Gothic idea; and though both demand for their full development a certain degree of magnitude, without which they sink into mere models, very ample magnitude is not demanded. York Minster and St. Paul's united would scarce cover one fourth the space occupied by the Crystal Palace, and yet they are both great buildings, and it is not. Hercules, the son of the most potent of the gods, was great; whereas the earthborn giants that he conquered and slew were simply bulky. And in works of art so much depends, in like manner, on lineage, that things of plebeian origin, however large they may eventually become, rarely if ever attain to greatness. Two or three centuries ago, some lover of flowers and shrubs bethought him of shielding his more delicate plants from the severity of the climate by a small glass-frame, consisting of a few panes. In course of time, the idea embodied in the frame expanded into a moderate-sized hot-house, then into a green-house of considerably larger size, then into a tall palm-house; and, last of all, an ingenious gardener, bred among groves of exotics protected by huge erections of glass and iron, and familiar with the necessity of adding to the size of a case as the objects which it had to contain multiplied or were enlarged, bethought him of expanding the idea yet further into the Crystal Palace of the Exhibition. And such seems to be the history and lineage of perhaps the largest of all buildings: it is simply an expansion of the first glass-frame that covered the first few delicate flowers transplanted from a warmer to a colder climate, and, notwithstanding its imposing proportions, is as much a mere *case* as it was. And were its size to be doubled, — if, instead of containing two hundred miles of sash-bars and nine hundred thousand superficial feet of glass, it were stretched out so as to contain four hundred miles of bars and eighteen hundred thousand feet of glass, — it would be of course a larger building than it is, but not a greater. Nay, I should perhaps rather remark that it would be impossible by addition to render it not only *more*, but even *less* great than it is, — in itself a mark of inferiority. To a truly great building it would be impossible to add; for unity, as a whole, forms the very soul of all great edifices. He would be a bold man who would attempt making a single addition to St. Paul's, — one tower more would ruin it; whereas the length or breadth of a railway terminus may be increased *ad infinitum*, without in the least affecting its unity or proportions; for the railway terminus is also a mere case, and its unity and proportions bear reference to but the rule of convenience, which directs that it should be made quite large enough to hold what it had been erected to shelter. It is ill, I may add, with an architecture, of what at least ought to be the higher kind, when it is found to come under this law of the lower. I was sorry to observe that the singularly ornate pile which contains the new Houses of Parliament, now nearly finished, could, like the glass palace or a railway terminus, and very unlike either St. Paul's or York Minster, be made either twice as long as it is now, or shorter by one half, without rendering it either a better building or a worse.

Once fairly entered within the edifice, the objects first

singled out by the eye were a few noble statues, such as the Greek Slave and the Amazon and Tiger; nor was it until these works of genius were scanned that the humbler works of mere talent and art succeeded in forcing themselves on the attention. And yet these last served to show much more definitely the actual stage of progress at which the nations that produced them had arrived than the higher order of works. When genius is the artist, the goal is soon reached; whereas talent labors slowly. Genius, too, can work very much alone, and bid, as Johnson expresses it, "help and hinderance alike vanish before it;" whereas talent requires the assistance of a thousand coadjutors. And so we find that genius, laboring in ancient Greece greatly more than two thousand years ago, produced its statues and its architecture, its orations, its history, and its poetry, to be models and patterns to the world throughout all after ages. But the world at the time was too much in its infancy to excel in works of talent; and so Greece, even when erecting the Parthenon, or sculpturing the figures on its frieze, could not have built and furnished a single ship of the line, or raised such a palace of glass and iron as that of Mr. Paxton. I found, however, that in works of genius, as certainly as of talent, what may be properly termed the civilized nations of the world march abreast of each other in nearly the same line, — not serially in file; and it is one of the advantages of the Exhibition that it should teach this lesson. The United States of America, France, Austria, Northern Germany, Denmark, Italy and Sardinia, England and Scotland, are all laboring in nearly the same arts, artistic and mechanical, and producing nearly the same results. Their inhabitants are intellectually of like stature, and similarly trained, — a fact which national pride, schooled in the Exhibition, will now scarce venture to deny, and which, we are disposed to think, the English people will be much the better for knowing; seeing that to undervalue a competitor or opponent is one of the most certain ways possible to secure defeat, and to form a correct estimate of him one of the most effectual means of avoiding it.

It was interesting enough to read, in the extent of some manufactures all over the world, as shown in the various departments of the Exhibition, a chronicle of their great antiquity. Tried by this test, the art of the weaver seemed to be the most ancient; it was, in at least this display of human industry, the most widely diffused. With the exception of a few barbarous islands, where a kind of coarse paper, or animal skins, or the layers of vegetable tissue, form an imperfect substitute for cloth, every nation presented for examination its textile fabrics, very diverse in pattern in most instances, but constructed on the same mechanical principles, and ornamented, if not in the same style, at least by the same arts. That quality of thread, for instance, of reflecting light according to the disposition of its fibres and to the angle in which it is viewed, which forms the foundation of the style of ornament employed in damask, and in so many other fabrics, seems to be known all over the world, — in China, with its insulated and far-distant centre of civilization on the one hand, as certainly as in America on the other; and in all countries the same arts have been employed to make this quality paint without color the surface of the fabric. It is now more than three thousand years since the patriarch Job compared the short life of man to the swift and brief flight of a weaver's shuttle. Judging from what appears in the Exhibition, it seems not improbable that weavers' shuttles, and this simple art of painting by light without the aid of chemistry, may have been spread all over the world at the dispersal of mankind from before the great tower. And it seemed quite curious enough to reflect, that in this world's other great building the nations should have assembled for the first time, to show whether and to what extent they had been improving the talent, or whether, like a few of the barbarous tribes, they had not sunk into utter degradation and buried it in the earth. In passing along through this textile department, — one of the most largely represented in the Exhibition, — it was interesting to mark the different ideas that had been superadded from certain countries to the original stock. The Jacquard idea seemed one of the most important; and we find that, with the potency of a true idea, it has spread all over Europe and America. The other great idea in this department is of such recent origin, that we found it well-nigh still restricted to its original centre of production. We refer to the invention of our fellow-citizen Mr. Richard Whytock, of barring the threads across in the state of yarn, according to a nice calculation, with varying stripes of color, and of then forming them, by simply committing them to the loom, into rich patterns, that grow up under the workman's hand he scarce knows how. The rich magnificence of the pieces exhibited — a magnificence that, in at least their immediate neighborhood, threw all competition into the shade— demonstrated the happiness of the idea, and its uniqueness in the Exhibition, though occurring in one of the oldest of arts, its great originality.

One of the next things that struck, in the general survey, was the tendency of all the merely ornamental ideas presented in the Exhibition to arrange themselves in the mind, irrespective of the dates of their production, into modern and ancient. The semi-barbarous and the civilized nations are contemporary; the workmen who produced the fabrics or jewellery of China and of Tunis are as certainly living men, still engaged in the production of more, as the workmen of France or of England; and yet *their* works bear the stamp of antiquity; while those of the civilized nations, save where, in a few cases, a false taste has led to a retrograde movement, bear the true modern air. They are things, not of the past, but of the present. Not a few of the ideas which they embody are absolutely old: their sculpture is formed on the old model of Greece, — their architectural ideas are either Grecian or Gothic; and yet, though associations and recollections of the ancient do mix themselves up with the later style, we feel that, unlike the semi-barbarous productions of the less civilized nations, they are not *old-fashioned*, but new. In one sense, *new* and *true*, *old* and *false*, are evidently convertible terms. A false idea in art always becomes old; while a true idea lives on, and bears about it the freshness of youth. The false idea is consigned to the keeping of but the antiquary; whereas civilized man, as such, becomes the depository of the true one; and in his countless reproductions it continues to bear about it the fresh gloss of youth. And at length, with even the recent, if false, we come to associate ideas of the obsolete and the old. I was much struck, in the mediæval department of the Exhibition, — a department which we owe to Puseyism, — by the large amount of the false in art which this superstition has been the means of calling back from its grave. The Gothic architecture is true, — one, as we have already said, of the two great architectural ideas of the world. But the Gothic sculpture and the Gothic painting are both false; and Puseyism has, with the nonsense and false doctrine of the middle ages, been restoring both the false painting and the false sculpture. The grotesque figures gaudily stained into glass, or grimly fretted into stone, harmonized well with tall candles of beeswax and cotton wick, to light which is worship, and with snug little cages of metal, into which priests put their god when they have made him out of a little dried batter. We are told that James VII. strove hard to convert his somewhat unscrupulous favorite, the semi-infidel Sheffield, to Popery. "Your Majesty must excuse me," said the courtier: "I have at length come to believe that God made man, which is something; but I cannot believe that man, to be quits with his Maker, turns round and discharges the obligation by making God." In such a display of human faculty as the Great Exhibition, the strangely-expressed feeling of Sheffield must surely have come upon many a visitor of the mediæval apartment. What man is, — how glorious in intellect, how rich in genius, and how powerful in his control over the blind forces of nature, — was manifested, in by much the larger part of the Exhibition, in a manner in which none present had ever seen it manifested before. And what then must be the character and standing of that Great Being by whom man was created? Under the ample roof, however, there were here and there grotesque corners filled with the false and the old-fashioned; and, curiously enough, there were posted in these grotesque corners, as specimens of human workmanship, false, old-fashioned gods, — gods with paunch bellies, and gods with bloated, negro-like faces, and gods with from fourteen to twenty hands and arms apiece; and here, in yet one other grotesque corner, amid a false sculpture, we found copes, and albs, and painted candles seven feet high, and little cages for holding what the early reformers termed the "bread-god," which priests manufacture. Here, as in the other idolatrous apartments, false, old-fashioned arts were associated with a false, old-fashioned religion, and both wore alike on their foreheads the stamp of mortality and decay.

Popery, however, had, I found, one grand advantage over Puseyism in its use of art. With Puseyism all was restoration from a barbarous age, that possessed only one true artistic idea among many false ones; whereas Popery, on the other hand, had availed itself of art in all its stages; and so all its artistic ideas were the best and truest of their respective ages. When a Michel Angelo appeared, it forthwith adopted the sculpture of a Michel Angelo; when a Raphael appeared, it forthwith adopted the painting of a Raphael. Instead of perpetuating an obsolete fashion in its trinkets and jewels, it set its Benvenuto Cellini to model and set them anew; nay, in Italy, surrounded by noble fragments of the old classic architecture, it broke off its associations with the Gothic, and erected its fairest temples in the old Vitruvian symmetry, under the eye of a Palladio. This great difference between the two churches was most instructively shown in the portion of the Exhibition devoted to the display of stained glass. The English contributions, manufactured for the Puseyite market, abounded in ugly saints and idiotical virgins, flaming in tasteless combinations of gaudy color; whereas in much of the stained glass contributed by the popish countries of the Continent the style is exquisitely Raphaelesque. But I cannot better describe the difference between the two schools than in the admirable pictures of Warton, with which, as representative of the wisdom of Popery in its generation, compared with the folly of Puseyism, we for the present conclude. It is of the mediæval style that the poet speaks: —

"Ye brawny prophets, that, in robes so rich,
At distance due possess the crisped niche;
Ye saints, who, clad in crimson's bright array,
More pride than humble poverty display;
Ye virgins meek, that wear the palmy crown
Of patient faith, and yet so fiercely frown;
Ye angels, that from clouds of gold recline,
But boast no semblance to a race divine;
Shapes that with one broad glare the gazer strike;
Kings, bishops, nuns, apostles, all alike; —
No more the sacred window's round disgrace,
But yeld to Grecian groups the opening space.
* * * *
And now I view, instead, the chaste design,
The just proportion, and the genuine line;
Those native portraitures of Attic art,
That from the lucid surface seem to start;
The doubtful radiance of contending dyes,
That faintly mingle, yet distinctly rise
'Twixt light and shade; the transitory strife,
The feature blooming with immortal life;
The stole, in causal foldings taught to flow,
Not with ambitious ornaments to glow;
The tread majestic, and the beaming eye,
That, lifted, speaks its commerce with the sky;
Heaven's golden emanation, gleaming mild
O'er the mean cradle of the virgin's child.
* * * *

Thy powerful hand has broke the Gothic chain,
And brought the bosom back to truth again, —
To truth, by no peculiar taste confined,
Whose universal pattern strikes mankind."

SECOND ARTICLE.

I found the various articles of the Exhibition ranged under the four great heads of raw materials, manufactures, machinery, and the fine arts. In the first department I saw the stuff, whether furnished by the bowels of the earth or produced on its surface, on which man has to work; in the second, that into which, for purposes of use or of ornament, he succeeds in fashioning it; in the third, his various most ingenious modes of making dead matter his fellow-laborer and slave in the task of moulding the stubborn materials into shape and form; and in the fourth, his strainings after something higher than mere utility, and his wonderful ability of creating a perfection in form and expression greater than that which he finds in living nature. I could have wished that into this last department fine pictures, as certainly as fine statues, had been admissible. The display of either was not properly the object of the Great Exhibition; and yet it would have been incomplete without them. From the two sister arts, — those of the painter and of the statuary, — all that imparted elegance and beauty to the labors of the manufacturer had been derived. The workers in wood, stone, and metal had borrowed their delicate sculpturings from the statuary; the workers in silk and thread, in clay and in glass, in dyes and in paints, in japans and in varnishes, had borrowed their choicest patterns from the painter; all that added beauty to comfort in the implements and appliances of a high civilization had been derived from the twin arts; they had thrown, as if by reflection, the flush of genius on the common and ordinary things of life; I saw their vivid impress at almost every stall; and as sculpture was present in some of her higher productions, I hold that painting in some of her higher productions should have been present also, as that other art which, in the staple productions of the Exhibition, had added beauty to comfort, and the exquisite and the ideal to the common and the ordinary.

In examining the raw materials furnished by the various countries of the world, some of them many thousand miles apart, what first struck was the great uniformity of character and appearance which prevailed among the sections devoted to mineral and mining products, and the great diversity which marked the animal and vegetable ones. Whatever was furnished by the primary rocks bore almost the same character all over the world. The granites and porphyries of the southern hemisphere differed in no respect from those of the northern one; or the iron, lead, and copper ores of the Old World from those of the New. Even specimens sent by one state or kingdom as marvels from their size or purity, were in most cases fully matched by specimens of the same kind sent by some other state or kingdom thousands of miles apart. Russia, for instance, furnished plates of mica a full foot across; but then the United States did the same; and a mass of virgin copper from Massachusetts, which weighed two thousand five hundred and forty-four pounds, was more than matched by a block of similar copper from Trenanze in Cornwall, which weighed only two thousand five hundred pounds, but was merely a portion of a mass fifty superficial feet in extent, of so much greater weight that it could not be raised entire out of the mine. The frequent occurrence of copper in a *virgin* — that is, pure and malleable — state, among the ores of the world, as presented to view in the Exhibition, threw light on the place which the bronze age holds in the chronology of the antiquary. Its place is always second to the age of stone. All the iron ores exhibited existed as mere stones. If a bit of virgin iron be here and there occasionally found, the chemist ascertains that, unlike any of the iron of earth, it is mixed with nickel and chrome, and concludes that it came as a meteorolite from heaven; for it is still doubtful whether there be properly any virgin iron on earth which the earth itself has produced; at least, if it at all exists, it is a greatly rarer substance than gold. And iron in the stony state is a much less eligible substance for tool or weapon making than ordinary stone. But virgin copper is greatly superior to either flint or jasper in at least ductility; and such is its purity, that the savage who found the first mass of it in the rock could beat it out into a sword or spear-head, with simply one stone for his anvil and another for his hammer. In every country of the world in which copper is to be found at all, the copper or bronze age is found to have come immediately after that of stone, and in advance of that of iron. That resemblance borne among themselves by the mineral productions of the earth in all countries, which the Exhibition made so strikingly manifest, has been remarked both by Humboldt and by Captain Basil Hall. "Humboldt," says the latter writer, in his voyage to Loo Choo, "somewhere remarks the wonderful uniformity which obtains in the rocks forming the crust of the globe, and contrasts this regularity with the diversity prevailing in every other branch of natural history. The truth of this remark was often forcibly impressed upon our notice during the present voyage; for wherever we went, the vegetable, the animal, and the *moral* kingdom, if I may use such an expression, were discovered to be infinitely varied: even the aspect of the skies was changed; and new constellations and new climates coöperated to make us sensible that we were far from home. But on turning our eyes to the rocks upon which we were standing, we instantly discovered the most exact resemblance to what we had seen elsewhere."

There were, however, a few centres to be found in this Exhibition of the world's industry, where the production of some mineral in larger and finer masses than it had been detected elsewhere, among at least the civilized nations, had originated some branch of art or manufacture unique in the show. Of this, perhaps the most striking example was furnished by the Russian department, where the malachite furniture and ornaments, wholly unlike aught displayed in any other section, were of the most gorgeous and impressive beauty. A few specimens of the material in its rude state lay on a table beside the wrought articles, and were certainly of much greater size and mass than any specimens of the mineral which I had hitherto seen in any collection. One fragment seemed about a foot square on its larger surface, and from six to eight inches in depth. Malachite is one of the ores of copper. It consists of from fifty to sixty per cent. of that metal, combined with oxygen, carbonic acid gas, and water, in the solid form; it may, in fine, be regarded as a green verdigris, hardened by its union with the gases into a compact marble, susceptible of a fine polish, and occurring usually in cavities in the stalactitical or botryoidal form. Its color internally is found to vary from darker to lighter, as in most stalactites, in graceful lines parallel to its lines of surface, and that speak, in those flowing curves, of aqueous deposition. The worker in malachite cuts it up into thin veneers, which, according to the nature of his work, he lays down upon a ground either of stone or of metal, taking care that the curve of one fragment merges gracefully into the curves of the neighboring ones; and thus large and apparently continuous planes of the substance are formed, as in tables, chimney-pieces, and doors; or it is curved and hollowed so as to wrap round noble vases bordered with gold, or even wrought into ornately carved chairs. The beauty of the articles thus produced is so great that they formed one of the centres of the Exhibition, upon which the living tide constantly set in; but their great cost must restrict their use to what their exquisite beauty peculiarly fits them to grace, — the palaces of princes and the mansions of nobles. One magnificent door of this substance, which from top to bottom looked like an opaque emerald, was valued, we understood, at about ten thousand pounds sterling.

The vegetable and animal substances exhibited under the head of raw materials formed a marked contrast, in their great diversity, to the mining and mineral products. In the colonial department, almost every climate and zone sent specimens of its plants and trees, its mammals, reptiles, fishes, and birds; and the variety was of course very great. There were, however, a few of the mineral products of the latter geological ages that came under the same law of diversity as that which obtained among the plants and animals. Coal and the coal plants, judging from the specimens, seem to bear well-nigh the same character all over the world, and to be spread very widely in each hemisphere; but amber, the fossilized resin of an extinct species of pine, seems very much restricted, like some of our existing pines, to a European centre; and though there were specimens in the Exhibition furnished from various European localities, — among the rest, from the Norfolk coast, — all the finer and larger specimens came, we found, from Northern Germany, and in especial from the southern coast of the Baltic. In the glass case of one exhibitor, beautiful pieces of amber, dug out of the ground, lay side by side with fragments of the fossil pine (*pinus succinifer*, which had produced it; in another there were large masses which had been cast up by the sea, of a quality so fine that similar masses have been sold at the rate of a hundred dollars per pound weight; in yet another case there were specimens of the various implements and ornaments into which amber is formed, and which rendered it of old, and in some degree still, an important article of commerce; and in yet another and vastly more interesting case still, had one but the time and opportunity necessary to observation, there was a set of specimens of amber selected for the sake of the organisms, vegetable and animal, which they contained, and which had taken — so said the catalogue — twenty-five years to collect. It is a curious circumstance, that naturalists have now discovered in this substance fossilized fragments of forty-eight different species of shrubs and trees, and no fewer than eight hundred different species of insects. We looked with no little respect on the various specimens of amber furnished by the Exhibition, as of great interest to the geologist, from the circumstance that it has formed the best of all matrices for the preservation of the minuter organisms of the later tertiary periods; and of great interest to the historian, from the circumstance that it was the means of first awakening the commercial spirit in northern Europe, and of inducing the equalizing tides of civilization to set in from the shores of the genial Mediterranean to those of the frozen Baltic.

In the vegetable department, though the intertropical colonies sent their splendid exotics, and the woods, roots and plants of the New World contended with those of the Indian Archipelago and the southern hemisphere, I saw nothing that at all equalled in completeness the collection of the Messrs. Lawson of Edinburgh. It consists of all plants, seeds, and trees which are reared in Scotland for the use of man; and, interesting at all times, it would have formed, had it been made in the last age, one of the best possible apologetic defences for Scotland against the gibes of the English. "Do you ever bring the sloe to perfection in your country?" inquired Johnson, in one of his merrier moods, of the obsequious Boswell. The Messrs. Lawson show most conclusively that we bring to perfection a great deal more. We find it stated that the making of their collection cost them about two thousand pounds sterling, — evidence enough of itself that the vegetable productions of Scotland useful for food and in the arts cannot be few. There are many Scotchmen, and in especial Scotch women, who complain of the climate of their country. I dare say it must have occurred to some of them, amid the beautiful specimens of the Messrs. Lawsons' collection, that the wonder is, not that the climate of Scotland should be occasionally severe, but that in the average it should be so mild and genial. There is not another country in the world lying between the fifty-fifth and fifty-ninth degrees of latitude, whether in the northern or southern hemisphere, that could mature one-half the productions exhibited by the Messrs. Lawson. On the American coast, under the same degrees, the isothermal line is that of the north of Iceland; the ground always remains frozen hard as a rock to the depth of a few feet from the surface; and as the winter sets in, the sea forms into a continuous cake of ice along the shores. The lines of latitude fairly taken into account, we challenge for Scotland the finest climate and the most productive soil in the world. And yet, at a time comparatively recent to the geologist, though, of course, removed beyond the historic period, the case was widely different. The scratched and polished rocks of the Pleistocene period, its moraines and travelled stones, its drift gravels, its boulder clays, and its semi-arctic shells, testify to an age of ice and snow, of local glaciers and drifting icebergs, in which not a tithe of the vegetable productions exhibited by the Messrs. Lawson could have been reared in Scotland. I am glad to learn that this interesting collection, so honorable to the skill and industry of the collectors, and which so thoroughly bears out the deductions of science regarding the isothermal conditions of Scotland, is to be transferred entire to the horticultural museum at Kew Gardens.

In the exhibition of birds and beasts, which came in part under the head of materials derived from the animal kingdom, and in part illustrated the art of the animal-stuffer, I saw some cabinets of rare interest; but I could fain have wished that the general section had been more complete. Such a collection of the birds, fishes, and quadrupeds of Scotland as that which the Messrs. Lawson exhibited of its plants would have well repaid the study of days. Nor, of course, would less of interest have attached to the animals of other countries, with their rivers and seas. I saw one tastefully-arranged case of stuffed birds from the wild west coast of Assynt, and recognized in the name of the exhibitor, Mr. W. Dunbar, an intelligent naturalist resident at Loch Inver, whose freely communicated stores of knowledge occupy, though not always with the due acknowledgment, a large space in a popular work on Sutherlandshire. His case contained chiefly the game-birds of the county, which might be regarded either as the raw material which our sporting gentlemen convert into food at the very moderate cost, when they are eminently successful in the process, of about thirty pounds sterling per stone; or, a more pleasing view, as adequately representative of an important portion of the natural history of the county. Nothing could be more perfectly life-like or natural than these stuffed birds of Mr. Dunbar. The great achievement presented by the Exhibition, however, in this department, was furnished by a German State. On no one object under the vast crystal roof, not even on the Kohinoor itself, did a greater tide of visitors set in, whether on shilling or on half-crown days, than on what were known, though not so entered in the official catalogue, as "The Comical Creatures of Wurtemburg." The catalogue simply bore that Herman Ploucquet, preserver of objects of natural history at the Royal Museum of Stuttgardt, had contributed to the show "groups of stuffed animals and birds, nests of birds of prey, hawks pouncing upon owls," etc., etc.; and certainly nothing could be more natural and true than these groups. They were made to represent, with all the energy of life, the scenes so frequently enacted in the animal world. It was not, however, to the purely natural that the Exhibition owed its interest, but to the introduction of an idea long familiar to the poet and the fabulist, and which painting and sculpture, in at least some of their humble departments, have borrowed from literature, but which, to at least the bird and animal stuffer, seems to be new. Most of Mr. Ploucquet's groups, though animals are the actors, represent scenes, not of animal, but of human life. The "Batrachomaomachia" of Homer, in which frogs and mice enact the part of the heroes of the Trojan war, and "make an Iliad of a day's campaign," furnished merriment to the old Greeks. Æsop and his numerous imitators followed in the same wake; until at length the representation of men under the forms, and bearing the characters of animals, became one of the commonest of literary ideas. And from literature it found its way, as we have said, into painting and sculpture. But the introduction of the animals themselves into such scenes seems to be a new, and, judging from the great popularity of Ploucquet's figures, a most successful idea. It is interesting, and really not uninstructive, to mark how thoroughly the animal physiognomy can be made to express at least the lower passions and more earthy moods of the human subject. One of the stories illustrated by the ingenious German is an eminently popular one on the Continent, — that of Reynard the Fox. "Among the people," says Carlyle, "it was long a house-book and universal best companion. It has been lectured on in universities, quoted in imperial council-halls, lain on the toilets of princes, and been thumbed to pieces on the bench of artisans." Reynard bears, of course, in the story, his character of consummate cunning and address; and in the opening scene, where a *bona fide* fox is introduced, lolling at his ease on a sofa, with his hind legs set across, his tail issuing from between them and curled jauntily round his left fore-paw, and his head reclining upon his right, there is an expression of cool, calculating cunning, as strongly, we had almost said as artistically marked, as in the Lovat or the John Wilkes of Hogarth.

II.

CRITICISM FOR THE UNINITIATED.

FIRST ARTICLE.

We have just been spending a few hours for the first time among the pictures of the Exhibition of the Royal Scottish Academy, and spending them very agreeably. A good picture is inferior in value to only a good book; and in one important respect at least bad ones are better than inferior books, seeing one can determine their true character at scarce any expense of time. There are no second and third pages to turn after perusing the first; and if there be nothing to strike or nothing to please, this negative quality of the piece, as fatal surely to a picture as to a book, is discovered at a cost proportioned to its value. The connoisseur, like the critic, has his rules of art and his vocabulary; but though some eyes are doubtlessly more practised than others, and some judgments better informed, I do not deem the art itself of very difficult attainment. To please is the grand end of the painter; and he can attain his object in only two different ways, — by either a close imitation of the objects he represents, or by the choice of objects interesting in themselves. Now, it needs no art whatever to decide whether or no he has succeeded in the first and simpler department, — the faithful representation of what he intended to delineate. The birds that pecked at the grapes of the ancient painter; the countryman who attempted to scale the painted flight of stairs; the artist who stretched his hand to draw aside the well-simulated curtain which seemed to half-conceal the work of his rival, — all these were equally skilful judges. Even the decision of the birds themselves was such a decision as no connoisseur would have dared dispute; and many an ingenious piece of criticism has the memory of it survived. In the same way, the mastiff who came running up to his master's portrait wagging his tail was a perfectly qualified judge of its fidelity. The other department of the art — the choice of subjects — requires higher qualities in the connoisseur; but it is not exclusively in picture-galleries that his skill is to be acquired. Nay, I am mistaken if it may not be acquired outside of the picture-gallery altogether, and in utter ignorance of the technicalities of the art. Take landscape, for instance. Who can doubt that Shenstone, who had of all men the most exquisite eye for the real scenes of nature, must have had an eye equally exquisite for those very scenes when transferred to canvas? He was more than a great connoisseur: he was also a great artist, — an artist who dealt in realities exclusively, and planted his thickets and formed his waterfalls with all the exquisite perception and inventive originality of high genius. No one can suppose that Shenstone's taste and skill would not have served him in as good stead amid a collection of pictures as at Hagley or in the Leasowes; or that, however unskilled in the connoisseur's vocabulary, he would have proved other than a first-rate connoisseur.

The "poet's lyre," says Cowper, "must be the poet's heart;" he must feel warmly before he can express strongly. I suppose nearly the same remark may be applied both to the painter and the men best qualified to appreciate the painter's productions. An intense feeling of the beautiful and a nice perception of it invariably go together; and unless a person has experienced this feeling, in the first instance, amid the delights of the original nature, there is no virtue in rules or phrases to convey it to him from the painter's copy. I am not aware that Professor Wilson knows anything of these rules or phrases. Certain I am, however, that this master of gorgeous description, who makes the reader more than see the scene he paints, for he makes him feel it too, must have an exquisite eye for landscape, whether it be on or off canvas. He is one of the born connoisseurs. And what this man of genius possesses in so great a degree is possessed as really, though in immensely varied gradations, by almost all. Akenside describes the untaught peasant lingering delighted amid the glories of a splendid sunset, intensely happy, and yet scarcely able to say why. Assuredly that same peasant would be quite qualified to distinguish between a daub and a fine picture. Imagine him passing homeward, after "his long day's labor," in one of those exquisite evenings of early June that live with a "sunshiny freshness in memory," as Shelley finely expresses it, long after they have passed. There is a splendid drapery of clouds in the west, tinted by those hues of heaven which can be fully expressed by neither the words nor the colors of earth, — those hues of exquisite glory — of gold, and flame, and pearl, and amber — which the prophets describe as encircling the chariot of Deity. The sun rests in the midst, less fiercely bright than when he looked down from the middle heavens, but dilated apparently in size, and more glorious to the conception, because more accessible to the eye. The landscape below is soft and pastoral. There is a dim undulating line of blue hills on the one hand, and the far-off sea on the other. A light, fleecy cloud hangs over the distant village, and seems a bar of pale silver relieved against the wooded hill behind. A lonely burying-ground, surrounded by ancient trees, and with the remains of an old time-shattered edifice rising in the midst, occupies the foreground. We see the white tombstones glittering to the sun, and the alternate bars of light and shadow that mark more dimly the sepulchral ridges of yellow moss which rise so thickly over the sward; while beyond, on the side of a wide-spreading acclivity, there is a quiet scene of fields, and hedgerows, and clumps of wood, with here and there a group of white cottages, all basking in the red light. And mark the loiterer, — one of the intellectual peasants of our own country, — a well-selected specimen of the class which, in at least thought, feeling, and power, has found its meet type and representative in

> "Him who walked in glory and in joy,
> Following his plough upon the mountain-side."

How his steps become gradually fewer and more slow! and how at length, unconscious of aught except what Akenside exquisitely describes as the "form of beauty smiling at his heart," he stands still, to lose, in the happiness of the present, every gloomier recollection of the past, and every darker anticipation of the future! Undoubtedly that untaught peasant is a connoisseur of the higher class. The birds peck the grapes, the mastiff recognizes the portrait; but the peasant can judge of more than mere likeness, — he can exquisitely feel the beautiful; and he is perfectly qualified to say that the work of art which can reawaken in him this feeling is assuredly a work of genius. But why all this wild radicalism, this lowering of the aristocracy of criticism, this breaking down of the fictitious distinctions of connoisseurship! In the first place, I am merely making my apology for having derived very exquisite pleasure from even a first visit to the pictures of the Academy; and, in the second, for daring to do what I am just on the eve of doing, — for daring to assure the reader, that, if he has an eye and a heart for nature, he may go there, however unskilled in the rules of the vocabulary of criticism, and derive much pleasure from them too. I am merely standing up, as Earl Grey and Cobbett have expressed it, *for my order*, — the uninitiated.

I have spent some of my happiest hours amid exhibitions of a different kind from the Exhibition in the Academy; and some of my most vivid recollections refer to scenes redolent of the wild and the sublime of nature, and to the emotions which these have awakened. May I venture to describe the feeling in connection with one sweet scene — a wooded dell in the far north — in which I have perhaps oftenest experienced it, and which came rushing into my mind as I lingered in front of one of the richest landscapes of the Exhibition? It is a recess of deepest solitude; but the sweet Highland stream that comes winding through it, passing alternately from light to shadow and from motion to repose, imparts to it an air of life and animation, and we do not feel that it is lonely. Man is so little an animal, says Rousseau, that he is as effectually sheltered by a tree twenty feet in height as by one of sixty. True; but his ideas are much larger than himself, and he has too close a sympathy with nature not to experience an ampler expansion of feeling under the loftier than under the lower cover. In this solitary dell, the banks, which on either hand, at every angle and indentation, advance their grassy ridges or retire in long, sloping hollows, partake perhaps rather of the picturesque than of the magnificent; but the trees which rise along their sides, and which for the last century have been slowly lifting themselves to the freer air of the upper region, look down from more than the higher altitude instanced by Rousseau. Often, when the evening sun was casting its slant red beams athwart their topmost branches, and all beneath was brown in the shade, I have sauntered along this little stream, lost in delicious musings, whose intermingled train of thought and feeling I have no language to convey. I have felt that the cogitative faculty in these moods had not much of activity; but then, though it wrought slowly, it wrought willingly and unbidden; and around every minute thought there would swell and expand an atmosphere of delightful feeling, which somehow seemed to owe its origin as much to the magnitude as to the quiet beauty of the surrounding objects, and which has reminded me fancifully, but strongly, of that minutest of all the planets, — of the asteroids rather, — whose atmosphere rises over it to more than ten times the height of the atmosphere of our own planet; I have looked up to the branches that twisted and interlaced themselves so high over head, and the leaves that seemed sleeping in the light; I have seen the deep blue sky far beyond; I have caught glimpses through the chance vistas of little open spaces, shaggy with a rank vegetation, and which I have loved to deem the haunts of a solitude still deeper than that which surrounded me; I have marked the varieties of beauty which distinguish the several denizens of the forest, — the ash, with his long massy arms, that shoot off from the trunk at such acute angles, and his sooty blossoms spread over him as if he wore mourning; the elm, with his trunk gnarled and furrowed like an Egyptian column, and his flake-like foliage laid on in strips that lie nearly parallel to the horizon; the plane, with his dark green leaves and dense, heavy outline, like that of a thunder-cloud; the birch, too, a tree evidently of the gentler sex, with her long flowing tresses falling down to her knee; — and as I looked above and around, I felt my heart swelling with an exquisite emotion, that feasts on the grand and the beautiful as its proper food; and surely that mind must be chilled and darkened by the pall of a death-like scepticism, that does not expand with love and gratitude, under the influence of so exquisite a feeling, to the great and wonderful Being who has imparted so much of good and fair to the forms of inanimate nature, and has bestowed on the creature such a capacity of enjoying them.

SECOND ARTICLE.

In the middle of the second exhibition-room, on the west side, there is a picture of Allan's which almost every visitor stands to study and admire; and we observed not a few who, like ourselves, came back a second and a third time to look at it again and again. Let criticism say what it please, this is praise of the very highest order. The piece represents one of the first heroes and greatest men of Scotland, — Robert the Bruce; and represents him when greatest and noblest, — uniting to a courage truly heroic the tenderness and compassion of a gentle and affectionate nature. It embodies with exquisite truth Barbour's affecting story of the king and the "poore lavender."

The scene, as all our readers must remember, is laid in Ireland. The redoubtable hero of Bannockburn had been compelled to retreat before the immensely superior forces of the English and their Irish allies. Both the retreating and the pursuing army had been resting for the night, — the one in a valley, the other on an adjoining hill; but the pursuers were early astir, and their long array had been seen from the Scottish encampment stretching far into the background on the ridge of the neighboring height, and all in full advance. The Scotch, too, had been preparing for a hasty retreat; Edward Bruce and the Black Douglas had mounted their war-horses, and the warriors behind were all on foot and in marching column, when they were suddenly arrested by the voice of the king. He had heard a woman shrieking in despair when just on the eve of mounting his horse, and had been told by his attendants, in reply to a hurried inquiry, that one of the female followers of the army, a "poore lavender" (that is, laundress), mother of an infant who had just been born, was about to be left behind, as being too weak to travel, and that she was shrieking in utter terror at thoughts of falling into the hands of the Irish, who were accounted very cruel. We quote the words of Sir Walter, who softens, with a tact and delicacy worthy of study, the less tasteful, though scarcely less powerful, narrative of the metrical historian. "King Robert was silent for a moment when he heard the story, being divided betwixt the feeling of humanity occasioned by the poor woman's distress, and the danger to which a halt would expose his army. At last he looked round his officers with eyes which kindled like fire. 'Ah, gentlemen, never let it be said that a man who was born of a woman, and nursed by a woman's tenderness, should leave a mother and an infant to the mercy of barbarians. *In the name of God, let the odds and the risk be what they will, I will fight Edmund Butler rather than leave these poor creatures behind me.'* "

The painter has chosen the moment of this noble exclamation for fixing the scene on his canvas. King Robert occupies the centre, — a wonderfully perfect transcript of Sir Walter's exquisite description in the "Lord of the Isles," and one of the most commanding figures we have ever seen. There is a strength more than Herculean in the deep broad chest and the uplifted arm, — the very arm which clave Sir Henry Bohun to the teeth through the steel headpiece; but, to employ the language of Lavater, "it is not the inert strength of the rock, but the elastic strength of the spring." The ease is admirable as the force; the figure possesses the blended power of an Achilles, alike unmatched in the race and the combat. His look is raised to heaven, — a look intensely eloquent, for it unites the indomitable resolution of the unmatched warrior with a devout awe for the Being in whose strength he has determined to abide the battle. The features, too, grave and rugged like those of his countrymen, possess that beauty of expression, far surpassing the beauty of mere form, which a mind conversant with high thoughts and noble emotions can alone impart to the countenance. The painter has drawn the Bruce, mind and body, — the master-spirit of the time, and through whom, under Providence, Scotland at this day is a country of free men, not of degraded helots, like at least two thirds of the unfortunate Irish.

On the left of the warrior-king is the new-made mother with her infant; she is a poor young creature, of simple beauty, — such a one as the Mary of Burns or the Jessie of Tannahill. It would really have been a great pity to have left her to the barbarous, pitiless Irish, — the ruthless savages who, even in the times of the first Charles, could so cruelly destroy the Protestant females of the country,

—quite as unable to resist, and quite as unoffending, as the "poore lavender." There is something very admirable in the air of lassitude which invests the whole figure: one hand barely sustains the infant, which, in the midst of danger and extreme weakness, she evidently regards with all the intense, though but newly-awakened, affection of the mother; the other finely-formed arm I had almost said supports her in her half-reclining position; but it is by much too weak for that, and tells eloquently its story of utter exhaustion and recent suffering. There is much good taste, too, shown in the painter's selection of the surrounding attendants; in the old woman, and in the girl, who half-compassionates the mother, half-admires the child; in the aged monk, too, evidently a good, benevolent man, who in all probability directed the devotions of his countrymen when they knelt at Bannockburn, and who is particularly well pleased that the Bruce has determined rather to fight Edmund Butler than to desert the "poore lavender."

On the king's right are his brother Edward Bruce, and James, Lord of Douglas, mounted, as we have said, on their war-steeds. Edward is well-nigh as perfect a conception as his brother the king. It needs no Lavater to tell us, from the speaking countenance, that the warrior on the right cannot be other than the frank, fearless, rashly-spoken, affectionate man, who hastily wished Bannockburn unfought because his friend had been killed in the battle. His whole figure is instinct with character. There he stands, a capital man-at-arms, first in the charge and last in the retreat; especially good at a light joke too, particularly when matters come to the worst; but not at all to be trusted as a leader. He is right well pleased on this occasion with brother Robert. "Fight Edmund Butler! ay, ten Edmund Butlers, if they choose to come; but we can't leave the poor woman." Possibly enough, however, the poor woman would have been left had Edward been first in command,—not certainly from any indifference, but out of sheer thoughtlessness. Edward would never have thought of asking what the cry meant.

We are not quite so satisfied with the Black Douglas. He is a stalwart warrior, keen and true in the hour of danger as his steel battleaxe; but the tenderness of the character is wanting. The painter has given us rather the Black Douglas of Sir Walter as drawn in his last melancholy production, "Castle Dangerous," when the mind of our greatest master of character was more than half gone, than the good Lord James of Barbour. Barbour devotes an entire page to the personal appearance of the Douglas, and certifies his description by assuring the reader that he had derived his information solely from men who had seen him with their own eyes. His metrical history was given to the country rather less than half a century after the death of his hero. He describes him as tall and immensely powerful, and with a "visage some dele gray;" and the painter, true to the description, has made him just gray enough. The expression, however, was peculiarly soft, modest, and pleasing; and, in accordance with his appearance, he spoke with a slight lisp, "which set him wonder well." He was a mighty favorite, too, we are told, with the ladies of King Robert's company, the Queen, and her attendants, — he was so gentle and so amusing; and when, early in the king's career, they were hard beset among the mountains, no one exerted himself half so much as the Douglas in supplying all their little and all their great wants, — in providing them with venison from the hillside and fish from the river, or, as the Arch-Dean quite as well expresses it, "in getting them meit." After dwelling, however, on all his amiabilities of character and expression, and particularly the latter, the historian tells us in his happiest manner, —

> "But who in baittle mocht him see,
> Another countenance had hee."

Old James Melville gives us nearly a similar description of

Kircaldy of Grange, "Ane lyon in the feild, and ane lambe in the hous;" and what does not quite please us in the Douglas of the picture, because it runs somewhat counter to our associations, is, that, though the spectator of a scene so moving, he should yet have got on his battle countenance. We have the lion, not the lamb. This, however, is not intended for criticism. The picker of minute faults in works of great genius reminds us always of the philosopher in Wordsworth's epitaph, — the "man who could peep and botanize upon his mother's grave."

There is another point in the picture of great interest, and very admirably brought out. It is at once exquisitely true to nature, and illustrates finely one of the most masterly strokes in Barbour. We are told by the ancient poet, that when the king, single-handed, had defended the rocky pass beside the ford against the troop of Galloway men, and had succeeded in beating them back, after "dotting the upgang with slain horse and men," his followers, just awakened from the slumbers in which he had been watching them so sedulously, came rushing up to him. They found him sitting bareheaded beside the ford, "for he was het," and had taken off his helmet, to breathe the more freely after his hard exercise. The exploit had gone far beyond all they had ever seen him accomplish before. He had defended them against "a hail troop, him alone;" and they came crowding round to get a glimpse of him. The very men who were with him every day, and who saw him almost every minute, were actually jostling one another, that they might look at him. Now, this is surely exquisite nature; and the idea is as happily brought out by Allan as by Barbour himself. The men are crowding to see their king; and never were there countenances more eloquent. There is love and admiration in every feature; and we feel that such a general with such followers could be in no imminent danger of defeat, after all, from the multitudes of Edmund Butler. The minor details of the picture seem to be finely managed. There is a clear, gray light; the sun has not yet risen, but it is on the eve of rising; all is seen clearly that any one wishes to see, and the rest is thrown into the soft, bluish, tinted shade peculiar to the hour. Randolph appears in the middle distance; and no person acquainted with the strictly just but stern-hearted warrior would desire to see him brought a step nearer. He would merely have come to say, with that severe face of his, that he really thought there was too much ado about a poor washerwoman; but that, if Edmund Butler was to be met with, why, he would just meet with him.

Edmund Butler, however, was not met with on this occasion. The wary leader knew that Robert the Bruce was the first general of his age; and that when he halted to offer battle, it could not be without some hidden reason, which rendered it no safe matter to accept the challenge which the halt implied. And so the English leader halted too, until the king resumed his march; and thus the "poore lavender" was saved at no actual expense to her countrymen. The story is one of those which deserve to live; nor is it probable that what Allan has painted, and Sir Walter described, "the country will willingly let die." We felt, when standing in front of this admirable picture, that the art of the painter, all unfitted as it is for serving devotional purposes, may yet be well employed in giving effect to a moral one.

THIRD ARTICLE.

In estimating the real strength of a country, one has always to take into account its past history. The statistics of its existing condition are no doubt very important. It is well to know the exact amount of its population, and the extent of its resources. It is a great deal more important, however, to ascertain what its people were doing a century or two ago, — what the nature of their contests and their success in them, and what the issue of their battles. It is not enough to count heads, or to calculate on the mere physical power of a certain quantum of thews and sinews. If the country's history be that of an enslaved and degraded race, who took their law from every new invader, neither its physical strength nor the greatness of its revenues matters anything: it is utterly weak and powerless. If, on the contrary, its battles were hard-fought, and terminated either in signal victory on the part of its people, or in a defeat that led merely to another battle, — if, in all its struggles, however protracted, the enemy was eventually borne down, and the object of the struggle secured, — depend upon it, that country, whether it reckon its population by thousands or by millions, is rich in the elements of power. The national history in these cases is more than a test of character; — it is also an ingredient of strength. The past breathes its invigorating influences upon the present; the battles won centuries before become direct guarantees, through the enthusiasm which they awaken, for the issue of battles to be fought in the future; the names of the brave and the good among the ancestors become watchwords of tremendous efficacy to the descendants; the children "honor their fathers," and "their days, therefore, shall be long in the land."

But what has all this to do with criticism? A great deal. As you enter the second exhibition-room, turn just two steps to the left, and examine the large picture before you. It is one of the masterpieces of Harvey, — "The Covenanters' Communion;" and very rarely has the same extent of canvas borne the impress of an equal amount of thought or feeling. The Covenanters themselves are before us, and we return to the times of which, according to Wordsworth, the "echo rings through Scotland till this hour." Not in vain did these devoted people assemble to worship God among the hills; not in vain did these venerable men, these delicate women and tender maidens, unhesitatingly lay down their lives for the cause of Christ and his church. Their solitary graves form no small portion of the strength and riches of the country. They retain a vivifying power, like the grave of Elisha, into which when the dead man was thrown he straightway revived. Those opponents of the church who assert, in the present struggle, that the cherished memory of our martyrs serves only to foster a spirit of fanatical pride among the people, are as opposed to right reason as devoid of true feeling. It fosters a truly conservative spirit, which it is well and wise to cherish; and one of the eminently wholesome effects of the present struggle is the reciprocity of feeling, if we may so express ourselves, which it awakens between the past and the present. The determination of the present revives the memory of the past, and the memory of the past gives tenfold force and effect to the determination of the present. Martrys never die in vain. We doubt not there is a time coming when even the memory of the noble Spaniards of the sixteenth century who perished unseen, for their adherence to Protestantism, in the dungeons of the Inquisition, and that of the noble Venetians of the same dark period who were consigned at midnight, and in chains, for the same sacred cause, to the depths of the Adriatic, will yet awaken among their countrymen, as an animating spirit to urge them on with double vigor to the attack, when Babylon is to be utterly destroyed. Most assuredly, Scotland at least has not yet reaped the entire benefit which she is to derive from the blood of her martyrs. The commonest seeds retain their vitality for centuries; the seed of the church retains its vitality for centuries too.

I shall attempt a description of Harvey's exquisite picture, for the sake of such of my readers as live at a distance. The *locale* of the scene represents one of those wild upland solitudes so common among our lower mountain ranges, — one of those hollows amid the hills known only to the shepherd and the huntsman, which are shut out by the surrounding summits from the view of the neighboring country, and which, rising high over the reign of corn, and almost over that of wood, presents only a widespread barrenness. There is a solitary fir bush in the background, which at a lower elevation would have been a tree; and its stunted and dwarf-like appearance tells of the ungenial climate and the unproductive soil. All else up to the very hill-tops is dark with heath; and there is a sky well-nigh as dark beyond; for there is scarce transparency enough in the accumulated masses of heavy clouds, that betoken a night of tempest, to relieve the outline. But there is a light in the foreground. The previous service of the day has been protracted for many hours; there has been a long "action sermon" on the wrestlings of the Kirk, and a long, impressive prayer; and the sun at his setting is throwing his last red gleam on the group, with one of those striking fire-light effects which only nature and genius ever succeed in producing. The rays reach not beyond, but are absorbed in the heath; and there is truth in this too: one of the most striking effects of the moon when just rising, or the sun when just setting, is, that the light seems to be looking at darkness, and the darkness abiding the look. These, however, are but the minor features of the picture.

The congregation is but a small one; the fierce persecution has been long protracted, and all the chaff has blown off. The battle of Bothwell has been fought and lost: many have laid down their lives on the scaffold, and many on the hillside. The flower of the country is wasting in dungeons, or toiling in chains in the colonies. There is no hope of deliverance from man; and we have in the little group before us a mere remnant, tried in the very extremity of suffering, and found faithful and true. There is more than a Sabbath-day sacredness impressed upon the scene; and the utter poverty in which the solemn feast is celebrated adds powerfully to the effect. A cottage bench, barely large enough to bear the "communion elements," serves for the long, low table; but, in the recollection of other days, they have covered it with a white linen cloth.

The flagon is evidently not of silver, nor yet the plate which bears the bread; but the cups are: they have been carefully secreted from the spoiler, and devoutly reserved in the midst of extreme want, and though the fines of Middleton and Lauderdale have fallen ruinously heavy on the recusants, for the service of the sanctuary. The communicants are ranged on the heath on both sides. Three reverend elders are standing in front of the table, — grave, strong-featured men, well stricken in years, with high, thoughtful foreheads, and in both form and countenance so thoroughly Scotch that the spectator is convinced at a glance they could belong to no other country in the world except our own. Had I met them in the north of Scotland, I would have said they were three of *the men*, and that I was very sure they could all speak judiciously *to the question*. There is an air of reasoning sagacity about them. Their very type of forehead is metaphysical, high, full, erect. They could not have stopped short of Calvinism, even had they wished it. The clergyman stands alone on the opposite side, with his back to the setting sun, and the pale reflected light from the linen cloth thrown upon his face. I have striven to read the expression. The spare figure and the attenuated hands tell at once their story; but the countenance yields its full meaning more slowly, and, I would almost say, more doubtfully. But it has evidently much to tell. What was the character of the latter divines of the covenant, — its Camerons, Pedens, Renwicks, and Cargills, — the men who excommunicated in the Torwood that "man of blood, Charles Stuart," for his "cruel slaughter of the saints of God," — the men who, when the persecution waxed hotter and hotter, became only the more determined to resist, but who, though the will remained unsubdued and unshaken, experienced, in the intensity of their distress, something approaching to aberration in the other faculties, and in their more unsettled moods did battle in lonely caves with shades of darkness from the abyss, or saw in their waking visions the events of the future rising up thick before them. Well did Solomon say that persecution maketh even wise men mad. The spectator has but to think of the character which the countenance really should express, and he will find it no easy matter to conceive how the painter could have expressed it differently. There is an air of intense melancholy that tells almost of a weariness of life, mingled with what, for want of a better word, I must term a ghostly expression. There is the appearance, too, of fatigue and exhaustion, and the impression of a strangely-mixed feeling, that hovers, as it were, between the visible and the spiritual world. The whole figure and countenance, in short, gives us the idea of human nature tried over-severely, and the "willing spirit" failing through the "weakness of the flesh."

On the spectator's left hand there is a group of the communicants thrown much into the shade. There are two stern-looking men among the others, who have evidently perused with great satisfaction the chapter in the "Hind Let Loose" "Concerning owning tyrants' authority," and the other equally emphatic chapter, — "Defensive arms vindicated." The one rests upon his broadsword; and there is a powder-horn and carabine lying beside the other. The group on the right is decidedly the most exquisite I ever saw, either on or off canvas. It is instinct with character, and rich in beauty. The communicants have just partaken of the bread; and never was the devotional feeling — the awe and reverence proper to the occasion — more truthfully expressed. One of the men, young in years but old in sufferings, still retains the bread in his hand. His air has all the solemnity of prayer. A young girl sits beside him, the very beau-ideal of a beautiful Scotch female in humble life, — simple, modest, devout, — a very Jeanie Deans, too, in quiet good sense, only a great deal handsomer than Jeanie. I could not look at her without thinking of the young and delicate female, her contemporary and countrywoman, whom the cruel dragoons bound to a stake below flood-mark, while the tide was rising, and whom they urged, as the water rose inch by inch, to abjure her church and close with "black Prelacy," but who, faithful to the last, chose rather to perish amid the waves of the sea. There is a still younger girl beside her, who has evidently not yet been admitted into full communion with the church, and with whose deep seriousness there mingles an air of dejection. An old woman, on the extreme edge of life, is seated in the middle of the group; and there is perhaps some exaggeration in the figure, but the mind and the feeling with which it is animated triumphs over the defects. It is not the thin, sharp features, and the almost skeleton arm, that attract our attention; it is the all-pervading intensity of the devotional feeling. The old man who sits beside her with his face covered is admirably in keeping with the rest. Such is an imperfect description of a picture which must not only be seen, but also carefully perused, ere its excellence can be adequately appreciated. The gentleman who criticized it in our last, rates it considerably lower than I have done; and there are other pictures which he estimates highly that lie perhaps beyond the reach of my sympathy. I am unable to understand them. I therefore again remind the reader that I pretend to no critical skill, and that my only criterion of merit in a picture is simply the amount of pleasure which I derive from it, and the quantum of thought which I find embodied in it. I have literally to *feel* my way along the canvas.

Allan's picture of the Bruce reads a high moral lesson. What is the moral taught by Harvey's Communion? It is a controversial picture on the side of the church. It sets before us, with all the truth of impartial history, the rebels and outlaws of the bloody and dissolute reign of Charles II., and teaches powerfully the useful truth that these offenders against the majesty of the law were in reality the preserving salt of the age, — that these dwellers in dens and caves were the meet representatives for the time of the dwellers in dens and caves described by the apostle, and of whom the "world was not worthy." The dissolute Middleton, the crafty Rothes, the brutal Lauderdale, the bloody Mackenzie, were the judges and law authorities of the time. A gross and profligate atheist, bribed against his own people by foreign gold, sat upon the throne. His court was a sty of licentiousness and impurity. Wickedness had broken loose in those "evil days;" and for twenty-eight years together the people of God were hunted upon the hills. But a time of retribution came; the wicked died "even as the beast dieth," and went to their place leaving names behind them that sound like curses in the ears of posterity. The reigning family — those infatuated and low-thoughted Stuarts, who, in their short-sighted and debasing policy, would have rendered men faithful to their princes by making them untrue to their God — were driven from their high places and their country to wander homeless under the curse of Cain, — to bring disaster on every nation that sheltered them, and death and ruin on every adherent that espoused their cause. And at length, when the spectacle of their misery and degradation was fully shown to the kingdoms of the earth, the last vial of wrath was poured upon their heads, and they passed into utter extinction. But the names of the persecuted survive in a different savor; their sufferings have met with a different reward; the noble constancy of the persecuted, the high fortitude of the martyr, still live; a halo encircles their sepulchres; and from many a solitary grave and many a lonely battle-field there come voices like those which issued from behind the veil, — voices that tell us how this world, with all its little interests, must pass away, but that for those who fight the good fight there abideth a rest that is eternal. I heartily thank this man of genius and right feeling for the lesson which his pencil has taught. Such pictures more than please, — they powerfully instruct.

FOURTH ARTICLE.

At the further end of the first exhibition-room, on the left hand, there is a moonlight scene by M'Culloch, — "Deer Startled," — which only a man of genius could have transferred from nature to the canvas. It is actually what it professes to be, — a landscape lighted up by the moon; and the scene itself, a deep Highland solitude, is full of a wild and yet quiet poetry.

The mind of every man has its picture-gallery, — scenes of beauty or magnificence, or of quiet comfort, stamped indelibly upon his memory. More than half the exile's recollections of home are a series of landscapes. The poor untaught Highlander carries with him to Canada pictures enough in the style of M'Culloch to store an exhibition-room, — pictures of brown solitary moors, with here and there a gray cairn, and here and there a sepulchral stone, — pictures, too, of narrow, secluded glens, each with its own mossy stream that sparkles to the light like amber, and its shaggy double strip of hazel and birch, — of hills, too, that close around the valleys, and vary their tints, as they retire, from brown to purple, and from purple to blue. He carries them all with him to the distant country. The gloomy forest rises thick as a hedge on every side of his wooden hut; the huge stumps stand up abrupt and black from amid his corn, in the little angular patch which his labor has laid open to the air and the sunshine. These are the objects which strike the sense; but the others fill the mind; and when year after year has gone by, and he sits among his children's children a wornout old man, full of narratives about the brown moors and the running streams of his own Scotland, his eyes moisten as the scenes rise up before him in more than their original freshness; and he tells the little folk, as they press around him, that there is no place in the world that can be at all compared with the Highlands, and that no plant equals the heather.

One of Wordsworth's earliest lyrics, a sweet little poem which he gave to the world at a time when the world thought very little of it, though it has become wiser since, embodies a similar thought. The poet represents a poor girl — originally from a rural district, who had been both happier and better ere she had come to form a unit in the million of London — passing in the morning along Cheapside, when a bird, caged against the sunny wall, breaks out in a sudden burst of song. Her old recollections are awakened at the sound; the street disappears, and the dingy houses; she sees the meadow tract, with the overhanging trees, where she used to milk her cows; she sees, too, the cattle themselves waiting her coming; and, in the words of the lyric, "a river flows down through the breadth of Cheapside." Poor Susan! "Her heart is stirred," and her eyes fill.

Every human mind has its pictures. Were it otherwise, who would care anything for the art of the painter? When standing in front of M'Culloch's exquisite landscape, I was enabled to call up some of my own, — moonlight scenes of quiet and soothing beauty, or of wild and lonely grandeur. I stood on a solitary seashore. A broken wall of cliffs, more than a hundred yards in height, rose abruptly behind, — here advancing in huge craggy towers, tapestried with ivy and crowned with wood, there receding into deep, gloomy hollows. The sea, calm and dark, stretched away league after league in front to the far horizon. The moon had just risen, and threw its long fiery gleam of red light across the waters to the shore. A solitary vessel lay far away, becalmed in its wake. I could see the sail flapping idly against the mast, as she slowly rose and sank to the swell. The light gradually strengthened; the dark bars of cloud, that had shown like the grate of a dungeon, wore slowly away; the white sea birds, perched on the shelves, became visible along the cliffs; the advancing crags stood out from the darkness; the recesses within seemed, from the force of contrast, to deepen their shades; the isolated spire-like crags that rise thick along the coast, half on the shore, half in the sea, flung each its line of darkness inwards along the beach. A wide cavern yawned behind me, rugged with spiracles of stalactites, that hung bristling from the roof like icicles at the edge of a waterfall; and a long rule of light that penetrated to the innermost wall, leaving the sides enveloped in thick obscurity, fell full on what seemed an ancient tomb and a reclining figure in white, — sports of nature in this lonely cave. There was an awful grandeur in the scene: the deep solitude, the calm still night, the huge cliffs, the vast sea, the sublime heavens, the slowly rising moon, with its broad, cold face! I felt a half-superstitious feeling creep over me, mingled with a too oppressive sense of the weakness and littleness of man. Pride is not one of the vices of solitude. It grows upon us among our fellows; but alone, and at midnight, amid the sublime of nature, we must feel, if we feel at all, that we ourselves are little, and that God only is great.

The scene passed, and there straightway arose another. I stood high in an open space, on a thickly-wooded terrace, that stretched into an undulating plain, bounded with hills. The moon at full looked down from the middle heavens, undimmed by a single cloud; but far to the west there was a gathering wreath of vapor, and a lunar rainbow stretched its arch in pale beauty across a secluded Highland valley. A wide river rolled at the foot of the wooded terrace; but a low silvery fog had risen over it, bounded on both sides by the line of water and bank; and I could see it stretching its huge snake-like length adown the hollow, winding with the stream, and diminishing in the distance. The frosts of autumn had dyed the foliage of the wood; the trees rose around me in their winding-sheets of brown and crimson and yellow, or stretched, in the more exposed openings, their naked arms to the sky. There was a dark moor beyond the fog-covered river, that seemed to absorb the light; but directly under the nearest hill, which rose like a pyramid, there was a tall solitary ruin standing out from the darkness, like the sheeted spectre of a giant. The distant glens glimmered indistinct to the eye; but the first snows of the season had tipped the upper eminences with white, and they stood out in bold and prominent relief, nearer, apparently, than even the middle ground of the landscape. The whole was exquisitely beautiful, — a scene to be once seen and ever remembered.

I must attempt a description of the picture of M'Culloch. The moon is riding high over head in a cloudy and yet a quiet sky. There is a greenish transparency in the piled and rounded masses. Even where most dense, the thinner edges are light and fleecy; and the whole betokens what White of Selborne would have termed a mild and *delicate* evening. There is a lonely moor in front, a piece of water, and a stunted fir tree. The light falls strongly both upon the water and where the heathy bank shelves gradually toward it on the right, while the middle ground of the picture, with its scattered trees, lies more in the shade. The clouded sky tells us, however, that the whole country on such an evening cannot be other than checkered with a carpeting of alternate light and shadow. There is a screen of hills behind, dim and yet distinct; and a few startled deer — startled we know not why — are grouped in front. Such are the main features of the picture; but it is one thing merely to tell these over as in a catalogue, and quite another to convey an adequate idea of the wild and yet simple poetry which they express. The extreme loneliness of the scene, the calm beauty of the evening, the unknown cause of fright among these untamed denizens of the moors and mountains, — what can they have seen? what can they have heard? It is night, and deep solitude. Are the spirits of the dead abroad?

M'Culloch has another very sweet picture in the exhibition of this year, — "A Highland Solitude with Druidical Stones." We find it in the large middle room, on the left hand as we pass inwards. It is, though equally Highland, an entirely different scene from the other; and yet, in describing it, — for the pen has no such variety of shades as the pencil, and no such pliant flexibility of outline, — I must employ some of the same words. I must repeat, for instance, that there is a heathy moor in the foreground, and a screen of hills behind, and that a sky checkered with clouds has dappled the landscape with sunshine and shadow. There is a transient shower sweeping gloomily along a narrow glen, while the hills to the right are smiling in purple to the sun. The Druidical stones rise gray in the mid-ground; and the smoke, apparently of a shepherd's fire, is ascending slantways from among them, before a light breeze. It is, as I have said, a sweet picture, but inferior in feeling to the other, and perhaps not altogether what its name would have led us to expect. I question, however, whether that blended feeling of the sublime and the solemn, with which it is natural to contemplate the monuments of an antiquity so remote that they lie wholly beyond the reach of history, and which form the sole and yet most doubtful memorials of unknown rites and usages, and of tribes long passed away, can be reawakened by the imitations of the painter. I have felt it strongly on the scene of some forgotten battle sprinkled with cairns and tumuli, and where the stone-axe and the flint-arrow are occasionally turned up to the light, to testify of a period when the aborigines of the country were making their first rude essays in art, and when the *man* had not yet risen over the *savage*. I have felt it when, standing where some ancient burial mound had been just laid open, I saw the rude unglazed sepulchral urn filled with half-burned fragments of bone, or with rudely-formed ornaments of jet or amber, fashioned evidently ere the discovery of iron. I have felt it, too, amid the Druidical circle, and beside the tall unshapen obelisk. But I did not feel it when standing before M'Culloch's second picture; and I questioned whether in what he had failed any other could have succeeded. With what Johnson terms the "honest desire of giving pleasure," I shall briefly attempt a description of the scene in which I have felt it most strongly, a scene to be visited in the gray of the evening, or by the light of the moon.

There is a soft pastoral valley, formed by the river Nairn, not much more than a mile to the southwest of the field of Culloden. Low-swelling eminences rise on either hand. The view is terminated, as we look downward, by a prominent rounded hill, on which are the remains of one of those ancient earthen forts or duns — combinations of green mounds and deep angular fosses — which seem to have constituted in our own country, like the hill-forts of New Zealand in the present day, the very first efforts of ingenuity in defensive warfare, — the very first inventions of the weaker party in their attempts to withstand the stronger. As we look up the glen towards the west, we see the view shut in by another rounded hill, and it also bears its ancient stronghold, — one of those puzzles of the antiquary, — a vitrified fort. The low rude wall all around the top of the eminence has been fixed into one solid mass by the force of fire; and we marvel how the rude savage who applied the consolidating agent, all unacquainted as he was with mortar, and unfurnished with tools, should have been so expert a chemist. He was a glassmaker on a large scale, probably before the discovery of the Phœnician merchants. It is in the valley below, however, on a level meadow-plain beside the winding Nairn, known as the plain of Clava, that we find most to interest and to astonish. It is a city of the ancient dead, thickly mottled in its whole extent with sepulchral cairns, standing stones, and Druidical temples. Detached columns of undressed stone, shaggy with moss and spotted with lichens, rise at wide intervals apparently in lines, as if to unite the other structures in one general design. There are cairns beside cairns, and circles within circles; and there rose high over the rest only a few years ago, but they have since been injured by some curious excavator, three accumulations of stone, immensely more huge than the others, and more artificially constructed, that seemed to mark out the resting-place of the kings or chieftains of the tribe. The bases of these larger cairns were hemmed in by circular rings of upright stones; and a wider ring, of larger masses, encircled the outside. A dark, low-roofed, circular chamber occupied the space within. Its walls were constructed of upright stones; and uncemented flags, overlapping each other until they closed atop, formed the rude, dome-like roof. In the fat, unctuous earth which composed the floor there were found unglazed earthen urns, as rudely fashioned as the surrounding building, and filled with ashes and half-calcined bones. It is a curious fact, that, even so late as the close of the last century, Highlanders in the neighborhood buried amid these ancient tombs such of their children as died before baptism. For, according to a superstition derived from the Church of Rome, and in some remote localities not yet worn out, unbaptized children were deemed unholy; and in this belief their remains were consigned to the same unconsecrated ground which contained the dust of their remote pagan ancestors. It is another striking fact, — a fact full of poetry, — that near the western end of the plain of Clava there are the remains of an ancient Christian chapel, which still bears the name of the *clachan*, or church; and a traditional belief survives in the district that it was planted in this citadel of idolatry by the first Christian missionaries. Would that we were acquainted with its story! and yet it would probably be merely another illustration of the fact that the religion that most inculcates humility and self-denial is of all animating principles the most daring and heroic.

FIFTH ARTICLE.

What sort of painters, think you, do the Scotch promise to become? Why, painters equal to any the world ever produced, if the national mind be only suffered to get into a national track, and our artists have sense and spirit enough, however much they may admire the pictures of other countries, not to imitate them. The genius of our countrymen, as shown in their literature, is eminently of a pictorial character. The national feeling is vividly descriptive. As early even as the days of James IV., old Gavin Douglas, and his contemporary Will Dunbar, could fill page after page with splendid descriptions, as minutely faithful as the descriptions of Cowper in his "Task," and scarcely less poetical. The "Seasons" of Thomson form a series of landscapes; and never, surely, were there landscapes more felicitously conceived or more exquisitely finished. It has become the fashion of late to decry M'Pherson; but rarely has Europe seen a mightier master of description. The scenery of Burns is nature itself. Who ever excelled Grahame in pictures of quiet beauty, or Professor Wilson in the wild and the sublime of Alpine landscape? And, last and greatest, we stake Sir Walter Scott for the vividly graphic, for strength of outline and beauty of color, against every painter of every school, and all the writers of the world. The people whose literature exhibits such powers have, if they wish to become painters, only to try. But let them beware of imitation. The straight-nosed beauties of Greece were no doubt very great beauties, and its historical characters very fine characters indeed. There is something very admirable, too, in the genius of Italy. No people ever excelled the Italians in drawing legendary saints, with glories of yellow ochre round their heads, or angels mounted on the wings of pigeons. But what of all that? It is not by painting the straight-nosed beauties of Greece or the winged angels of

Italy that the Scotch artist need expect to confer honor on either Scotland or himself. Let him do what was done by Thomson and Burns and Sir Walter Scott, and what Wilkie and Allan and Harvey are employed in doing, — let him walk abroad into nature, and study the history of his country. The mere imitative faculty is one of the lowest; the Chinese possess it in perfection, and so does the chimpanzee.

But am I not evincing a barbarous and Gothic disregard of the classical? Very far from it. I have read all Cowper's "Homer" and Dryden's "Virgil" again and again. I could almost repeat that portion of the Odyssey in which the wanderer of Ithaca is described sitting apart in his own hall, a poor, despised beggar, when his enemies are expending their strength in vain attempts to bend his bow; and I have felt my heart leap within me, when, scorning reply to their rude taunts, he leaned easily forward on the well-remembered weapon, and, bending it with scarce more of effort than the musician employs in straightening the strings of his harp, sent the well-aimed arrow through all the rings and the double planks of the oaken gate beyond. I have luxuriated, too, over the exquisite descriptions of the Æneid, — amid the horrors of the burning town, for instance, till I almost saw the pointed flames shooting far aloft into the darkness, and almost heard the tramplings and shouts of the enemy in the streets, — amid the terrors, too, of the tempest, when the fierce surge rolled resistless over the foundering vessel, and the scattered fleet labored heavily amid the loud dash of the billows and the wild howl of the wind. And when I looked for the first time on Laocoon and his children crushed in the ruthless coil of the serpent, — a too faithful allegory of the human race, — the story of Virgil rose at once before me, and I felt the blended genius of the poet and the sculptor breathing in an intense human interest from the group. But what classical artists and authors were born to accomplish has been accomplished already; and no man ever became great, nor ever will, by servilely following in their track. The more an author or artist copies them, the less is he like them; for the imitative turn, which delights in catching their manner, is altogether incompatible with the originality of their genius. And hence it is that our modern classics, whether painters or sculptors, or manufacturers of unreadable epics, rank invariably among the men of neglected merit. They overshoot those sympathies of a common humanity to which their masters could so powerfully appeal in the past, and which their contemporaries are scarcely less successful in awakening in the present, each in a track of his own opening. The sculptors of Great Britain were classical and imitative for a whole century; and all they produced in that time, in consequence, was a lumbering mass of unreadable allegories in stone, which no one cares for; groups of Prudences with fine necks; of Mercies, too, with well-turned ankles; and of Cupids looking sly; and, had they been employed in cutting them in white-sugar or gingerbread, all would have now agreed that the choice of the material mightily heightened the value of the work.

Among the rising painters of our country, I know no artist whose productions better serve to corroborate the truth of remarks such as these than the pictures of Thomas Duncan. Brown justly reckons the principle of contrast, or contrariety, among the causes which suggest and connect ideas. One of Duncan's living pictures — "Prince Charles and the Highlanders entering Edinburgh after the battle of Preston," a picture exquisitely Scotch, instinct with character and rich in interest — shows more powerfully, on this principle, the folly of toiling in the dead school of classical imitation, than even the *effete* of the artists who irrecoverably lose themselves within its precincts of death. I spent two full hours before his picture, and regretted I could not spend four.

The morning sun has risen high over the Old Town of Edinburgh, and the beams fall clear and bright through a cloudless autumn sky, on half the high-piled, picturesque tenements of the Canongate, and half the street below. The other half lies gray in the shade. I saw, just in front, on the sunny side, the castellated jail of the burgh, with its blackened turrets and its Flemish-looking clock-house. The barred windows are thronged with faces; and a few disarmed, half-stripped, forlorn-looking soldiers, huddled together on an outer staircase, show that the incarcerated crowd are military prisoners from the field of Preston. The street lies in long perspective beyond, house rising over house, and balcony projecting beyond balcony. Every flaw and weather-stain has the mark of truth; every peculiarity of the architecture reminded me of the scene and the age. A dense crowd occupies the foreground. The Highlanders, after totally routing the superior numbers of Cope, have entered the city with their Prince at their head, and have advanced thus far on their march to Holyrood House. The apparently living mass seems bearing down upon the spectator. There is a mischievous-looking, ragged urchin, half-extinguished by the cap of some luckless grenadier, who has possibly no further use for it, scampering out of the way; and an unfortunate barber, the very type of Smollett's Strap, has got himself fast jambed between a projecting outside stair and the brandished war-axe of a half-naked and more than half-savage gillie, who is exerting himself with tremendous vigor in clearing a passage, and who, as if to add to the poor barber's distress and peril, is looking in another direction. There are other strokes of the comic in the piece. In one corner a Jacobite laird, *blin' fou*, is threatening destruction with unsheathed whinyard to all and sundry who will not drink the Prince's health. In another, two pipers are marching side by side. The one, a long-winded young fellow, cast in the Herculean mould of his country, and proud of his strength and his music, is adjusting the drone of his pipe with a degree of self-complacency that might serve even the Dean of Faculty himself. The other, an old man of at least seventy-five, with features fiercely Celtic, and an expression like a thunder-cloud, is evidently enraged at the better breath of his opponent; but, collecting his strength for another effort, he seems determined rather to die than give in. The Prince rides in the centre on a noble steed, that seems starting out of the canvas. We recognize him at once, not only from his prominent place and princely bearing, but from the striking truth of the portrait, — one of the most spirited, perhaps, that has yet appeared, and most like the man when at his best. Has the reader never noticed the striking resemblance which the better portraits of Prince Charles bear to those of his remote ancestress, Queen Mary? I was first struck by it when, in glancing my eye over a bookseller's window, I saw side by side the frontispieces of "Chambers' History of the Rebellion" and the "Life of Mary Queen of Scots," — both numbers of "Constable's Miscellany;" and I have had since repeated opportunities of verifying the remark. It is, I believe, no uncommon matter for resemblances of this kind to reappear in families at distant intervals. Sir Walter, no ordinary observer of whatever pertained to the nature of man, whether physical or intellectual, has repeatedly embodied the fact in his inventions; but I do not know a more striking instance of it in real history than the one adduced.

All the more celebrated heroes of the rebellion are grouped round the Prince, full, evidently, of a generous enthusiasm, in which the spectator can hardly avoid sympathizing. There was little of moral worth or of true kingly dignity in the latter Stuarts; and I could not forget that the "gallant adventurer," who, with at least all the courage of his ancestors, threw himself upon the generosity of the devoted and warm-hearted Highlanders, was in reality a cold, selfish man, who sunk in after life into a domestic tyrant and a besotted debauchee. And yet I could not avoid sharing in the well-expressed excitement of the Prince's gallant adherents, as they drink in his looks with all the intense and rapturous exultation of a loyalty which has passed from the earth with the generation that cherished it. No such pervading love or deep devotion awaits the kings or princes of the present time. Behind the Prince rides Clanranald, the chief of Clan-Colla. His Highlanders take precedence of the other clans, for the Bruce had assigned them their place of honor in the right when they fought at Bannockburn. Young Clanranald, a tall, handsome youth, and his cousin Kinloch Moidart, have advanced in front; old Hugh Stewart, a rugged, deep-chested veteran of the Black Watch, who fought in all the battles of Charles, and whose portrait is still preserved, presses on behind them; and the gigantic miller of Inverrahayle's Mill, a tremendous specimen of the wild mountaineer, is still more conspicuous among a group of clansmen on the left. There is a dense crowd behind, and what seems a thick wood of spears and axes, with here and there a banner, — among the rest, an English standard taken from the dragoons at Preston. A heap of other trophies lies in front, over which Hamish M'Gregor, the son of the celebrated outlaw Rob Roy, keeps watch.

An intensely interesting group occupies the left. There we see Lord George Murray, the cool-headed, far-seeing statesman of the expedition, who dared honestly to tell his Prince disagreeable truths, and who was liked none the better because he did so; the gallant Lochiel, too, who in his devoted loyalty joined in the enterprise with his brave Camerons, even though he had anticipated from the first that the result would be disastrous. There also is the Marquis of Tullibarden, the original of Sir Walter's Baron of Bradwardine, a fine old Lowland cavalier, dressed, in honor of the Prince, in a birthday suit, half-covered with lace, and of a fashion at least twenty years earlier than the time. There is a galaxy of high-born dames beside him, relatives of the family, — one of them at least of exquisite beauty, and all of them what clever artists do not invariably succeed in painting, even when they try most — ladies. Their countenances seem lighted up with the triumph of the occasion; and the children of the family, sweet little things, worth all the cupids that the imitators ever chiselled or painted, are employed in strewing white roses in the path of the Prince. The opposite side of the picture is occupied by a group of a different but not less interesting character.

On an outer stone stair, on the shady side of the street, — one of those appendages characteristic of the Scoto-Flemish style of domestic architecture, — there is a group of citizens. Professor Maclaurin, the celebrated mathematician, the man who first brought down the philosophy of Newton to the level of common minds, and whose simple, unpretending style rises in some passages to the dignity of the sublime, purely from the force and magnitude of his thoughts, leans calmly over the rail. The good zealous Whig had proposed to the magistrates his well-laid scheme for fortifying and defending the city, and had exerted himself in carrying it into effect; but the necessary courage to carry out his measures was lacking on the part of the people, and so he has had just to fall back and rest him on his philosophy. John Home, the author of "Douglas," and one of the first historians of the Rebellion, stands beside him. He, too, though a mere youth at the time, had bestirred himself vigorously in the same cause, and is now evidently bearing the reverse of his party as he best can. But the figure behind them, one of the most masterly in the picture, is instinct with a sterner spirit. Had there been five hundred such men in the city to back the philosopher, the Highlanders, with all their valor, would have been kept outside the wall. He stands at the stair-head, scowling at the enemy and all their array of spears and battleaxes, — one of the followers of Richard Cameron, girt with a buff belt, from which his Andrea Ferrara hangs suspended, and bearing a heavy Bible. Depend on it, had that man fought at Preston, he would have stood beside the good and gallant Colonel Gardiner unmoved in the midst of rout and panic, and have left, like him, a gashed and mangled corpse to mark where the tide of the battle had turned. Such is a meagre outline of Duncan's exquisite picture. It is said to have cost the almost continuous labor of two years; and the anticipated expense of multiplying it by the graver — and never was there a picture more worthy — is calculated at about three thousand pounds. The pictorial history of Scotland promises to excel all its other histories, and it does not contain a more brilliant page than that contributed by Duncan.

Gallant Highlanders, men of warm hearts and tender feelings, and spirits that kindle as the danger comes, the phantom of mistaken loyalty deludes you no longer; you have closed with a better faith; and, while the strength of the character still remains unbroken, all its fierceness is gone. I have lived amid the quiet solitude of your hills, and, as I have passed your cottages at the close of evening, have heard the voice of psalms from within. I have sat with you at the humble board, to share your proffered hospitality, — the hospitality of willing hearts, that thought not of the scanty store whence the supply was derived. I have marked your untaught courtesy, ever ready to yield to the stranger, and have laid me down in security at night amid your hamlets, with only the latch on the door. I have seen you pouring forth your thousands from brown distant moors and narrow glens, to listen with devout attention to the words of life from the lips of your much-loved pastors, and to worship God among your mountains in the open air. I know, too, the might that slumbers amid your gentleness of nature; and that, when the day of battle comes, "and level for the charge your arms are laid," desperate indeed must that enemy be, and much in love with death, that awaits the onset. A day may yet arrive, should Socialism and Chartism, with their coward cruelty, inundate society in the plains, when we may look to your hills for succor; but that day has not yet come. You tell us that, though little able to assist the church with the pen or on the platform in her present troubles, your hearts are all with us; and that, should the worst come to the worst, we may reckon on the Highlanders of Scotland as thirty thousand fighting men. And we know what sort of fighting men you are, and what sort of hearts you bear. But reserve your strength, brave countrymen, for another day and a different quarrel. Should the church which you love fall prostrate before her adversaries, and wickedness rush unchecked over the land to trample and destroy, your swords may be required, not to protect her friends from her enemies, but to protect both her friends and her enemies too.

SIXTH ARTICLE.

Immediately below one of Wilkie's admirable pictures, — "The Spanish Posado," — there is a painting, not particularly showy, and which might possibly enough come to be overlooked among productions of less merit and more glitter, but which is at once so simple, unaffected, and true to nature, that it bears the formidable neighborhood wonderfully well. It is the work of a young and rising artist, Tavernor Knott, — a gentleman who, at the age of twenty-two, has learned to compress a large amount of just thought and fine feeling within a few square feet of canvas, and who, I am convinced, will be better known to his country-folks in the future than he is at present. I do not know whether his subject might not have prejudiced me in his favor, — "A Scotch Family Emigrating;" but I have certainly derived much pleasure from an attentive perusal of his picture, and it has served to recall to my recollection a good many similar scenes from real life, of a half-pleasing, half-melancholy character. I have never yet seen a party of emigrants quitting their country forever, half-broken-hearted, as they almost always are, without forgetting all my political economy, and sympathizing with them in their regret. Hazlitt says, very truly though somewhat quaintly, that when men compassionate themselves, other men compassionate them too. We admire the fortitude of the stoic, but we never pity his sufferings. But a kindly, manly Scot, proud of his country, and attached to his friends, and yet compelled by stern necessity to part from both, and parting from them with a swelling heart and wet eyes,—we must pity the poor fellow, and feel sorry that he is leaving us, let population increase as it may. I know of scenes which have taken place in the Highlands of Scotland which I hope neither Malthus nor M'Culloch could have contemplated with a dry eye; and I shall instance one of them. All the Highlanders of an inland district in Sutherlandshire were ejected from their homes by the late Duke a good many years ago, to make way for a few *sheep-farmers.* The poor people, a moral and religious race, bound to their rugged hills with a strength of attachment hardly equalled in any other country, could not be made to believe the summonses of removal real. Their fathers had lived and died among these very hills for thousands of years. They had spent their blood, and had laid down their lives of old, for the good Earls of Sutherland. Nay, when their Countess, in her maiden years, had expressed a wish to raise a regiment among them for the service of the country, a regiment had risen at the bidding of their chief's daughter, and had marched off to the war. Every man among them brought his Bible with him, and the enemy never bore them down in the charge. And now could it be possible that they were to be forced out of their own country! They at first thought of resistance; and, had they carried the thought into action, it would have afforded perilous employment to a thousand armed men to have ejected every eight hundred of them; but they had read their New Testaments, and they knew that the Duke had become proprietor of the soil; and so the design dropped. Shall we write it? — some of their houses were actually fired over their heads, and yet there was no bloodshed! Convinced at length that no other alternative remained for them, they gathered in a body in the churchyard of the district to take leave of their country for ever, and of the dust of their fathers last. And there, seated among the graves, men and women, the old and the young, with one accord, and under the influence of one feeling, they all "lifted up their voices and wept." This tract of the Highlands is now inhabited by sheep.

Mr. Knott's picture represents rather a Lowland than a Highland scene. There is a humble cottage, half overshadowed by trees, in the foreground, surrounded by a level country. The sea spreads beyond. We see the ship in the distance which is to bear away the emigrants; and the loaded wagon in the middle ground is evidently conveying their effects to the shore. The group stands in front of the cottage. There are a few supplementary figures introduced into the scene, partly for the sake of heightening the effect by the force of contrast, — for they have no direct interest in it, — and partly to bring out its minor details; for, though little moved by it, they are yet all employed in it. One, an elderly man, with spectacles on, is painfully scrawling out a direction-card for a box; there is a rough, thick-set, sun-burned sailor from the beach, who is leaning over him, evidently criticising the penmanship, but satisfied, apparently, that it may just pass; and a tall stripling stands directly in front, prepared with a coil of cord to bear the box away. In an opposite corner there is a boy of the family parting with a favorite dog, which he is handing over, bound in a string, to a companion. The poor little fellow is much dejected, and not at all likely soon to forget Scotland, nor his dog either. The stroke is a fine one; but there is a still finer stroke in the same part of the group. A barefooted, simple-looking lassie, of about fifteen, who has been living with the family, taking care of the child, a sweet, chubby thing, is kissing her charge, not dry-eyed, and bidding it farewell; and baby, though it does not exactly know what is the matter, is quite disposed to return the caress.

A vigorous man, in the prime of early manhood, — the father of the boy and the infant, and of two little girls in the foreground, — has turned round in a half-absent mood to the shut door. He has been bearing up, with apparent fortitude, for the sake of the others, and under a high sense of what constitutes the firm and the manly in character. The present, however, is a moment of partial forgetfulness; the assumed firmness is laid down, and his thoughts are hovering in sadness, as he looks back on his humble dwelling, between the enjoyments of the past and the uncertainties of the future. His wife, a woman of great beauty, — not merely that of feature and complexion, which may exist wholly disjoined from all that we most value in the sex, but that of expression and character also, — is leaning on the arm of her father-in-law, a venerable old man. Unlike her husband, she has had no part to act on the occasion, nor has she simulated the fortitude or the indifference which she does not possess nor feel. She is drowned in tears. The sweet little girl who holds on by her gown, and the girl beside grandpapa, are both too young to participate in the general regret; and yet they, too, have an air of absence and unhappiness about them, caught, as it were, by sympathy from the others. The old man, the patriarch of the family, is one of the most striking figures in the picture. Wilkie himself has rarely produced anything more characteristically Scotch. There is a deep seriousness impressed on the somewhat rugged features, blent with a dash of sadness; for he, too, feels that he is leaving his home and the country of his fathers. But he has thought of another and more certain home; and the consolations which he is pressing on his daughter-in-law, whose hand he is affectionately grasping in his own, are evidently of the highest character. Venerable old man!

Divested of hopes and beliefs such as yours, the aged emigrant would be of all men the most unhappy. It has been well said by Goldsmith, that "a mind long habituated to a certain set of objects insensibly becomes fond of seeing them, visits them from habit, and parts from them with reluctance;" and it is chiefly from such objects that age derives its pleasures. It cannot give to novelty the feelings appropriated by recollection; and must fare ill, therefore, in a foreign land, in the midst of what is strange, and what, from its very nature, cannot become otherwise, — in the midst, too, of hardships and privation. The old man in such circumstances must be either like the cottar of Burns, — the "priest-like father" of the family, — or he must be by much the unhappiest member of it.

Such is an imperfect description of Mr. Knott's picture, as I have been enabled to read it. It has no doubt its faults, like every other; but these seem mostly to be mere faults of execution, from which no young artist can be wholly free, whatever his genius, — not faults of conception. The foliage of the trees which half-embosom the cottage does not repose in the softened sunshine with perhaps all the grace of nature, and the tiled cottage does not strike as characteristically Scottish. A roof of heath, or fern, or straw, with here and there a patch of stone-crop, and here and there a tuft of grass or a cluster of houseleek, would better repay the painter's study. But these are very minute matters; and he would be a connoisseur worth looking at who would place such things in the balance against the large amount of thought and feeling displayed in the group. The painter who can impart character to men and women, both national and individual, can well afford to leave a tree or a cottage without much to distinguish them, and be a superior painter still.

Of all the figures of the piece, the old man pleases me the best, though the female, his daughter-in-law, is also very exquisite. I have perused with deep interest the letters of an aged emigrant, who quitted the north of Scot land for Upper Canada about eight years ago. He was one of the excellent though now fast diminishing body known in Ross-shire and the neighboring districts by the name of *the men*; and, though marked perhaps by a few eccentricities, he was by no means a low specimen of the class. He settled among some of the outer townships, — I forget which, — where there were no ministers and no churches; and he saw for the first time, in his seventieth year, the Sabbath rise over the wild and trackless woods of America, all unmarked from the other days of the week. But John Clark had brought his Bible with him, and no superficial knowledge of its contents; and, regularly as the day came round, he assembled his family, like one of the Pilgrim fathers of old, for the purpose of religious worship, and to press upon them the importance of religious truth. Some of the neighbors learned to drop in. His fervent prayers, and his homely but forcible expositions, full of masculine thought, had the true popular germ in them; and John's log cottage became the meetinghouse of the thinly-peopled district; until at length the accumulating infirmities of a period of life greatly advanced interfered with his self-imposed duties, and set him aside. He is still alive, however, at least he was so a few months ago; and at that time, in the midst of great bodily debility, far removed from all his Christian friends of the same stamp or standing with himself, and with the near prospect of laying down his worn-out frame, to mingle with the soil in some gloomy recess of the wild forest, thousands of miles from the lonely Highland churchyard where the remains of his fathers and of some of his children are laid, with those of the wife of his youth, John was yet more than resigned; he was rejoicing, — will our readers guess for what? He had just heard of the revival at Kilsyth, and of the attitude assumed by the Church of Scotland in behalf of the rights of the Christian people and of the Headship of her Divine Master. What, I marvel, does infidelity propose giving to such men in exchange for their religion?

I am impressed by the absolute necessity which exists for emigration. Circumstances have settled the point. Whatever the sacrifice of feeling, it has ceased to be an open question whether or no our countrymen should leave us for other fields of exertion. The population of the country is already redundant in a degree which occasions much distress among the working classes, and much consequent bad feeling; for the true cause of the evil is misunderstood; and this already redundant population is increasing at the portentous rate of nearly a thousand per day. Besides, it is according to the design of Providence that the human race should spread forth as they multiply. The Scotch are only doing for Canada and the insular regions of the far south what the Celtæ and the Scandinavians did for Scotland three thousand years ago; and is it not well that the process should be so different now from what it was when the Goths and the Vandals overwhelmed the Roman empire? It is civilization and the arts that are advancing on the regions of barbarism, and sending out their pickets and their advanced guards far into the waste, — not barbarism that is bursting in, as of old, to bear down civilization and the arts. But we can at once recognize these principles, — principles, indeed, too obvious not to be recognized, — and yet regret cases of what we may term wholesome emigration none the less. Nothing can be more healthy than the drain on a redundant town or country population: it is blood-letting to an apoplectic patient; and the emigrating thousands are as little missed as water withdrawn from the ocean. "The crowds close in, and all's forgotten." Very different is the case, however, when the population of upland districts have been torn up root and branch, and uninhabited wildernesses formed where a simple-hearted but surely noble race lived contented in times of quiet, and constituted the strength of their country in the day of war. There have been cottages on many a hillside emptied of their inhabitants within the last twenty years, which shall never again be gladdened by the domestic circle ; and the heath is creeping slowly in lonely dells and sweeping acclivities, over many a narrow range of meadow, and many a little field whose flattened and sinking furrows shall never again yield to the plough. The contemplation of such scenes amid the depopulated solitudes of the Highlands has always inclined me to sadness, especially in the inland districts which, as they nad no dependence on the fluctuations of trade, were little exposed to those extreme depressions which have borne so heavily of late years on the inhabitants of the islands and the sea-coasts, and in which, I know from experience, much happiness has been enjoyed, and an intense love of country cherished.

Rather more than twelve years ago I was led into the central Highlands of the north. I first left behind me the comparatively level fields of the low country, with their hedgerows and intervening belts of planting, and then the upper skirting of forest, which waved mile after mile on the lower declivities of the hills. I next passed on a half-obliterated path along the upper ridges, rising and descending alternately, — now shut out from the widening landscape in some brown moory hollow, roughened with huge fragments of rock, now on a swelling eminence that, overtopping the previously surmounted height, blended in one vast prospect the region of moor, of forest, and of corn, and, far beyond, the widely extended sea. The last eminence was at length surmounted, and a broad tract of table-land, slightly depressed toward the middle, bounded on the opposite side by low craggy hills, with here and there an inky pool and here and there a gloomy morass, spread out for miles before me in black and unvaried sterility. I toiled drearily across, and reached the opposite boundary of hill. It overlooked a deep pastoral valley of considerable extent. A wild Highland stream, skirted on either bank by a straggling row of alders, went winding through the midst. On either side there were patches of vivid green, encircled by the brown heath, like islands by the ocean, which had once been furrowed by the plough. As I advanced I saw the ruins of deserted cottages. All was solitary and desolate. Roof-trees were decaying within mouldering walls. A rank vegetation had covered the silent floors, and was waving over hearths the fires of which had been forever extinguished. A solitary lapwing was screaming over the ruins, rising and falling in sudden starts, darting off along the ground, now to the right, now to the left, and then turning abruptly round in mid air, and almost brushing me as she passed. She had built her nest within some deserted cottage, and was employing her every instinct to lure me away. A melancholy raven was croaking on a neighboring eminence. There was the faint murmur of the stream, and the low moan of the breeze; but every sound of man had long passed from the air; and the bright sunshine seemed to fall idly on the brown slopes and greener levels of this uninhabited and desolate valley. I have rarely been more impressed. I was reminded of what I had read of eastern armies, whose track may be followed years after their march by ruined villages and a depopulated country, — of scenes, too, described by the prophets, — lands once populous "grown places where no man dwelleth, or son of man passeth through."

III.

GEOLOGY VERSUS ASTRONOMY.

It was remarked early in the last century by a French wit, who was also an astronomer, that when the potentates of earth ceased to quarrel about their sublunary territories, they would in all likelihood begin to dispute about the plains and mountain ranges of the moon. They would give, he said, their own names to its peaks and craters, and fall to blows for the nominal possession of some of its more prominent eminences or profounder hollows. The prediction, however, seems to be as far from its fulfilment as ever. The present war with Russia shows that the quarrels of rulers respecting their earthly territories, so far from being at an end, or nearly so, are as serious and irreconcilable as at any former period; and hitherto, at least, kings and princes have left all disputes about the nomenclature of the moon's geography to be settled by the moon's geographers. The celestial map-makers have already had their quarrels on the subject. One of them named the places on the moon's surface after philosophers eminent in all the various departments of mind; another named them after the terrestrial seas and mountains which they seemed to resemble; a third, interposing, strove to give them back to the philosophers again, but struck off the former list all philosophers save the astronomical ones; and now the moon's surface bears, in the maps at least, marks of all the three combatants. It has its Alps and its Apennines and its Caucasus, its Sea of Serenity and its Sea of Storms, its Aristarchus and its Plato, its Tycho and its Copernicus. There is, as we may perceive, no danger of a too unbroken peace on earth regarding the condition of the moon, or of any of the other heavenly bodies, even though neither Napoleon nor Nicholas should interfere in the quarrel.

In fine, every department of science has its controversies; and it is well that it should be so. It saves the world from all danger of connivance to deceive it, on the part of scientific men, — a thing which the world is somewhat prone to suspect, — and proves, on the whole, the best mode of eliciting truth. There are certain stages, too, in the course of discovery, when controversy becomes inevitable. "Tempests in the state are commonly greatest," says Bacon, "when things grow to equality, as natural tempests are greatest about the equinoctia." And we find that it is so in science also. When comparatively new sciences rise, in certain departments specially their own, to assert an equality with old ones, that, when they stood alone, had been extended beyond their just limits, controversies almost always result from the new-born equality in the disputed province. In the middle ages, for instance, there existed but one great science, — theology; and, pressed far beyond its just limits, it impinged on almost every province of physical research and every department of mind. In the fifteenth and sixteenth centuries — peculiarly the ages of maritime discovery — geography rose into importance; and after a prolonged controversy, which at one time had well-nigh crushed Columbus, it was finally established, in opposition to the findings of St. Augustine and Lactantius, that the world is round, not flat, and that it has antipodes. In the sixteenth and seventeenth centuries astronomy became a great and solid science; and, after a still fiercer controversy than that of the geographers, it asserted a supremacy in its own special walk against popish theologians such as Caccini and Bellarmine, and against Protestants such as Turretine. We have seen a similar controversy carried on in the present century — which has witnessed the rise of geology, just as the fifteenth, sixteenth, and seventeenth centuries witnessed that of geography and astronomy, — between theologians who were also geologists, such as Chalmers, Sedgwick, and Sumner, and theologians who were wholly ignorant of geology, such as Granville Pen, Eleazor Lord, and Moses Stuart. And, as in astronomy and geography, the controversy may now be regarded as ultimately settled in favor of the new science, within at least the new science's own proper province. There are, however, other controversies than theological ones, wich rise when, according to Bacon, "things grow to equality;" and that equality to which geology has attained with astronomy during the last fifty years may be properly regarded as the real cause of the very interesting controversy carried on at the present time between the author of the "Essay on the Plurality of Worlds," understood to be one of the distinguished ornaments of English science, and our great countryman Sir David Brewster, — a philosopher who, while supreme in his own special walk, is perhaps of all living men the most extensively acquainted with the general domain of physical science. The English writer, though he presses his argument by much too far, may be regarded as representative of the geological side; Sir David of the astronomical.

There are, we have said, certain stages in the course of discovery at which controversy becomes inevitable; and it seems demonstrative of the fact that the new arguments in which these controversies originate arise much about the same time, without concert or communication, in minds engaged in the same or similar pursuits. Had they not been originated by the man who first made them known, they would have been originated almost contemporaneously by some one else. Almost all discovery has a similar course. Adams and Le Verrier were engaged at the same time in calculating the irregularities of Uranus, and inferred from them the existence and position of the great planet, actually discovered almost simultaneously, shortly after, by Dr. Galle and Professor Challis; and it is a known fact that Mr. Lassel and Professor Bond discovered on the same evening the eighth moon of Saturn, though the Atlantic flowed between them at the time. And we find a resembling simultaneousness of inference and conclusion exemplified by the work which has given occasion to the present controversy. The argument which it amplifies and expands, and, as we think, carries by much too far, and into conclusions not legitimate, was first given to the world seven years ere the appearance of this English volume, in the columns of a Scotch newspaper, and full six years in a separate work, published and rather extensively circulated both in Britain and America. And in glancing over the first edition of the "Essay on the Plurality of Worlds," we had expected — not, perhaps, taking sufficiently into account that simultaneity of thought at certain stages of acquirement to which we refer — that some acknowledgment ought to have been made to the writer who had originated the argument so long before. We ascertain, however, from the second edition of the English work now before us, that its author had framed his argument for himself, independently altogether of the previously-published one. "I have no wish," he says, "to lay any stress upon the originality of the views presented in the Essay. I now know that, several years ago (in 1849), Hugh Miller, in his 'First Impressions of England' (chap. xvii), presented *an argument from geology very much of the nature of that which I have employed;* and that the Rev. Mr. Banks, in a little tract published in 1850, urged the very insecure character of the doctrine that the planets and stars are inhabited. These coincidences with my views I did not know till my Essay was not only written but printed. As to myself, the views which I have at length committed to paper have long been in my mind." There is an error in the date given here. The argument to which the author of the Essay refers as "much of the nature" of his own was first published, not in 1849, but in October, 1846, when it appeared in the columns of the "Witness" as part of one of the chapters of "First Impressions,"— a work which was published in the collected form as a volume early in the following year. Essentially, however, the reference is perfectly satisfactory and, mayhap, not wholly uninteresting, as corroborative of our position, that at certain periods, after a certain amount of fact in some new department has been acquired, inferences never drawn before come to be drawn simultaneously by minds cut off by circumstances from all intercourse with each other. The argument, as originally stated in the "Witness," we shall take the liberty to repeat, slightly abridged, not only from its bearing on one of the most curious controversies of modern times, but as it may also serve to indicate what we deem the just degree in which the inferences of astronomers regarding the inhabitability of the planets are to be qualified by the facts of the geologist.

"There is a sad oppressiveness in that sense of human littleness which the great truths of astronomy have so direct a tendency to inspire. Man feels himself lost amid the sublime magnitudes of creation, — a mere atom in the midst of infinity; and trembles lest the scheme of revelation should be found too large a manifestation of the divine care for so tiny an ephemera. Now, I am much mistaken if the truths of geology have not a direct tendency to restore him to his true place. When engaged some time since in perusing one of the sublimest philosophic poems of modern times, — the 'Astronomical Discourses' of Dr. Chalmers, — there occurred to me a new argument that might be employed against the infidel objection which the work was expressly written to remove. The infidel points to the planets; and, reasoning from an analogy which on other than geologic data the Christian cannot challenge, asks whether it be not more probable that each of these is, like our own earth, not only a scene of creation, but also a home of rational, accountable creatures. And then follows the objection, as fully stated by Dr. Chalmers, 'Does not the largeness of that field which astronomy lays open to the view of modern science throw a suspicion over the truth of the gospel history? and how shall we reconcile the greatness of that wonderful movement which was made in heaven for the redemption of fallen man with the comparative meanness and obscurity of our species?'

Geology, when the Doctor wrote, was in a state of comparative infancy. It has since been largely developed; and we have been introduced, in consequence, to the knowledge of some five or six different creations of which this globe was the successive scene ere the present creation was called into being. At the time the 'Astronomical Discourses' were published, the infidel could base his analogy on his knowledge of but one creation; whereas we can now base our analogy on the knowledge of at least six creations, the various productions of which we can handle, examine, and compare. And how, it may be asked, does this immense extent of basis affect the objection with which Dr. Chalmers has grappled so vigorously? It annihilates it completely. You argue, may not the geologist say to the infidel, that yonder planet, because apparently a scene of creation like our own, is also a home of accountable creatures like ourselves. But the extended analogy furnished by geologic science is full against you. Exactly so might it have been argued regarding the earth during the early creation represented by the Silurian system, and yet the master-existence of that extended period was a crustacean. Exactly so might it have been argued regarding the earth during the term of the creation represented by the Old Red Sandstone; and yet the master-existence of that scarce less extended period was a fish. During the creation represented by the Carboniferous period, with all its rank vegetation and green-reflected light, the master-existence was a fish still. During the creation represented by the Oolite, the master-existence was a reptile, a bird, or a marsupial animal. During the creation of the Cretaceous period, there was no further advance. During the creation of the Tertiary formation, the master-existence was a mammiferous quadruped. It was not until the creation to which we ourselves belong was called into existence that a rational being, born to anticipate a hereafter, was ushered upon the scene. Suppositions such as yours would have been false in at least five out of six instances; and if in five out of six *consecutive* creations there existed no accountable agent, what shadow of reason can there be for holding that a different arrangement obtains in five out of six *contemporary* creations? Mercury, Venus, Mars, Jupiter, Saturn, and Uranus may have all their plants and animals, and yet they may be as devoid of rational, accountable creatures, as were the creations of the Silurian, Old Red Sandstone, Carboniferous, Oolitic, Cretaceous, and Tertiary periods. They may be merely some of the 'many mansions' prepared in the 'Father's house' for the immortal existence of kingly destiny made in the Father's own image, to whom this little world forms but the cradle and the nursery.

"But the effect of this extended geologic basis may be neutralized, the infidel may urge, by extending it yet a little further. Why, he may ask, since we draw our analogies regarding what obtains in the other planets from what obtains in our own, — why not conclude that each one of them has also had its geologic eras and revolutions, — its Silurian, Old Red Sandstone, Carboniferous, Oolitic, Cretaceous, and Tertiary periods; and that now, contemporary with the creation of which man constitutes the master-existence, they have all their fully-matured creations, headed by rationality? Why not carry the analogy thus far? Simply, it may be unhesitatingly urged in reply, because to carry it so far would be to carry it beyond the legitimate bounds of analogy; and because analogy pursued but a single step beyond the limits of its proper province is sure always to land the pursuer in error. Analogy is not identity..... A sagacious guide in its own legitimate field, it is utterly blind and senseless in the precincts that lie beyond. It is nicely correct in its *generals*, perversely erroneous in its *particulars;* and no sooner does it quit its proper province — the general for the particular — than there start up around it a multitude of solid objections, sternly to challenge it as a trespasser on grounds not its own. How infer, we may well ask the infidel, — admitting, for the argument's sake, that all the planets come under the law of geologic revolution, — how infer that they have all, or any of them save our own earth, arrived at the stage of stability and ripeness essential to a fully developed creation, with a reasoning creature as its master-existence? Look at the immense mass of Jupiter, and at that mysterious mantle of cloud, barred and streaked in the direction of his *trade*-winds, that forever conceals his face. May not that dense robe of cloud be the ever-ascending steam of a globe that, in consequence of its vast bulk, has not sufficiently cooled down to be a scene of life at all? Even the analogue of our Silurian creation may not yet have begun in Jupiter. Look, again, at Mercury, where it bathes in a flood of light, enveloped within the sun's halo, like some forlorn smelter sweltering beside the furnace mouth. A similar state of things may obtain on the surface of that planet, from a different though not less adequate cause. But it is unnecessary to deal further with an analogy so palpably overstrained, and whose aggressive place and position in a province not its own so many unanswerable objections start up to elucidate and fix."

Such, virtually, is the argument which has been reproduced and greatly expanded in the "Essay on the Plurality of Worlds." We think, however, that the ingenious and accomplished author of that work has pressed it too far, and forgotten that, though it introduces into the reasonings of the astronomer, regarding the existence of rational inhabitants in the planets, the *modifying element of time*, it does not affect his general conclusions. It merely shows, from the extended experience of the earth's history which geology furnishes, that these conclusions may not refer to the *now* of the planetary universe, but to some period in a perhaps very remote future. For the argument of the astronomer, in a condensed form, let us draw on Fontenelle, — a man who wrote ere geology had yet any existence as a science. It is thus he makes his philosopher reason with his lady friend the Marchioness, in a general summary: "We cannot pretend to make you *see* them [the inhabitants of the planets]; and you cannot insist upon demonstration here, as you would in a mathematical question; but you have all the proofs you could desire in our world, — the entire resemblance of the planets with the earth which is inhabited, the impossibility of conceiving any other use for which they were created, the fecundity and magnificence of nature, the certain regards which she seems to have had to the necessities of the inhabitants, as in giving moons to those planets remote from the sun, and more moons still to those yet more remote; and, what is still very material, there are all things to be said on one side, and nothing on the other. In short, supposing that these inhabitants of the planets really exist, they could not declare themselves by more marks, or by marks more sensible." Such is the statement of Fontenelle; and, though it can be no longer affirmed that nothing can be said on the *opposite* side, seeing that we have now a very ingenious volume written on the opposite side, by not merely a clever, but also a highly scientific man, it will be found that in the course of discovery the argument has rather strengthened than weakened. Let us take, for instance, the portion of it founded on the existence and distribution of moons. It was known when Fontenelle wrote his "Conversations on the Plurality of Worlds," that the earth had one moon, Jupiter four moons, and Saturn five. It is now further known that Saturn has eight moons, and Uranus also eight; and if only one has yet been detected revolving round Neptune, it must be taken into account that the latter planet is twice further distant from our earth than Saturn, and so dimly discernible that it is still a question whether it possesses a ring or no, — that our earliest acquaintance with it is not yet more than eight years old, — that even Saturn's eighth moon was discovered only six years ago, — and that not only not a few of the moons of Neptune, but even some of the moons of Uranus, may be still to find. The general fact still holds good, that in proportion as the larger planets most distant from the sun require, in consequence, moons to light them, the necessary moons they have got; just as on our own earth the animals who live most distant from the sun, and require, in consequence, thicker protective coverings to keep them warm, have got these necessary protective coverings, whether of fatty matter or of fur. But the argument derivable from the light and heat of the sun himself seems scarce less strong. Let us avail ourselves of it, as condensed by Sir David Brewster, from Sir Isaac Newton's first letter to Dr. Bentley. "He [Sir Isaac] thought it inexplicable by natural causes, and to be ascribed to the counsel and contrivance of a voluntary agent, that the matter [of which the solar system is formed] should divide itself into two sorts, part of it composing a shining body like the sun, and part an opaque body like the planets. Had a natural and blind cause, without contrivance and design, placed the earth in the centre of the moon's orbit, and Jupiter in the centre of his system of satellites, and the sun in the centre of the planetary system, the sun would have been a body like Jupiter, and the earth that is, without light and heat; and, consequently, he [Sir Isaac] knew no reason why there is only one body qualified to give light and heat to all the rest, but because the Author of the system thought it convenient, and because one was sufficient to warm and enlighten all the rest." "To warm and enlighten all the rest!" Newton recognizes the hand of the Divine Designer in that peculiar collocation of matter through which the lamp and furnace of the system is placed in its centre, and the opaque objects to be warmed and heated arranged at certain distances around it. But why the application of light and heat to masses of dead matter? Light and heat, in a lesser or greater degree, are necessary to the existence of all organisms, plant and animal, but not to the existence of matter not organized. A lamp is necessary in a railway carriage that travels by night, if there be passengers within, but not in the least necessary to the carriage itself, if there be only the empty seats to shine upon. And if, of all the planets that not only revolve round the central lamp and furnace, but have also special lamps of their own, the earth be the only inhabited one, not only is the waste most enormous, but the argument of design, so profoundly deduced by Sir Isaac, must be pronounced to be of no force in more than thirty cases for one, that is, in the cases of all the supposed uninhabited planets in which there exists nothing capable of being benefited by being either lighted or warmed. Or, to avail ourselves of Sir David's happy illustration, the Creator of a solar system with many uninhabited planets, and only a single inhabited one, would resemble some "mighty autocrat who should establish a railway round the coasts of Europe and Asia, and place upon it an enormous train of first-class carriages, impelled year after year by tremendous steam-power, while there was a philosopher and a culprit in a humble van, attended by hundreds of unoccupied carriages and empty trucks." And, of course, were the unoccupied carriages to be lighted up with lamps apparently for the benefit of the passengers which they had not, and were these lamps to be fewer or more numerous in each case in meet proportion with the degree of darkness to be encountered, and as the necessities of actual passengers would require, the puzzle involved in the why and wherefore of the whole concern would be still increased. The old argument for the inhabitancy of the planets, regarded as an argument of *ultimate* design, still remains unaffected by the discoveries of the geologist.

But, on the other hand, let not the *modifying* influence of these discoveries be denied. Such is their effect on the argument, that, though we may receive it in full as truly solid, we may yet, in perfect consistency with its conclusions, deem it a moot point whether there be *at the present time* a single inhabited world in the system save our own. We cannot express, either by figures or by algebraic signs, save by the signs that express unknown quantity, the geologic periods. We only know that they were of enormous extent. Let us, however, for the argument's sake, represent the period during which man has been upon earth by the sum 5000, the periods during which the successive plant-and-animal-bearing systems of the geologist were in being by the sum 1,000,000, and the earlier death periods, during which the gneiss, the older quartz rock, the mica schist, and the non-fossiliferous clay slate were formed, by the sum 500,000; and let us then suppose that some intellectual being, wise as a Newton, and reasoning on exactly his principles and those of Sir David Brewster, had existed during all these terms, converted into years, at a distance from the earth as great as that which separates the earth from the planets Mars or Venus; further, let us suppose that once in every five thousand years for the first half-million, the query had been propounded to him by the Creator, as the Creator questioned Job of old, — "Intellectual being, is yonder planet inhabited, or no?" and that during the million of years that followed, the query should be repeated after the same intervals in the modified form,

—"Is yonder planet inhabited by rational, accountable creatures, or no?" Now, nothing can be more clear than that, reasoning on Sir Isaac's and Sir David's premises, the reply would be given in each instance in the affirmative. It would be seen by the reasoning creature that the distant earth-planet was lighted up and heated by the great central furnace and lamp, the sun; that it had its clouds, and therefore its atmosphere; that it had its grateful interchange of day and night, of summer and winter, autumn and spring; and, further, that it had its attendant moon, to stir up its seas with purifying tides, and to light up its nights. And yet most probable it is that the first hundred answers to the query — those which related to the existence of mere animal being — would have been false ones; and most certain it is that the next two hundred answers to the query — those which related to the existence of natural life — would be false also. Not until after the lapse of a million and a half of years, when the question would come to be put for the three hundred and first time, would it elicit the true response. And let us remember that whatever was may be; and that what were the first states of our own planet may be the present states of the various planets that revolve with it round the central furnace and lamp. Here again we cannot cast our argument into an exact geometrical or arithmetical shape. We cannot even say, founding on the assumption of proportionate periods already given, that as our earth was for three hundred periods of five thousand years each without rational inhabitants, and possessed of such an inhabitant during only the three hundred and first period of that length, so it is probable that of three hundred and one contemporary planets only one is a scene of rational existence, and the others either not inhabited at all, or inhabited by but sentient irrationality. We cannot give the argument any such exact form, seeing that an unreckoned but possible, nay, probable element, comes in to destroy its exactitude. The other planets *may*, nay, in all likelihood, *have* been ripening as certainly as our own, and the period of rational inhabitancy may have arrived in not a few of them. Quite as perilous, however, would it be to argue from the particular analogy furnished by the history of the earth, that all, or even the greater part of them, had so ripened. Why, even the fruit of one season, whether apples or apricots, does not all ripen at the same time on the same tree; far less do the fruits of different trees ripen at the same time. And we are sufficiently acquainted with the planets to know that, with certain general resemblances, they are very different fruit indeed from our own earth. Even supposing Jupiter, for instance, to be in every respect save size a second earth (which, by the way, demonstrably he is not), he would take, on the soberest calculations of the geologist, many hundred times more time to ripen than our small planet. And so may it be predicted of Saturn and Uranus, and Neptune also, and most probably, from the different circumstances in which they are placed, of the smaller planets Mercury and Venus. But while this geological question, in relation to the present time of *ripe* or *unripe*, must be now brought in to qualify the reasonings of the astronomer, let us not forget that these reasonings have, with reference to *ultimate* results, a value as positive as ever. From the crustaceous eyes of many facets that existed during the times of the Silurian period, and the ichthyic eyes of but one facet or capsule that existed during the times of the Old Red Sandstone, the geologist infers that during these periods there existed light; while the astronomer, taking up the converse of the argument, infers that where there is light (joined, of course, to the other necessary conditions of life, such as planetary matter existing in the twofold form of solid nucleus and surrounding atmosphere) there must be eyes, — eyes, *therefore* light, solar or lunar, etc., — light, solar or lunar, *therefore* eyes. And just as the geologic argument is noways invalidated by the fact that there are animals in the fœtal state furnished with eyes darkly veiled in the womb, for which *light* does not *yet* exist, it in no degree invalidates the astronomical argument that there have been, and most probably now are, *fœtal* planets furnished with light, solar or lunar, for which *eyes* do not yet exist. Such, in this controversy, seems to be the due balance and adjustment of the opposite arguments, — astronomic and geologic arguments that modify, but in no degree destroy, each other.

We can of course do little more, within the limits of a single article, than just touch at a few points, on a subject upon which men such as Sir David Brewster, and, shall we say, Professor Whewell, fill each a volume apiece. Let us, however, submit to them, as very admirable, both in form and substance, the claims of geology, as stated by the English Professor: —

"Astronomy claims a sort of dignity over other sciences, from her *antiquity*, her *certainty*, and the *vastness* of her discoveries. But the *antiquity* of astronomy as a science had no share in such speculations as we are discussing; and if it had had, new truths are better than old conjectures; new discoveries must rectify old errors; new answers must remove old difficulties. The vigorous youth of geology makes her fearless of the age of astronomy. And as to the *certainty* of astronomy, it has just as little to do with these speculations. The certainty stops just where these speculations begin. There may, indeed, be some danger of delusion on this subject. Men have been so long accustomed to look upon astronomical science as the mother of certainty, that they may possibly confound astronomical discoveries with cosmological conjectures, though these be slightly and illogically connected with those. And then, as to the *vastness* of astronomical discoveries, — granting that character, inasmuch as it is to a certain degree a matter of measurement, — we must observe that the discoveries of geology are no less vast; they extend through time, as those of astronomy do through space; they carry us through millions of years, that is, of the earth's revolutions, as those of astronomy do through millions of the earth's diameter, or of diameters of the earth's orbit. Geology fills the regions of duration with events, as astronomy fills the regions of the universe with objects. She carries us backward by the relation of cause and effect, as astronomy carries us upward by the relation of geometry. As astronomy steps on from point to point of the universe by a chain of triangles, so geology steps from epoch to epoch of the earth's history by a chain of mechanical and organical laws. If the one depends on the axioms of geometry, the other depends on the axioms of causation.

"So far, then, geology has no need to regard astronomy as her superior, and least of all when they apply themselves together to speculations like these. But, in truth, in such speculations geology has an immeasurable superiority. She has the command of an implement in addition to all that astronomy can use, and one, for the purpose of such speculations, adapted far beyond any astronomical element of discovery. She has for one of her studies, — one of her means of dealing with her problems, — the knowledge of life, animal and vegetable. Vital organization is a subject of attention which has in modern times been forced upon her. It is now one of the main points of her discipline. The geologist must study the traces of life in every form; must learn to decipher its faintest indications and its fullest development. On the question, then, whether there be in this or that quarter evidence of life, he can speak with the confidence derived from familiar knowledge; while the astronomer, to whom such studies are utterly foreign, beause he has no facts that bear upon them, can offer on such questions only the loosest and most arbitrary conjectures; which, as we have had to remark, have been rebuked by eminent men as being altogether inconsistent with the acknowledged maxims of his science.

"When, therefore, geology tells us that the earth, which has been the seat of *human* life for a few thousand years only, has been the seat of *animal* life for myriads, it may be millions, of years, she has a right to offer this as an answer to any difficulty which astronomy, or the readers of astronomical books, may suggest, derived from the consideration that the earth, the seat of human life, is but one globe of a few thousand miles in diameter, among millions of other globes at distances millions of times as great. Let the difficulty be put in any way the objector pleases. Is it that it is unworthy of the greatness and majesty of God, according to our conception of him, to bestow such peculiar care on so small a part of his creation? But we know from geology that he has bestowed upon this small part of his creation — mankind — this special care. He has made their period, though only a moment in the ages of animal life, the only period of intelligence, morality, religion. If, then, to suppose that he had done this is contrary to our conceptions of his greatness and majesty, it is plain that our conceptions are erroneous: they have taken a wrong direction. God has not judged as to what is worthy of Him as we have judged. He has found it worthy of him to bestow upon man his special care, though he occupies so small a portion of time; and why not, then, although he occupies so small a portion of space?

"Or is the objection this, — that if we suppose the earth only to be occupied by inhabitants, all the other globes of the universe are wasted, — turned to no purpose? Is waste of this kind considered as unsuited to the character of the Creator? But here again we have the like waste in the occupation of the earth. All its previous ages, its seas, and its continents, have been wasted upon mere brute life, often, so far as we can see, for myriads of years upon the lowest, the least conscious form of life, — upon shell-fishes, crabs, sponges. Why, then, should not the seas and continents of other planets be occupied at present with a life no higher than this, or with no life at all?"

IV.

THE SPACES AND THE PERIODS.

That vast development of natural science which forms a leading characteristic of the present age gives an importance to questions such as that which it involves which they did not possess at any former period; and must, we doubt not, materially affect in the future the entire front of that ever-fresh controversy which has been maintained since the earliest ages of the church around the Christian evidences. Let us address ourselves to the present portion of our subject, — the great extent of the geologic periods, through the medium of a simple illustration.

Let us suppose that shortly after the arrival of the Mayflower at the shores of New England, and just as the Pilgrim Fathers are preparing to begin their labors among the deep primeval forests which cover the country, there occurs a friendly controversy between two of the party regarding the age of these vast woods. All the trees are of kinds unknown at home; and though loftier, many of them, than the great oaks of England, and not a few of them not less bulky, it is maintained by one of the disputants that they may yet have come under very different laws of growth, and may not be one twentieth part so old. These hoary forests, he argues, though it would require some three or four centuries to form such on the eastern shores of the Atlantic, may on its western shores be less than fifty years old; nay, not only may the woods of the country be as of yesterday compared with those of England, but even its animals may be of such rapid growth that the mouse-deer, though of ponderous bulk and size, may be in reality only a few months old; and the oyster, which on the English beds takes from five to seven years, as shown by its annual *shoots*, to be fit for market, may in the greatly larger American species be equally mature in as many weeks. The disputant contends — and at this stage of the controversy contends truly — that they are furnished with no correct *unit* by which to measure the age of either the unknown plants or unfamiliar animals of the new country. Let us yet further suppose that in the immediate neighborhood of the infant settlement there is a small lake, which the settlers find it necessary for sanitary purposes to drain, and that they cut through, in the work, one of those deep mosses of northern America in which the gigantic bones, and not unfrequently the entire skeletons, of the mastodon occur. Let us suppose that they first cut through several yards of solid peat; that they then reach a tier of rather small tree-stumps sticking in the soil; that a second tier of somewhat larger tree-stumps lies beneath; that they then reach a third tier of still larger stumps; that under the stratum of earth which underlies these they find a thick bed of marl composed chiefly of very minute shells; and that embedded in the marl they find the skeleton of a mastodon. Judging from data furnished on the eastern side of the Atlantic, the Pilgrim who has been asserting, in opposition to his neighbor, the antiquity of the American woods, argues from these appearances that the moss deposit must be of great age, and the underlying skeleton of an age greater still. Mosses in Old England, containing three tiers of stumps, are demonstrably as old as the times of the Roman invasion. Even the Roman axe has in some instances been found sticking in the lower trunks; and at least the huge unknown skeleton just found in the moss must, he urges, be quite as ancient as the times of Agricola or Julius Cæsar. His antagonist, however, challenges the inference. The previous question has, he asserts, first to be settled. The rate of growth of the American wood and the American shells has to be determined ere any calculation can be founded on either the three tiers of stumps or the overlying or intervening deposits of vegetable matter, or yet on the thickness of the shell-marl which underlies the whole. For if, as he contends, the growth of animals and vegetables be, as is possible, very rapid in the new world, the moss and shells, instead of being at least sixteen or seventeen hundred years old, may not be above sixty or seventy years old, and the huge animal beneath may have been living only eighty or a hundred years ago. At length, however, the required *unit* of measurement turns up. In cutting a tree for the erection of his hut, the Pilgrim who maintains the opposite side of the argument finds it strongly marked by the annual rings. And there can be no doubt that the rings *are* annual ones. Between the tropics, when rings occur at all, they may indicate the checks given to vegetation by the dry seasons; and as the year has in certain localities two of these, each twelvemonth may be represented in the tree, not by one, but by two rings. But in the latitude of New England, where winter presses his iron signet on the soil with much firmness, one strongly-marked ring represents the year; and so, if it be found that a tree of some eighteen or twenty inches in diameter has its hundred concentric rings, it may be safely predicated that it has stood its century. And such, in the supposed case, is the inference of the Pilgrim. He has at length got a unit, in reality fixed by the great, never-varying astronomic movements which give to the world its seed-time and its winter; and finding, as he cuts tree after tree, the same evidence repeated, — ring answering to ring, here larger and there smaller, but in their average proportions corresponding with those of the English woods, — he is constrained definitively to conclude that the trees of the new country grow as slowly, or nearly so, as those of the old one; and he confidently challenges his antagonist to test the data on which he founds. Nor can he hold that his newly-found unit, though, strictly speaking, only a measure of the age of the various forest trees in which it occurs, has bearing only on them. If trees grow as slowly in the new country as in the old, can he rationally hold that its other classes of vegetables — its ferns, equisetacæ, club-mosses, grasses, and herbaceous plants generally — grow much faster than their cogeners at home? Further, though his unit does not enable him to measure exactly the age by the mossy deposit, with its three tiers of stumps and its underlying mastodon, it at least enables him to determine that it must be very old. It gives him in succession the age of each tier; and when he infers respecting the intervening and overlying deposits of vegetable matter, that, as the trees grow slowly, the deposits must have been formed correspondingly slow in about the average ratio of similar formations on the other side of the Atlantic, it justifies the inference; nay, it is not without its bearing on the probable growth of the animals of the country also. It would be utterly wild to hold that in a country in which an ordinary-sized pine was the slow growth of a century, a mouse-deer or a grizzly bear shot up to its full size in a few weeks or months. And if in the foliaceous shells of the coast, such as its oysters, he finds exactly such layers of growth, or *shoots*, as those from which the oyster-fisher at home computes the age of the animals, each "*shoot*" being the work of a year, can he avoid the conclusion that here also he has got a unit by which to measure the time during which the organisms have lived, and from which he may conclude, in all sobriety, that if the bed of shell-marl which contains the remains of the mastodon be very thick, it must of necessity be very old? If he cannot, in strictness, apply his units to every plant or every shell, or yet to every deposit of vegetable or animal origin, they at least tell him that the same general laws of growth obtain on the one side of the Atlantic as on the other, and warn him against inferring, like his antagonist, that the cases in which he has not yet been able to apply them are in any degree anomalous, or under laws that are different.

We have but to apply to the geological periods of at

least the Secondary and Tertiary divisions the reasoning of our illustration here, in order to determine that they must have been immensely prolonged. In no degree is the argument more affected by the portion of time which separates our age from the ages of the Oolite, than by the portion of space which separates our country from the eastern shores of America. In the woods of the great palæozoic division the lines of growth are uncertain and capricious. Many of the trees furnish no trace of them whatever, just as there are recent intertropical trees in which they do not occur; and in some of the others they appear capriciously and irregularly, as in those intertropical trees in which the growth is checked from time to time by intense heats and occasional droughts. But in the woods of the Lias and Oolite winter has set his seal; the annual rings of *Peuce Eiggensis* and *Peuce Lindleiana* are as regularly and strongly marked as those of the Scotch fir or Swiss pine; nor, be it added, are they of larger size. In one specimen of our collection, but in one only, the rings average nearly a quarter of an inch in breadth; the tree added in a single twelvemonth almost half an inch to its diameter; but the specimen is an exceptional one. In the others they average from about a line to an eighth part; and in one specimen no fewer than twenty-eight rings occur in the space of an inch. The slow-growing tree, of which it formed a portion, — sluggish in its progress as a Norwegian pine on some exposed mountain-side, — added only half an inch to its diameter in *seven* years. The unit here tells certainly of no rapid development of life, but, on the contrary, of a development quite as tardy as that of the present age of the world in latitudes as high as our own; and, though we cannot decide with the same certainty respecting the rate of growth in the animals contemporary with those trees, we may surely most naturally infer that ostrea of some ten or twelve layers, or gryphites (extinct members of the same family) of some fifteen or twenty, could not have been very young; that as the ammonite, though thinly walled, was as solid in its substance as the nautilus, and had a great many more chambers, which were added to it peacemeal, one at a time, it could not have been of much quicker growth; and that, as the internal shell of the belemnite was much more ponderous than that of its successor the cuttle-fish, it must have attained to maturity quite as slowly. Further, not only can it be demonstrated that ivory teeth were every whit as dense in those ages as they are now, — a remark that applies equally to the later palæozoic periods, — but it can be shown also that some of these teeth were as sorely worn as in existing animals when very old. In short, the evidence that life, animal and vegetable, existed on the further side of the Tertiary geologic periods under the same laws as now, is as conclusive as that it exists under the same laws on the further side of the Atlantic. And these laws cast much light, as in the case of the peat-moss of our illustration, on the rate at which many of the mechanical deposits must have gone on. The Lias of Eathie, for instance, consist, for about four hundred feet in vertical extent, of an almost impalpable shale, divided into layers scarce thicker than pasteboard. It might well be predicated, from the merely mechanical character of the deposit, that its formation could not have been rapid. But how greatly is the argument for the lapse of a vast period of time for its growth strengthened by the fact that each one of these many thousand layers formed a crowded platform of animal life, and that so thickly are they covered with the remains of not only free shells, such as ammonites, but also of sedentary shells, such as the ostrea, that the organisms of but two of the more crowded platforms could not find room on a single one! And these shells were the contemporaries of slow-growing pines, that on the average increased in diameter little more than the fifth of an inch yearly.

Nor, though we lack the regulating unit, is the evidence of the lapse of vast periods during the deposition of the palæozoic systems much less complete. The oldest wood that presents its structure to the microscope — a fossil of the Lower Old Red Sandstone — exhibits no annual rings; but it presents as dense a structure as the Norfolk Island pine. The huge araucarian of Granton has a structure nearly as dense. We have already incidentally referred to the solid ivory and much-worn teeth of the reptile fishes of the Coal Measures. In the Mid-Lothian basin there are thirty seams of workable coal intercalated among deposits of various character, whose united thickness amounts to nearly three thousand feet, and under most of these seams the original soil may still be detected on which the plants that formed their coal flourished and decayed. Whole beds of the Mountain Limestone are composed almost exclusively of marine shells and the stems of lily encrinites. In the Old Red Sandstone there are three different formations abounding in fishes; and yet, so far as is yet known, there is not a single species of fish common to any two of them. And who shall tell us that the life-term of a *creation* is a brief period? In the Upper Silurian system we have examined a deposit more than fifty feet thick, every fragment of which had once been united to animal life, crustaceous, molluscan, or radiated. And how wonderfully, too, the further geologists explore, and the more carefully they examine, are their formations found to expand! Phillips estimated the thickness of the Coal Measures at ten thousand feet. Sir Charles Lyell, in one of his recent visits to America, found that the Coal Measures of Nova Scotia had a thickness of more than *fourteen* thousand six hundred feet. Phillips estimated all the deposits beneath the Old Red Sandstone at twenty thousand feet. The geologists of the Government survey find that the Silurians alone amount to about *thirty* thousand feet; and under these, in Scotland at least, lie the clay-slates, the mica-schists, and the enormous deposits of the gneisses. On the Continent, the remains of whole creations have been found intercalated between what had been deemed contiguous systems. An entire system, the Permian, has been detected between the Coal Measures and the Trias; and that shell-deposit that extends between the Gironde and the Pyrenees, once regarded as of the same age with the Coraline Crag, has yielded seven hundred species of shells — nearly twice the number of all the species found on the coasts of Britain — that belong neither to the Crag nor to the older Eocene. It is yet another creation that has appeared, for which fitting space must be found in the record. The more thoroughly the field-geologist examines, the larger become his demands on the eternity of the past for periods which it is certainly very competent to supply. His sibyl ever returns upon him; but, unlike her of old, it is with an increased, not a diminished store of volumes; and she ever demands for them a larger and yet larger price.

And why should the tale of years be refused her? Let year be heaped upon year, until the numerals that represent them, consisting all of nines, would extend in a close line from the sun to the planet Neptune, and they would still form but an inappreciable item in the lifetime of the Creator. We see nothing to regret in the truth, destined to become greatly more evident in the future than it is now, that there is nothing in all history, or in all creation, vast enough to be measured off against the periods of the geologist, save the spaces of the astronomer; or that, with relation to at least our own planet, *rational* existence is still in its immature infancy. Could we wish it to be otherwise? The world is still sowing its wild oats; and, though somewhat better, on the whole, than it has been, there is surely nothing in its present aspect to reconcile any one to the belief that it has attained to its ultimate development. Its present most prominent features, if we may so express ourselves, are the horrible sufferings of war and the lies of stock-jobbers.

V.

UNITY OF THE HUMAN RACES.[1]

THERE are certain typical forms of error that never die, though their details alter, and the facts and analogies on which they purport to be based vary with the increase of knowledge and the progress of the human mind. And it is of great importance that these should be studied, not only in their essential, and, if we may so express ourselves, generic character, but also historically, in the various modifications of shape and color which have marked them at their several periods of revival, and which will almost always be found to depend on some peculiarity of pursuit or opinion prevalent at the time, or, if connected with the physical sciences, on some newly-opened course of discovery. The various *species* of error once thoroughly mastered, the student will find ever after that it is with but its *varieties* he has to deal. Nay, by thoroughly knowing the *species*, and the history of the changes through which they passed at their several appearances, he may be able to anticipate the exact course which they would have to run should they reappear in his own times, when men worse taught, and unacquainted with this cycloidal character of error, will neither know whence they come nor whither they are going. The native sagacity of the late Dr. M'Crie was greatly sharpened by a knowledge of this kind, derived from his profound acquaintance with church history; and he is said to have predicted, while Rowism was

[1] The Unity of the Human Races proved to be the Doctrine of Scripture, Reason, Science, etc. By the Rev. Thomas Smyth, D.D., Member of the American Association for the Advancement of Science.

yet a howling enthusiasm, gibbering the untranslatable tongues, and stretching forth its hand to work miracles, that it was to end at no very remote period as a decrepit superstition. The fluttering butterfly was destined, agreeably to its previously-determined constitution, to produce a brood of creeping caterpillars, though only the laborious student who had acquainted himself with its specific character, as exhibited in former manifestations, knew that such was to be the case. This perception of the specific essentials and consequents of both truth and error constitutes, too, at once the charm and the value of such a mastery over the controversies which have arisen within the church, or in which, in self-defence, the church has been compelled to engage, as that possessed by the Principal of our Free Church College, Dr. Cunningham; and there are not a few opposed to college extension on the principle that, even in the Free Church, Professors of Church History of similar calibre and acquirement are not to be had in every district of country, and that yet such are imperatively demanded by the emergencies of the time. To distinguish between the permanent forms and the accidental circumstances, — between the ever-recurring cycloidal types and those mere varieties which belong to but one phase or period in the appearance of these, — must ever form no inconsiderable portion of the science of ecclesiastical history. Nay, save for this tendency in the typical forms of error to return upon the world altered in their features but unchanged in their framework, at least two thirds of all ecclesiastical history would be but a profitless record of the nonsense and errors of the past; and the beau ideal of a church history would be a work such as that of Milner, which is little else than a record of the better thoughts and deeds of Christian men chronologically arranged, and useless for the most important ends served by ecclesiastical history of the better type. It sounds no note of warning, and furnishes no armor of defence, against the cycloidal errors.

There are two of these returning errors of a diametrically opposite character, which arise out of natural science, and of which the last century has seen several revivals, and the centuries to come must witness many more. The one — that of Maillet and Lamarck — sees no impassable line between species, or even genera, families, and classes, and so holds that all animals — the human race as certainly as the others — may have commenced in the lowest forms, and developed during the course of ages to what they now are. The other — that of Kames and Voltaire — recognizes in even the varieties of the species impassable lines, and holds, in consequence, that the human race cannot have sprung from a single pair. And both beliefs are as incompatible with the fundamental truths of revelation as they are with one another. The Lamarckian form of error has been laid on the shelf for a time; nor will it be very efficiently revived until some new accumulation of fact, gleaned from the yet unexplored portions of the geologic field, or the obscurer fields of natural history, and pregnant with those analogical resemblances between the course of creation and the progress of embryology with which nature is full, will give it new footing, by associating it with novel and interesting truth. The antagonist error is at present all alive and active in America, where it has been espoused by naturalists of high name and standing; and it has already produced volumes of controversy. Nor is there a country in the world where, from purely political causes, there must exist a predisposition equally strong to receive as true the hypothesis of Voltaire. The existence of slavery in the Southern States, and the strong dislike with which the black population are regarded by the whites throughout the country generally, must dispose the men who hate or enslave them to receive with favor whatever plausibilities go to show that they are not of one blood with themselves, and that they owe to them none of the duties of brotherhood. We have perused with interest and instruction a very learned and able volume on this subject by the Rev. Dr. Thomas Smyth of Charleston, one of the most accomplished Presbyterian ministers of the United States, with whose works on the "Apostolical Succession" and the "Claims of the Free Church of Scotland" many of our readers must be already acquainted, and who, though residing in the centre of a slave district, and exposed to much odium on the part of the abolitionists, has been the first to come forward in this controversy to assert in behalf of the black man the "unity of the human races," and that all men have fallen in one common father, the first Adam, "created a living soul," and that there is salvation to all in one common Saviour, the "last Adam," "made a quickening Spirit." Much of the volume is taken up in dealing with the question in its older form. Voltaire held that there were "as well-marked species of men as of apes." Kames was more unhappy in his illustration. "If the only rule afforded by nature for classing animals can be depended upon," we find him saying, "there are different species of men *as well as of dogs*." Gibbon, though his remark on the subject takes the characteristic form of an ironical sneer, in which he says the contrary of what he means, deemed it more natural to hold that the various races of men originated in those tracts of the globe which they inhabit, than that they had all proceeded from a common centre and a single pair of progenitors. To the view, however, taken by these distinguished sceptics, — men eminent in the literary world, but of little weight in that of science, — all the greater naturalists of the last century were opposed. Kames, in the chapter of his "Sketches" specially devoted to the question, had to combat both Linnæus and Buffon; and the later naturalists who have specially concentrated themselves on the subject, such as Pritchard, Bachman, and Lawrence, have irrefragably shown that, tried by the marks which are regarded as constituting specific differences among the lower animals, the family of man consists of but one species. But the question raised in the modern form, without disputing this conclusion, eludes it by a new statement; and we could fain wish that Dr. Smyth had devoted a larger portion of his valuable volume to the controversy in its new phase. The fact is, while in its old form the greater naturalists were on the side of the orthodox theologian, some very distinguished naturalists take in its new form the opposite side. The difference in the statement may be summed up in a few words. It was held by Voltaire and his coadjutors that there are several *species* of men, who must of necessity have originated from several pairs; whereas, what is held by Professor Agassiz and several of the American naturalists is, that though the species be properly but one, it is according to the known analogies both of plants and animals that it should have originated in various centres, — a conclusion which the strongly-marked varieties of the race which occur in certain well-defined geographic areas serve, it is held, to substantiate, or at least to render the most probable.

It will be seen that against this restatement of the question many of the old facts and arguments do not bear. Theologically, however, — in every instance in which it assumes the positive form, and in which, building on its presumed analogies, and the extreme character and remote appearance of the several varieties of the species to which it points, it asserts that the beginnings of the race must be diverse, and its *Adams* and *Eves* many, — it is in effect the same. On the consequences of the result it can be scarce necessary to insist. The second Adam died for but the descendants of the first. Nay, so thoroughly is revelation pledged to the unity of the species, that if all nations be not "made of one blood," there is, in the theological sense, neither first nor second "Adam;" "Christ," according to the apostle, "hath not risen;" conversion is an idle fiction; and all men are yet in their sins. Further, that kind of brotherhood which unites the species by those ties of neighborhood illustrated by our Saviour is broken; and there are races of men reckoned up by millions and tens of millions in which we may recognize our slaves and victims, but not our brothers and neighbors. Nay, why should we respect the life of creatures not of our own blood? Bill Sykes tells Fagin the Jew, in "Oliver Twist," that he wished he was his dog; "for," said he, "the government that cares for the lives of men like you lets a man kill a dog how he likes." But if these tribes be men not of our own blood, — men who did not spring from the same source with ourselves, and for whom therefore Christianity can make no provision, — why the distinction? It is only to those whom we believe to be of our own blood that the distinction extends. It is as lawful to shoot an orang-outang or a chimpanzee as a dog or a cat; and with but mere expediency to regulate the matter, it might become quite as necessary to hunt down and destroy wild men as to hunt down and destroy wild dogs. Nay, we are not sure whether a somewhat mysterious admission to this effect may not be found in a passage quoted by Dr. Smyth from the writings of one of the American assertors of the diversity of races, Dr. Nott. "The time must come," says this latter gentleman, "when the blacks will be worse than useless to us. What then? Emancipation must follow, which, from the lights before us, is but another name for *extermination*." But though the remark, viewed in connection with such a doctrine, seems strangely ominous, we do not profess fully to understand it.

Within the limits of a newspaper article — narrow for such a subject when amplest — we can scarce be expected even to indicate the line which we think ought to be taken up in this controversy by the churches. To the historic evidence we find ample justice done by Dr. Smyth; and the historic evidence, so far as it goes, is, be it remembered, *positive*, — not merely inferential. We are less sure, however, of the line specially adopted against Agassiz in the field of natural history. The analogies *may* be on the side of the naturalist, as he says they are, and he may be quite right in holding that varieties of the race so extreme as that of the negro on the one side, and the blue-eyed, fair-haired, diaphanous Goth on the other, could not have originated *naturally* in a species possessed of a common origin during the brief period limited by authentic history on the one hand, and the first beginnings of a family so recent as that of *man* on the other. But though he may possibly be right as a naturalist, — though we think that matter admits of being tried, for it is far from settled, — he may be none the less wrong on that account as a theologian. His inferences may be right and legitimate in themselves, and yet the main deduction founded upon them be false in fact. Let us illustrate. There is nothing more certain than that the human species is of comparatively recent origin. All geological science testifies that man is but of yesterday; and the profound yet exquisitely simple argument of Sir Isaac Newton, as reported by Mr. Conduit, bears with singular effect on the same truth. Almost all the great discoveries and inventions, argued the philosopher, are of comparatively recent origin. Perhaps the only great invention or discovery that occurs in the fabulous ages of history is the invention of letters. All the others — such as the mariner's compass, printing, gunpowder, the telescope, the discovery of the New World and Southern Africa, and of the true position and relations of the earth in the solar system — lie within the province of the authentic annalist; which, man being the inquisitive, constructive creature that he is, would not be the case were the species of any very high antiquity. We have seen, since the death of Sir Isaac, steam, gas, and electricity introduced as new forces into the world; the race, in consequence, has in less than a century and a half grown greatly in knowledge and in power; and by the rapid rate of the increase, we argue with the philosopher that it can by no means be very ancient. Had it been on the earth twenty, fifty, or a hundred thousand years ago, steam, gas, and electricity would have been discovered hundreds of ages since, and it would at this date have no such room to grow. And the only very ancient history which has a claim to be authentic — that of Moses — confirms, we find, the shrewd inference of Sir Isaac. Now, with this fact of the recent origin of the race on the one hand, and the other fact, that the many various languages of the race so differ that there are some of them which have scarce a dozen words in common, a linguist who confined himself to the consideration of natural causes would be quite justified in arguing that these languages could not possibly have changed to be what they are, from any such tongue, in the some five or six thousand years to which he finds himself restricted by history, geology, and the inference of Sir Isaac. It takes many centuries thoroughly to change a language, even in the present state of things, in which divers languages exist, and in which commerce and conquest, and the demands of literature, are ever incorporating the vocables of one people with those of another. After the lapse of nearly three thousand years, the language of modern Greece is essentially that in which Homer wrote; and by much the larger part of the words in which we ourselves express our ideas are those which Alfred employed when he propounded his scheme of legislative assemblies and of trial by jury. And were there but one language on earth, changes in words or structure would of necessity operate incalculably more slowly. Nor would it be illogical for the linguist to argue, that if, some five or six thousand years ago, the race, then in their extreme infancy, had not a common language, they could not have originated as one family, but as several, and so his conclusion would in effect be that of the American naturalist. But who does not see that, though right as a linguist, he would be wrong as a theologian, — wrong in fact? Reasoning on but the common and the natural, he would have failed to take into account, in his calculation, one main element, — the element of miracle, as manifested in the confusion of tongues at Babel; and his ultimate finding would, in consequence, be wholly erroneous. Now, it is perhaps equally possible for the naturalist to hold that two such extreme varieties of the human family as the negro and the Goth could not have originated from common parents in the course of a few centuries; and certainly the negro does appear in history not many centuries after the flood. He had assumed his deep black hue six hundred years before the Christian era, when Jeremiah used his well-known illustration, "Can the Ethiopian," etc.; and the negro head and features appear among the sculptures and paintings of Egypt several centuries earlier. Nay, negro skulls of a very high antiquity have been found among the mummies of the same ancient kingdom. But though, with distinguished naturalists on the other side, we would not venture authoritatively to determine that a variety so extreme could have originated in the ordinary course of nature in so brief a period, just as we would hesitate to determine that a new language could originate naturally in other than a very extended term, we would found little indeed upon such a circumstance, in the face of a general tradition that the negroid form and physiognomy were marks set upon an offending family, and were scarce less the results of miracle than the confusion of tongues. We are far from sure, however, that it is necessary to have recourse to miracle. The Goth is widely removed from the negro; but there are intermediate types of man that stand in such a midway relation to both, that each variety, taking these as the central type, is divested of half its extremeness. Did such of our Edinburgh readers as visited the Exhibition of this season mark with what scholar-like exactness and artistic beauty the late Sir William Allan restored, in his last great picture ("The Cup found in Benjamin's Sack"), the original Egyptian form, as exhibited in the messengers of Joseph? Had the first men, Adam and Noah, been of that mingled negroid and Caucasian type, — and who shall say that they were not? — neither the Goth nor the negro would be so extreme a variety of the species as to be beyond the power of natural causes to produce.

We had purposed referring at some length to that portion of the argument which is made to rest on analogy. We have, however, more than exhausted our space, and merely remark that it is not at all a settled point that the analogies are in favor of creation in a plurality of centres. Linnæus and his followers in the past, and men such as Edward Forbes in the present, assert exactly the contrary; and, though the question is doubtless an obscure and difficult one, — so much so that he who takes up either side, and incurs the *onus probandi* of what he asserts, will find he has but a doubtful case, — the doubt and obscurity lie quite as much on the one side as the other. Even, however, were the analogies with regard to vegetables and the lower animals in favor of creation in various centres, it would utterly fail to affect the argument. Though the dormouse and the Scotch fir had been created in fifty places at once, the fact would not yield us the slightest foundation for inferring that man had originated in more than a single centre. Ultimately, controversies of this character will not fail to be productive of good. They will leave the truth more firmly established, because more thoroughly tried, and the churches more learned. Nay, should such a controversy as the present at length convince the churches that those physical and natural sciences which, during the present century, have been changing the very face of the world and the entire region of human thought, must be sedulously studied by them, and that they can no more remain ignorant without sin than a shepherd can remain unharmed in a country infested by beasts of prey without breach of trust, it will be productive of much greater good than harm.

VI.

NORWAY AND ITS GLACIERS.[1]

There is a striking resemblance in form and aspect between the Scandinavian races of Denmark, Sweden, and Norway, and the people of the northeastern coasts of Scotland. The resemblance, however, is not restricted to the races, — it extends also to the countries which they inhabit. The general features of Denmark and Sweden are very much those of the southern districts of our own country, — mayhap rather tamer on the whole, from a less ample development of the trap-rocks. And in Norway we have, if we except a small portion of its southern extremity, simply a huge repitition of the Western Highlands of Scotland; it is a Highlands roughened by greater hills, and intersected by deeper and more extensive lochs, and prolonged far beyond the Arctic circle. In, however, their physical conditions, both Norway and the Highlands are wonderfully alike ; but with this interesting difference, that some of the great agents which modified, in the remote past, the form of the rougher portions of our country, and regarding which we can only speculate and theorize, are still in active operation in Norway. The loftier Norwegian mountains rise to nearly twice the height of Ben Macdhui and Ben Nevis; the country, too, stretches about twelve degrees further to the north than Cape Wrath, and runs more than three hundred miles within the Arctic circle. And so it has its permanent

[1] Norway and its Glaciers visited in 1851, etc. By James D. Forbes, D.C.L., F.R.S., Sec. R.S., Ed., etc. etc., and Professor of Natural Philosophy in the University of Edinburgh.

snow-fields and its great glaciers, that are in the present day casting up their moraines, lateral and transverse, and grooving and rounding the rocks beneath, just as our own country had them in some remote and dateless age, ere, mayhap, the introduction of man upon our planet. There are other respects in which it is representative rather of the past than of the present of Scotland. It still retains its original forests, and presents, over wide areas, an appearance similar to that which was presented by the more mountainous parts of our own country ere the formation of our great peat-mosses. The range of the Grampians, when first seen by Agricola, must have very much resembled in its woody covering the southern Highlands of Norway at the present day. Professor Forbes, on nearing the Norwegian coast, was struck, on first catching sight of the land, by the striking resemblance which it bore to some of the gneiss tracts of the mainland of Scotland and the Hebrides. The gneiss islands of Tyree and Coll first occurred to his mind; and "doubtless," he says, "the same causes have produced this similarity of character, acting in like circumstances. Both belong to that great gneiss formation so prevalent in Norway, and also in Scotland, with which few rocks can compare in their resistance to atmospheric action and mechanical force. In both cases they have been subjected for ages to the action of the most tremendous seas which wash any part of Europe; and they have probably been abraded by mechanical forces of another kind, which have given the rounded outlines to even their higher hills." As, however, the Professor approached the shore, he became sensible of a grand distinction between the mountain scenery of Norway and the Scotch Hebrides. It was the Scotland of eighteen hundred years ago on which he was looking. "On closer observation," he says, "I perceived that the low, rounded, and rocky hills which I had at first believed to be bare were almost everywhere covered, or at least dotted over, with woods of pine, which, descending almost to the shore, gave a peculiarity of character to the scenery, at the same time that it afforded a scale by which to estimate its magnitude." The low hills which had at first rather disappointed him were now, he found, a full thousand feet in height.

There are several respects in which Norway may be regarded as a country still in its *green* youth. These primeval forests are of themselves demonstrative of the fact. Humboldt well remarks, that "an early civilization of the human race sets bounds to the increase of forests;" for "nations," he says, "in their change-loving spirit, gradually destroy the decorations which rejoice our eye in the north, and which, more than the records of history, attest the youthfulness of our civilization." There are other evidences that at least the northern portions of both Norway and Sweden were unappropriated by man during the earlier ages of British and Continental history. It is a curious fact, adverted to by Mr. Robert Chambers in his "Tracings of the North of Europe," that in the great Mùseum of Antiquities at Copenhagen, the relics of the stone period have been furnished by only Denmark and the southern provinces of Sweden and Norway. They are not to be found in the far provinces of the north; and the only district beyond the Baltic in which they occur in the ordinary proportions of the south and middle portions of Europe, is the low-lying, comparatively temperate province of Scania. It is doubtless an advantage, in some respects, at least for a wild and mountainous country to be still in its youth. Large tracts of the more ancient Scottish Highlands lie sunk in the hopeless sterility of old age. In many of their so-called forests, that are forests without a living tree, — such as the Moin in Sutherlandshire, or that tract of desert waste which spreads out around Kingshouse in Argyleshire, — the traveller sees, in the sections opened by the winter torrents, two periods of death represented, with a comparatively brief period of life intervening between. There is first, reckoning from the rock upward, a stratum of gray angular gravel, formed of the barren primary rocks, and identical with the angular gravels still in the course of forming under the attrition of the glaciers of Norway and the Alps. And it speaks of the ice-period of death, when the country had its permanent snow-fields and its great glaciers. Next in order immediately over the dead gravel, there occurs usually a thin stratum of mossy soil, bearing its tier of buried stumps — the representatives of an age of vegetable life when the Highlands were what Norway is now, — a scene of wide-spreading forests. And then over all, to the depth often of six or eight feet, we find, as representative of a second and permanent period of death, a cold, spongy, ungenial peat-moss, in which nothing of value to man finds root, save, perhaps, a few scattered spikes of deer-grass, that, springing early, furnish the flocks of the shepherd with a week or two's provision, just as the summer begins. But for every agricultural purpose these mossy wastes are in their effete and sterile old age, and the yearly famines show how the poor settlers upon them fare. Man failed to appropriate them during their cheerful season of youth and life; and over wide tracts they are dead — past resuscitation now. In Norway, with all its bleakness, the chances in favor of the people are better. The Norwegians have escaped the curse of clanship; and the country, still in the vigor of youth, is parcelled out among many proprietors, who till the lands which they inherit. Even in its wild animals, Norway is a larger Scotland, post-dated some ten or fifteen centuries. It has the identical beaver, bear, and wolf still living in its forests, whose remains are occasionally found in our mosses and marl-pits.

In another respect, however, Norway resembles our country at a greatly earlier time than that of the primeval forests. Its long line of western coast, with its many islands and long-withdrawing fiords, presents everywhere the appearance of a land not yet fairly arisen out of the sea. The islands are simply the tops of great mountains, that at once sink sheer into deep water; and the fiords, great glens, like Glen Nevis and Glencoe, that have not yet raised themselves out of the sea. One may voyage for many miles along this bold coast without finding a bit of shore on which to land; and such must have been very much the appearance of our Western Highlands in the old ice-ages, when the sea stood from five hundred to a thousand feet higher along our steep hillsides than it does now, or rather the land sat from five hundred to a thousand feet lower. Both Professor Forbes and Mr. Chambers refer to the great freshness of the raised terraces which stretch at various heights along the coast, as if to show where the surf had beat during prolonged intervals in the course of upheaval; and the latter gentleman seems to have been particularly struck by the freshness of the sea-shells that occur at great heights, and by their identity with those which now live in the neighboring seas. Professor Keilhau showed Mr. Chambers serpulæ on a rock-face, scarce a mile from the busy city of Christiania, still firmly adhering to the spot on which the creatures that inhabited them had lived and died. And yet that rock is now one hundred and eighty-six feet over the level of the sea. The great abundance and freshness of the shells found on some of the raised beaches of the country is of itself an object of wonder. "Uddwalla," says Mr. Chambers, in his "Tracings," "is a name of no small interest in science, because of a great bed of ancient shells found near it. The effect was novel and startling, when, on the hill-face o'erlooking the fiord, and at the height of two hundred feet above its waters, I found something like a group of gravel-pits, but containing, instead of gravel, nothing but shells! It is a nook among the hills, with a surface which had originally been flat in the line of the fiord, though sloping forward toward it. We can see that the whole space is filled to a great depth with the exuviæ of marine molluscs, cockles, mussels, whelks, etc., — all of them species existing at this time in the Baltic, — with only a thin covering of vegetable mould on the surface. I feel sure that some of these excavations are twenty feet deep; yet that is not the whole thickness of the shell-bed." In the fact of the identity of these shells with those that still live in the neighboring sea, we have an evidence of the comparative recentness of the upheaval of the land. In our own country, it is only those shells that lie embedded in the terrace which underlies the old coast-line that are identical, in at least the group, with the existing ones of the littoral and laminarian zones beyond. The higher-lying shells, not yet extinct, which occur in Britain at various heights, from fifty to fourteen hundred feet over the present sea-line, are, as a group, sub-arctic, and belong to the ice-age.

In one important circumstance, however, Norway and our own country must have had an exactly similar history. In both, the climate has been greatly more mild since at least the historic ages began than it was in an earlier time. When Scotland had its glaciers and snow wastes, Norway seems to have been enveloped in ice; whereas its climate is now one of the finest in the world for the same lines of latitude. That great gulf-stream which casts so liberally on the northern shores of Europe the tepid water of the tropics, is no doubt one of the main causes of this superiority in the climate of both Norway and our own country over all other countries in the same parallels. "It has been calculated," says Professor Forbes, "that the heat thrown into the Atlantic Ocean by the gulf-stream in a winter's day would suffice to raise the temperature of the part of the atmosphere which rests upon France and Great Britain from the freezing point to summer heat." And such are the effects on the distant coast of Norway, that, under the Arctic circle, or at least the sea-coast, the mercury rarely sinks beneath zero. The absence of the great gulf-stream would of course leave both countries to the climatal conditions proper to their position; it would insure to Scotland the severe and wintry climate of Labrador, and to Norway the still severer climate of northern Greenland.

Nor, as has been shown of late by Professor Hopkins, would it require a very considerable depression of the central parts of North America to rob northern Europe of the signal advantages of the gulf-stream. A greatly less considerable sinking of what is now the vast valley of the Mississippi, and of the lake district beyond, than that of which we have the evidence in our own country, would divert its waters into Hudson's Bay and the arctic seas beyond; and both Great Britain and Norway would be left to the severe climatal conditions of their latitudinal position on the globe. Nor is it in the least improbable that such, during the glacial ages, was the actual state of things. North America, as certainly as our own country, gives evidence of extensive submergence during the period of the existing plants and shells.

We must add, that Professor Forbes's volume is remarkably well written, and not less rich in the picturesque and the poetic than in the severely scientific. There has been a mighty improvement in this respect in what may be termed the pure literature of science during the last century; and at the present time some of the severest thinkers of the age take their place also among its best writers. Humboldt, the late Arago, Sir David Brewster, and Sir John Herschel, far excel, in the purely artistic department of authorship, most of our mere *litterateurs.* We have exhausted our space; but, referring our reader to Professor Forbes's interesting volume for his more scientific facts and observations, we must be permitted to show by the following extract how graphically he describes: —

"We are at the head of the Naræedal, one of those singular clefts common in Norway, bounded on either side by cliffs usually perpendicular, to a height of perhaps fifteen hundred or even two thousand feet; the bottom flat and alluvial, and terminating abruptly at the head of a steep but not precipitous slope. Down the slope the road is conducted by a series of zigzags, or rather coils, in a masterly manner, through a vertical height of eight hundred feet, — a very striking waterfall rushing down on either hand, and rendering the view in the opposite direction wonderfully grand. It is generally agreed that no more genuine specimen exists of Norwegian scenery than the Naræedal. From the foot of the descent to Gudvangen, on the banks of the Naræ-fiord, the road is nearly level, the whole descent on several miles being little more than three hundred feet. The mountains, however, preserve all their absolute elevation on either side, so that the ravine, though not quite so narrow, is deeper. The masses of rock on the right rise to five thousand or six thousand feet, and a thread of water forming the Keel-foss descends a precipice estimated at two thousand feet. The arrival at Gudvangen takes one by surprise. The walls of the ravine are uninterrupted; only the alluvial flat gives place to the unruffled and nearly fresh waters of this arm of the sea, which reaches the door of the inn. After dining, and procuring a boat and three excellent rowers, we proceeded to the navigation of the extensive Sogne-fiord, of which the Naræ-fiord, on which we now were, is one of the many intricate ramifications. The weather, which had fortunately cleared up for a time, was now again menacing, and a slight rain had set in when we embarked. The clouds continued to descend, and settled at length on the summits of the unscalable precipice which for many miles bound this most desolate and even terrific scene. I do not know what accidental circumstances may have contributed to the impression, but I have seldom felt the sense of solitude and isolation so overwhelming. My companion had fallen into a deep sleep; the air was still damp and calm; the oars plashed with a slow measure into the deep, blank, fathomless abyss of water below, which was bounded on either side by absolute walls of rock, without, in general, the smallest slope of debris at the foot, or space enough anywhere for a goat to stand, and whose tops, high as they indeed are, seemed higher by being lost in clouds, which formed, as it were, a level roof over us, corresponding to the watery floor beneath. Thus shut in above, below, and on either hand, we rowed on amidst the increasing gloom and thickening rain, till it was a relief when we entered on the wider though still gloomy Aurlands-fiord."

VII.

THE AMENITIES OF LITERATURE.

The love of literature amounts, with those who entertain it most strongly, to an engrossing passion; and there are few men of cultivated minds, however much engaged with other pursuits, who do not derive from it a sensible pleasure. Even when politics ran highest, and first-class periodicals, such as the "Edinburgh Review" and the "Quarterly," were toiling in the front of their respective parties, none but the most zealous partisans could deem their literary articles second in interest to their political ones; and to the great bulk of their readers, however sincere as Whigs or hearty as Tories, the literary ones always took the first place. They were read with avidity immediately on the delivery of the numbers which contained them, while the more serious disquisitions had to wait. Literature, in fine, was the sweetened pabulum in which the political principle of these works was conveyed to the public; and had the pabulum been less palatable in itself, or less generally suited to the public taste, the medicine would have failed to take. It has the advantage, too, of being so general a pebulum, that men of all parties and professions, if of equal acquirement and cultivation, take an equal interest in it. It is the most catholic of predilections, and neutralizes, more than any other, the bias of caste, church, and party. The Protestant forgets, in his admiration of their writings, that Pope and Dryden were Papists; the High Churchman luxuriates over Milton; old Samuel Johnson is admired by the Liberal and the Scot; and the Tory forgets that Addison was a Whig. In this, as in other respects, a love of literature is one of the humanizing principles, and in ages of controversy and contention its tendencies are towards union. It gives to men who differ in other matters a common ground on which they can meet and agree, and has led to many friendships and acts of forbearance and good-will between men who, had they been devoid of it, would have been bitter antagonists and personal enemies. There have been mutual respect and admiration from this cause between partisans on the opposite sides of very important questions. Swift and Addison still called each other friend, at a time when the point at issue between their respective septs was virtually the Protestant Succession; and Scott and Jeffrey were on fair terms when Whigs and Tories were engaged in a death-grapple, with the Reform Bill looming in the distance. Doubtless one of the causes of the often-remarked circumstance that while fifth and sixth rate partisans are almost always bitter in the feelings with which they regard their opponents, and ungenerous towards them in their resentments, the leaders of parties are comparatively tolerant and humane, may be traced in part to a community of tastes and sentiments in this important department, and in part to that superior tone of thought and feeling which it is one of the great functions of literature to foster and develop. Many of our readers must have had opportunity of remarking how pleasant it is, after one has been shut up for months, mayhap, in some country solitude, or engaged in some over-busy scene, without intelligent companionship, to meet with an accomplished, well-read man, with whom to beat over all the literary topics, and settle the merits of the various schools and authors. It is not less pleasant to turn to one's books after some period of close-engrossing engagement, and to clear off, among the masters of thought and language, all trace of the homely cares and narrow thinking which the season of hard labor had imperatively demanded. And it is so with peoples as certainly as with individuals. During the war so happily terminated, the nation was too busy and too much engrossed to listen patiently to disquisitions, however ingenious, on literature and the belleslettres. Leaders and articles on the state of the army and the prospect of the campaign, or the narratives and descriptions of "correspondents" in the Crimea, formed the staple reading of the time; and some of our most respectable booksellers could tell very feelingly, on data furnished by their balance-sheets, how little, in comparison, was the interest that attended reading of any other kind. The roar of war drowned the voice of the muses. Now, however, the country has got a breathing time; its period of all-engrossing occupation is over for the present; and works of general literature will once more form the staple reading of its more cultivated intellects. Good books will begin to sell better, when, at least, the publishing season commences, than they have done for the last two years; and by their measure of success they will certify respecting the tastes and leisure-hour occupations of that great and influential portion of the people which constitutes the reading public. And we recognize in a work now before us — "Essays, Biographical and Critical, chiefly on English Poets," by Professor David Masson,[1] which has just issued from the Cambridge press — one of the class of books which, in the circumstances of the time, this portion of the public will delight to read, and be the better and happier for reading.

Professor Masson is a high representative of a class of literary men peculiar to the age, — men who a century ago would have stood prominently forward in the ranks of authorship as the writers of elaborate volumes, but who, in the altered circumstances of a more hurried age than any of those which preceded it, are engaged mainly in providing the reading public with its daily bread, and, for the sake of present influence and usefulness, are content in some degree, so far as they themselves are concerned, to subordinate the future to the passing time. Almost all the

1 Essays, Biographical and Critical, chiefly on English Poets. By David Masson, A.M., Professor of English Literature in University College, London.

writing produced in our first-class newspapers, however distinguished for ability or influential in directing opinion, passes away with the day, or at least with the week, in which it has been produced. Like those ephemeridæ which, born in the morning, deposit their eggs and die before night, it makes its nidus in the public mind, and then drops and disappears. Contributions, however, to the higher quarterlies and first-class magazines have a better chance of life; and we have already a class of works drawn from these sources which bid as fair to live as almost any of the more elaborate authorship of the age. Such are the collected critiques of Jeffrey and Sydney Smith, the philosophic papers of Macintosh, the brilliant essays of Macaulay, and the soberer contributions of Henry Rogers. And to this class the Essays of Professor Masson belong; nor are they unworthy of being ranked among the very foremost of their class. There are essays in this volume which, for the minute knowledge of English literature which they display, and their nice appreciation of the distinctive and characteristic in our higher writers, we would place side by side with the *chef d'œuvres* of Jeffrey. Though consisting chiefly of contributions to the quarterlies, written at various times, and published in different periodicals, the pieces which compose the work have been so arranged that they form, with but few gaps, — which are more than compensated for by at least as many happy episodes, — a history of English literature from the early days of Milton down to those of Wordsworth. Nor are there backward glances wanting, which bring before the reader the primeval English literature of the times of Chaucer and Spencer. There are just two blanks in the work, which we could wish to see filled in some future edition, — a blank representative of that period which intervened between the times of Swift and of Chatterton, during which old Samuel Johnson gave law to the world of letters, and was well-nigh all that Dryden had been for the decade that preceded and the decade which succeeded the Revolution; and a second, though lesser blank, representative of the times during which Burns and Cowper flourished, and in which the school of Pope gave place to a more national, natural, and less elaborate school. Among what may be termed the episodes of the work, we would specially instance a dissertation on what we may term the boundary limits of prose and poetry, which we deem by far the ablest and most satisfactory which we have yet seen on the subject. Much has been written on what may be termed the conterminous limits of the two provinces; and the *suits* have been many that have originated in an erroneous drawing of the line. As in the famous case between Dandy Dinmont and Jack Dawson of the Cleugh, one party affirms that "the march rins on the tap o' the hill, where the wind and water sheers;" while another "contravenes that, and says that it hands down by the auld drove road; and that makes an unco difference;" — some critics so draw the line, that, like Bowles in his controversy with Campbell, they almost wholly exclude poets such as Pope and Dryden from their own proper domains; while others affirm that there exists no line between the two domains at all, but that whatever in thought or feeling finds expression in verse, may with equal propriety be expressed in prose. Byron's terse couplet on Wordsworth, whom it describes as a writer

> "Who, both by precept and example, shows
> That prose is verse, and verse is only prose,"

has, though in a somewhat exaggerated form, made this special view better known than even the men who assert it. Certainly there are broad grounds common to both prose and verse; and such is the groundwork of truth in Byron's satirical couplet, though in a widely different sense from that which the satirist himself intended, that there is not much in even the highest flights of the poetry of Wordsworth to which prose might not attain. We know not, for instance, a single passage in his greatest poem, "The Excursion," that might not find adequate expression, not only in the magnificent prose of Milton, or Raleigh, or Jeremy Taylor, but, so far at least as the necessary expression is required, in even that of Dryden or of Cowley. The same may be said of the poetry of Scott. The flights in "Marmion" or the "Lady of the Lake" rise no higher than those in Waverley or Ivanhoe. And yet, as Professor Masson well shows, there certainly is verse under whose burden the highest prose would utterly sink. We have remarked, in travelling through the Highlands of Scotland, that almost all the first-class hills of the country take the character of hills of the average size, with other hills placed, as if by accident, on the top of them; and there is a very lofty poetry that attains to its greatest elevation on a similar principle. The imagination, in the plenitude of its power, is ever piling, like the giants of old, mountain on the top of mountain. Let us draw our illustration from Milton. After comparing the arch-fiend, as he "lay floating many a rood" on the burning lake, to the old Titanian monster that warred on Jupiter, the poet rushes into another and richer comparison: he compares him to

> "That sea-beast
> Leviathan, which God of all his works
> Created hugest that swim the ocean stream."

And here, on the ground common to prose and verse, the comparison should stop. But the imagination of the great poet has been aroused; the glimpse of the huge sea-beast so fascinates him that he must look again; and a picture is the consequence, invested with circumstances of poetic interest, and finished with a degree of elaboration far beyond the necessities of the comparison: —

> "Him haply slumbering on the Norway foam,
> The pilot of some small night-foundered skiff'
> Deeming some island, oft, as seamen tell,
> With fixed anchor in his scaly rind,
> Moors by his side, under the lee, while night
> Invests the sea, and wished morn delays."

What a pile of imagery! Mountain cast on the top of mountain, — a feat for the greatest of the giants, and far beyond the reach of the most poetic prose-man, or the capabilities of prose itself. Our other example, though of a more homely character, will be found scarce less illustrative of this piled-up style, peculiar to the higher poesy. Burns, in his decidedly anti-teetotal "Earnest Cry and Prayer," after adverting to the deteriorating effects of the wines of southern Europe on the nerves and framework of the Continental soldiery, describes a Scottish soldier animated for the contest by the inspiration of usquebagh: —

"But bring a Scotsman frae his hill,
Clap in his cheek a Highland gill,
Say, Such is royal George's will,
An' there's the foe:
He has nae thought but how to kill
Twa at a blow."

Now, here is a vigorous stanza, — terse, clear, epigrammatic, and charged with thought equally fitted to do service either as prose or verse. But the poet catches a glance of the Highland soldier, the poetic blood gets up, and it becomes impossible, for the time, to arrest in his career either soldier or poet: —

"Nae cauld faint-hearted doubtings tease him;
Death comes; wi' fearless eye he sees him;
Wi' bloody han' a welcome gi'es him;
An' when he fa's,
His latest draught o' breathin' lea's him
In faint huzzas."

Here, again, we find the hill piled on the top of the hill after a different manner, but as decidedly as in Milton, and alike beyond the necessities or the reach of prose. This peculiar region of poetry seems to have formed a sort of inextricable wilderness to the more prosaic class of critics. Lord Kames, though a coarse, was an eminently sensible man; and his "Elements of Criticism" is a work that contains many striking things. What, however, the French critic termed "comparisons with a tail," seem fairly to have puzzled him. He could no more understand why similes should have caudal appendages, than his brother Judge, Lord Monboddo, could understand why men should want them. And so he instances as a mere "phantom simile, that ought to have no quarter given it," the very exquisite one which Coriolanus employs in describing Valeria, —

"The noble sister of Pophlicola,
The moon of Rome; chaste as the icicle
That's curdled by the frost from purest snow,
And hangs on Dian's temple."

The shrewd magistrate, who, to employ the delicate periphrase of Hector in the "Antiquary," used to address his learned compeers on the bench by the name ordinarily used to designate "a female dog," could not understand why the temple of Dian should be introduced into this comparison, or what right the icicle had in it at all; and so he ruled that it was palpably illegal for Shakspeare to write what he, a judge and a critic, could not intelligently read. The conclusion of Professor Masson on the respective provinces of poesy and prose is worthy of being carefully pondered by the reader: —

"In the whole vast field of the speculative and the didactic," says the Professor, — "a field in which the soul of man may win triumphs nowise inferior, let illiterate poetasters babble as they will, to those of the mightiest sons of song, — prose is the legitimate monarch, receiving verse but as a visitor and guest, who will carry back bits of rich ore, and other specimens of the land's produce; in the great business of record also, prose is preëminent, verse but voluntarily assisting; in the expression of passion, and the work of moral stimulation, verse and prose meet as coequals, prose undertaking the rougher and harder duty, where passion intermingles with the storm of current doctrine, and with the play and conflict of social interests, — sometimes, when thus engaged, bursting into such strains of irregular music that verse takes up the echo and prolongs it in measured modulation, leaving prose rapt and listening to hear itself outdone ; and, lastly, in the noble realm of poetry or imagination, prose also is capable of all exquisite, beautiful, and magnificent effects, but that by reason of a greater ease with fancies when they come in crowds, and of a greater range and arbitrariness of combination, verse here moves with the more royal gait. And thus prose and verse *are presented as two circles or spheres, not entirely separate, as some would make them, but intersecting and interpenetrating through a large portion of both their bulks, and disconnected only in two crescents outstanding at the right and left, or, if you adjust them differently, at the upper and lower extremities.* The left or lower crescent, the peculiar and sole region of prose, is where we labor amid the sheer didactic, or the didactic combined with the practical and the stern. The right or upper crescent, the peculiar and sole region of verse, is where *pathēsis*, at its utmost thrill and ecstasy, interblends with the highest and most darting *poiēsis*."

This is vigorous thinking and writing; and the Professor's volume contains many such passages. We would in especial instance the Essays on the "Literature of the Restoration," on "Wordsworth," and on "Scottish Influence on British Literature." But the longest and finest composition of the work — a gem in literary biography — is its "Chatterton, a Story of the Year 1770." There is perhaps no name in British poetry, of the same frequency of occurrence, that is so purely a name, as that of

> "The marvellous boy, —
> The sleepless soul that perished in his pride."

Such of his poems as were written in modern English, and in his own proper name and character, are not pleasing, and, sooth to say, not more than clever; while his poems written in the character of Rowley are locked up in what is virtually a dead tongue, considerably different from that of Chaucer or the "King's Quair," or, in short, from any other tongue ever written by any other poet. And as there is but little temptation to master a language, and that, too, a language which never was spoken, for the sake of a few poems, however meritorious, most men are content to take the fame of the Rowley writings on trust, or at least to determine by brief specimens that they are in reality the wonderful compositions which the critics of the last age pronounced them to be. And so Chatterton is now very much a bright name associated with a dark story. Further, of the story, little more survived in the public mind than would have furnished materials for an ordinary newspaper paragraph. Chatterton had not been very fortunate in his biographers; and it was but known, in consequence, that, living in an age not unfamiliar with literary forgery, — it is unnecessary to give instances within sight of the great Highland mountains, — he had fabricated a volume of old English poems greatly superior to any old English poems ever written, with the single exception of those of Chaucer; that, quitting his native place, where he had succeeded in earning not more than the modicum of honor which prophets ordinarily achieve for themselves when at home, he had gone to force his upward way among the wits of London; and that there, in utter destitution and neglect, he had miserably destroyed himself. Such was all that was generally known of Chatterton, even by men of reading. Professor Masson's singularly interesting and powerful biography fills up this sad outline as it was never filled up before; and shows how deep a tragedy that of the poor boy was, who, after achieving immortality, "perished in his pride," at about the age when lads who purpose pursuing the more laborious mechanical professions are preparing to enter on their apprenticeships. Further, without aught approaching to formal apology for the offences and shortcomings of the hapless lad, it shows us what a mere boy he was, in all except genius, at the time of his death. Sir Walter, in referring, in his "Demonology," to the young rascals on whose extraordinary evidence so many old women were burnt as witches in Sweden, has some very striking, and, we think, very just remarks, on the obtuseness of the moral sense in most children, especially boys.

"The melancholy truth, that the 'human heart is deceitful above all things, and desperately wicked,' is by nothing proved so strongly," we find him saying, "as by the imperfect sense displayed by children of the sanctity of moral truth. Both gentlemen and the mass of the people, as they advance in years, learn to despise and avoid falsehood, — the former out of pride, and from a remaining feeling, derived from the days of chivalry, that the character of a liar is a deadly stain on their honor; the other, from some general reflection upon the necessity of preserving a character for integrity in the course of life, and a sense of the truth of the common adage that 'honesty is the best policy.' But these are acquired habits of thinking. The child has no natural love of truth, as is experienced by all who have the least acquaintance with early youth. If they are charged with a fault while they can hardly speak, the first words they stammer forth are a falsehood to excuse it. Nor is this all. The temptation of attracting attention, the pleasure of enjoying importance, the desire to escape from an unpleasing task, or accomplish a holiday, will at any time overcome the sentiment of truth, — so weak is it within them." A sad picture, but, we fear, a true one; and in reading the tragic story of Chatterton, we were oftener than once reminded of it. We see in almost every stage of his progress the unripe boy, — precocious in intellect, and in that only. But with the following powerful passage, taken from the closing scene in the sad drama, we must conclude, — meanwhile recommending Professor Masson's work to our readers as one of singular interest and ability: —

"'He called on me,' is Mr. Cross's statement, 'about half-past eleven in the morning. As usual, he talked about various matters; and at last, probably just as he was going away, he said he wanted some arsenic for an experiment.' Mr. Cross, — Mr. Cross, — before you go to your drawer for the arsenic, look at that boy's face! Look at it steadily; look till he quails; and then leap upon him and hold him! Mr. Cross does not look. He *sells* the arsenic (yes, *sells*, for somehow during that walk, in which he has disposed of the bundle [of manuscripts], he has procured the necessary pence), and lives to repent it. Chatterton, the arsenic in his pocket, does not return to his lodging immediately, but walks about, God only knows where, through the vast town. 'He returned,' continued Mrs. Angell, 'about seven in the evening, looking very pale and dejected, and would not eat anything, but sat moping by the fire with his chin on his knees, and muttering rhymes in some old language to her. After some hours he got up to go to bed, and he then kissed her, — a thing he had never done before.' Mrs. Angell, what can that kiss mean? Detain the boy; he is mad; he is not fit to be left alone; arouse the whole street rather than let him go. She does let him go, and lives to repent it. 'He went up stairs,' she says, 'stamping on every stair as he went slowly up, as if he would break it.' She hears him reach his room. He enters, and locks the door behind him.

"The devil was abroad that night in the sleeping city. Down narrow and squalid courts his presence was felt, where savage men clutched miserable women by the throat, and the neighborhood was roused by yells of murder, and the barking of dogs, and the shrieks of children. Up in wretched garrets his presence was felt, where solitary mothers gazed on their infants, and longed to kill them. He was in the niches of dark bridges, where outcasts lay huddled together, and some of them stood up from time to time, and looked over at the dim stream below. He was in the uneasy hearts of undiscovered forgers, and of ruined men plotting mischief. He was in prison cells, where condemned criminals condoled with each other in obscene songs and blasphemy. What he achieved that night in and about the vast city came duly out into light and history. But of all the spots over which the Black Shadow hung, the chief, for that night at least, was a certain undistinguished house in the narrow street, which thousands who now dwell in London pass and repass, scarce observing it, every day of their lives, as they go and come along the thoroughfare of Holborn. At the door of one house in that quiet street the Horrid Shape watched; through that door he passed in, towards midnight; and from that door, having done his work, he emerged before it was morning.

"On the morrow, Saturday the 25th August, Mrs. Angell noticed that her lodger did not come down at the time expected. As he had lain longer than usual, however, on the day before, she was not alarmed. But about eleven o'clock, her husband being then out, and Mrs. Wolfe having come in, she began to fear that something might be the matter; and she and Mrs. Wolfe went up stairs and knocked at the door. They listened awhile, but there was no answer. They then tried to open the door, but found it locked. Being then thoroughly alarmed, one of them ran down stairs, and called a man who chanced to be passing in the street to come and break the door open. The man did so; and on entering they found the floor littered with small pieces of paper, and Chatterton lying on the bed, with his legs hanging over, quite dead. The bed had not been lain in. The man took up some of the pieces of paper; and on one of them he read, in the deceased's own handwriting, the words, 'I leave my soul to its Maker, my body to my mother and sisters, and my curse to Bristol. If Mr. Ca——': the rest was torn off. 'The man then said,' relates Mrs. Angell, 'that he must have killed himself; which we did not think till then. Mrs. Wolfe ran immediately for Mr. Cross, who came, and was the first to point out a bottle on the window containing arsenic and water. Some of the bits of arsenic were between his teeth, so that there was no doubt that he had poisoned himself.'"

VIII.

A STRANGE STORY, BUT TRUE.[1]

It is now nearly forty years since an operative mason, somewhat dissipated in his habits, and a little boy, his son, who had completed his twelfth year only a few weeks previous, were engaged in repairing a tall, ancient domicile, in one of the humbler streets of Plymouth. The mason was employed in re-laying some of the roofing; the little boy, who acted as his laborer, was busied in carrying up slates and lime along a long ladder. The afternoon was slowly wearing through, and the sun hastening to its setting; in little more than half an hour, both father and son would have been set free from their labors for the evening, when the boy, in what promised to be one of his concluding journeys roofwards for the day, missed footing just as he was stepping on the eaves, and was precipitated on a stone pavement thirty-five feet below. Light and slim, he fared better than an adult would have done in the circumstances; but he was deprived of all sense and recollection by the fearful shock; and, save that he saw for a moment the gathering crowd, and found himself carried homewards in the arms of his father, a fortnight elapsed ere he awoke to consciousness. When he came to himself, in his father's house, it was his first impression that he had outslept his proper time for rising. It was broad daylight; and there were familiar forms round his bed. He next,

1 Memoirs of Dr. John Kitto, D.D., F.S.A., Editor of the "Pictorial Bible" and the "Cyclopædia of Biblical Literature," Author of "Daily Bible Illustrations," etc. Compiled chiefly from his Letters and Journals. By J. E. Ryland, M.A. With a Critical Estimate of Dr. Kitto's Life and Writings. By Professor Eadie, D.D., LL.D., Glasgow.

however, found himself grown so weak that he could scarce move his head on the pillow; and was then struck by the profound silence that prevailed around him, — a silence which seemed all the more extraordinary from the circumstance that he could see the lips of his friends in motion, and ascertain from their gestures that they were addressing him. But the riddle was soon read. The boy, in his terrible fall, had broken no bone, nor had any of the vital organs received serious injury; but his sense of hearing was gone forever; and for the remainder of the half-century which was to be his allotted term on earth he was never to hear more. Knowledge at one entrance was shut out forever. As is common, too, in such circumstances, the organs of speech become affected. His voice assumed a hollow, sepulchral tone, and his enunciation became less and less distinct, until at length he could scarce be understood by even his most familiar friends. For almost all practical purposes he became dumb as well as deaf.

Unable, too, any longer to assist in the labors of his dissipated father, he had a sore struggle for existence, which terminated in his admission into the poor-house of the place as a pauper. And in the workhouse he was set to make list-shoes, under the superintendence of the beadle. He was a well-conditioned, docile, diligent little mute, and made on the average about a pair and a half of shoes per week, for which he received from the manager, in recognition of his well-doing, a premium of a weekly penny, — a very important sum to the poor little deaf pauper. Darker days were, however, yet in store for him; he was not a little teased and persecuted by the idle children in the workhouse, who made sport of his infirmity; his grandmother, to whom he was devotedly attached, and with whom he had lived previous to his accident, was taken from him by death; and, to sum up his unhappiness at this time, he was apprenticed by the workhouse to a Plymouth shoemaker, — a brutal and barbarous wretch, who treated him with the most ruthless indignity and cruelty, — threw shoes at his head, boxed him on the ears, slapped him on the face, and even struck him with the broad-faced hammer used in the trade. Such of our readers as are acquainted with Crabbe's powerful but revolting picture of Peter Grimes, the ruffian master who murdered his apprentices by his piecemeal cruelties, would scarce fail to find the original of the sketch in this disreputable wretch,— with this aggravation, too, in the actual as set off against the fictitious case, that the apprentices of Peter Grimes were not poor, helpless mutes, already rendered objects of commiseration to all well-regulated minds "through the visitation of God." And who could anticipate a different end for the sadly-injured and sorely-misused boy than that which overtook Peter's apprentices as they dropped in succession into the grave? Were it to be seen, however, that the deaf little fellow, apparently so shut out from the world, could record his sufferings at this time in very admirable English, the hope might arise that there was some other fate in store for one who had mind and energy enough to triumph over circumstances so unprecedentedly depressed and depressing. The following are extracts from a journal which he kept while under the brute master: —

"O misery, thou art to be my only portion! Father of mercy, forgive me if I wish I had never been born! Oh that I were dead, if death were an annihilation of being; but as it is not, teach me to endure life: to enjoy it I never can. Mine is indeed a severe and cruel master. . . . . Threw this morning a shoe in my face: I had made a wrong stitch. . . . . Struck again. . . . . Again. I could not bear it: a box on the ear, — a slap on the face. I did not weep in April [when his grandmother died], but I did at this unkind usage. I did all in my power to suppress my inclination to weep, till I was almost suffocated: tears of bitter anguish and futile indignation fell upon my work, and blinded my eyes. I sobbed convulsively. I was half mad with myself for suffering him to see how much I was affected. Fool that I was! Oh that I were again in the workhouse! . . . . . He threw his pipe in my face, which I had accidentally broken; it hit me on the temple, and narrowly missed my eye. . . . . I held the thread too short: instead of telling me to hold it longer, he struck me on the hand with the hammer (the iron part). Mother can bear witness that it is much swelled; — not to mention many more indignities I have received, — many, many more. Again, this morning, I have wept. What's the matter with my eyes!"

Alas, poor boy! And all this took place in proud England, — the land of liberty and of equal rights and laws! Flogging is not a punishment for men, but a very suitable one for brutes; and had the brute master in this case been tied up to the halberts and subjected to a round hundred, he would be a squeamish reformer indeed who could have objected to so just and appropriate a use of the lash.

Suddenly, however, this dire tyranny came to a close. A few excellent men connected with the management of the workhouse had been struck by the docility and intelligence of the young mute. One of them, Mr. Burnard, a gentleman who still survives, struck by his powers of thought and expression, had furnished him with themes on which to write. He had shown him attention and kindness, and the lad naturally turned to him as a friend and protector; and, stating his case to him by letter, the good man not only got him relieved from the dire thraldom of his tyrannical master, but, by interesting a few friends in his behalf, secured for him the leisure necessary to prosecute his studies, — for, even when his circumstances were most deplorable, the little deaf and dumb boy had been dreaming of making himself a name in letters, by producing books which even the learned would not despise, — and, by means of a liberal subscription, he was now enabled to go on reading and writing, with — wonderful change for him whose premium pence used to be all spent in the purchase of little volumes! — the whole books of a subscription library at his command. It is customary to laugh at the conceit and egotism of the young, as indicative of a mere weakness, which it is the part of after years of sober experience to dissipate or cure. There are cases, however, in which the apparent weakness is real strength,

— a moving power, without which, in very depressing circumstances, there would be no upward progress, for there would be no hope and no motive to exertion; and so the poor mute boy's estimate of himself, while yet an inmate of the workhouse, though it may provoke a smile, may be deemed not uninteresting, as in reality representative of an undercurrent in the character, destined to produce great results.

"*Dec. 5th*, 1821. — Yesterday I completed my sixteenth year; and I shall take this opportunity of describing, to the best of my ability, my person. I am four feet eight inches high; my hair is stiff and coarse, of a dark brown color, almost black; my head is very large, and, *I believe, has a tolerable good lining of brain within;* my eyes are brown and large, and are the least exceptional part of my person; my forehead is high, eyebrows bushy, nose large, mouth very big, teeth well enough, and limbs not ill-shaped. . . . . You have asked me why I have in many places used the expression, 'When I am old enough in other people's opinion.' The customs of this country have declared that man is not competent to his own direction until he has attained the age of twenty-one. Not so I. I never was a lad. From the time of my fall, deprived of many external sources of occupation, I have been accustomed to find sources of occupation within myself, — to think as I read, as I worked, or as I walked. While other lads were employed with trifles, I have thought, felt, and acted as a man. At ignominious treatment, at blows, I have suppressed my indignation and my tears till I have felt myself almost choked. I have, however, felt also *the superiority of genius, which would not allow ignorance to triumph.* I have walked hours on hours in the most lonesome lanes I could find, abstracted in melancholy musing; or, with a book in my hand, I have sat for hours under a hedge or tree. Sometimes, too, sheltered from observation by a rock, I have sat in contemplation by the riverside. At such times I have felt such a melancholy pleasure as I have not known *since I have been in the hospital.* O Nature! why didst thou create me with feelings such as these? Why didst thou give such a mind to one in my condition? Why, O Heavens! didst thou enclose my proud soul within such a casket? Yet, pardon my murmurs; I will try to be convinced that 'whatever is is right.' Kind Heaven, endue me with resignation to thy will, and contentment with whatever situation it is thy pleasure I should fill."

Such was the estimate formed of himself by the deaf workhouse boy, and such his mode of expressing it. Depressed as his circumstances might at this time seem, and little favorable, apparently, to the development of mind, they were yet not without their peculiar balance of advantage. Lads born deaf and dumb rarely master in after life the grammar of the language; for, though they acquire a knowledge of the words which express qualities and sentiments, or which represent things, they seem unable to attain to the right use of those important particles, adverbs, conjunctions, and prepositions, which, as the smaller stones in a wall serve to keep the larger ones in their places, give in speech or writing order and coherency to the others. But the deaf lad had not been born deaf: he had read and conversed, and even attempted composition, previous to his accident; so that his grandmother could boast of the self-taught boy, not without some shadow of truth, that her "Johnnie was the best scholar in all Plymouth." And now, writing having become his easiest and most ready mode of communication, the *speech* by which he communicated his ideas, he had attained to a facility in the use of the pen, and a command of English, far from common among even university-bred youths, his seniors by several years. He had acquired, too, the ability of looking at things very intently. It has been well said by the poet,

> "That oft when one sense is suppressed,
> It but retires into the rest."

And it would seem as if the hearing of this deaf lad had retreated into his eyes, which were ever after to exercise a double portion of the seeing function. All this, however, could not be at once understood by his friends. There seemed to be but few openings through which the poor deaf and dumb lad could be expected to make his way to independence, and what is termed respectability; and it was suggested that he should set himself to acquire the art of the common printer, and attach himself to a mission of the English Church,—still, we believe, stationed in Malta,—that sends forth from its press many useful little books, chiefly for distribution in the East.

Accordingly, in a comparatively short time the deaf lad did acquire the art of the common printer,—nay, more, he became skilful in setting the Arabic character; and, having a decided turn for acquiring languages, though unable to speak them, he promised, judging from his mechanical and linguistic abilities, to be a useful operative to the mission. *Unfortunately*, however,—for such was the estimate of the mission's conductors,—he was not content to be a mere operative: his instincts drew him strongly towards literature; and, ere quitting England for Malta, he had such a quarrel on this score with some very excellent men, that he threw up his situation, which, however, through the mediation of kind friends, he was again induced and enabled to resume. But at Malta, where the poor deaf lad suffered much from illness, and much from wounded affections,—for, shut out though he was from his fellows, he had yet had his affair of the heart, and the course of true love did not run smooth in his case,—the quarrel was again resumed, and he received a reprimand from the committee of the mission in England, which was virtually a dismissal. "The habits of his mind," said the committee, "were likely to disqualify him from that steady and persevering discharge of his duties which they considered as indispensably requisite." And to this harsh resolution the late excellent Mr. Bickersteth, by whom it was forwarded, added the following remark:—"You are aware our first principles as Christians are the sacrifice of self-will and self-gratification. If you can rise to this, and steadily pursue your work, as you engaged to do, you may yet fill a most important station, and glorify our Great Master. But if you cannot do this, it is clear that the Society cannot continue in its service those who will not devote themselves to their engagements." The deaf, solitary man felt much aggrieved. He said, and said truly,

"I gave the Society a pledge, which there does not live a man who could prove to an impartial person that I have not redeemed. When, after the labors of eight or nine hours, the office was closed for the day, I felt that I was at liberty to partake of some mental refreshment. This is the ground of my dismissal. Even if my attachment to literature were an evil, it might be tolerated whilst it did not (and *it did not*) interfere with my defined duties."

It is not now difficult to adjudicate between the poor deaf man and this learned and influential Missionary Society. No ordinary master printer in Edinburgh, or elsewhere, would think of treating one of his journeymen, or even one of his apprentices, after this fashion. The limits of a printer's work are easily ascertained. Nine tenths of the printers of Great Britain and Ireland are employed by the *piece*, the others are placed on what is known as a *settlement;* and, under either scheme, there is a portion of their time which is not sold to their masters, and with which, therefore, a master cannot *honestly* interfere. But the grand mistake of the committee, and of worthy Mr. Bickersteth, in this not uninstructive case, seems to have been founded on a certain *goody* sentiment, from which missionaries such as the brethren of the Society of Jesus would have been saved by their sagacious discernment of the capabilities and spirits of men, and the ordinary master printer, by his knowledge of the proper tale of work which an operative ought to furnish, and his full recognition of the common business rule, that the time is not the master's, but the operative's own, for which the master does not pay. The committee and Mr. Bickersteth evidently held, on the other hand, that the deaf lad, being a missionary printer, ought to have his heart and soul in the missionary printing, and in nothing else; that the work of writing and translating was a work to be done by other heads and hands than his, — heads and hands trained, mayhap, at Cambridge or Oxford; and that the literary studies pursued by the lad after office-hours were over were mere works of "self-will" and "self-gratification," and not suited to "glorify the Great Master." In order to glorify the Great Master, it was necessary, they held, that the deaf lad should give his heart exclusively to the printing of the mission. Alas! the good men were strangely in error. The Great Master had, we now know, quite other work for the deaf lad. We are ignorant of what the Oxford and Cambridge men of the Malta Mission have done, — what they could, we dare say, and we are sure they think it all too little; but their labors will scarce ever be brought into competition with those of the greatest Biblical illustrator of modern times. What Dr. Chalmers used to term his Biblical library consisted of four great standard works; and of these select four, Dr. Kitto's "Pictorial Bible" was much a favorite. "I feel quite sure," we find him saying, in his "Daily Scripture Readings," "that the use of the sacred dialogues as a school-book, and the pictures of Scripture scenes which interested my boyhood, still cleave to me, and impart a peculiar tinge and charm to the same representations when brought within my notice. Perhaps when I am mouldering in my coffin, the eye of my dear Tommy [his grandson] may light upon this page; and it is possible that his recollections may accord with my present anticipations of the effect that his delight in the 'Pictorial Bible' may have in endearing still more to him the holy Word of God." In the peculiar walk in which Dr. John Kitto specially excelled all other writers, the great Chalmers was content to accept him as his teacher, and to sit at his feet; and the poor, friendless, deaf lad, who so offended the committee of the Maltese Mission by devoting to literature the time which was indisputably his own, not theirs, was this same John Kitto, — a name now scarce less widely known, though in a different walk, than that of Chalmers himself.

Dismissed from his situation, he returned to England with but forlorn prospects. There was, however, work for him to do; and an unexpected opening, which providentially occurred shortly after his arrival, served greatly to fit him for it. A missionary friend, bound for central Persia, engaged him to accompany him on the journey as tutor to his two boys, — a charge for which his previous studies, pursued under the direst disadvantages, adequately fitted him; and, with his eyes all the more widely open from the circumstance that his ears were shut, he travelled through Russian Europe into Persia, saw the greater and lesser Ararats, passed through the Caucasian range of mountains, loitered amid the earlier seats of the human family, forded the Euphrates near its source, resided for about two years in Bagdad, witnessed the infliction of war, famine, and pestilence, and then — his task of tuition completed — journeyed homeward by Teheran, Tabreez, Trebizond, and Constantinople, to engage in his great work. His quiet life was not without its due share of striking incident. We have referred to a story of wounded affection. On his return to England, he found that she who had deceived and forsaken him had deeply regretted the part she had acted, and was now no more; and for years after, he bore about with him a sad and widowed heart. In his second return he had a companion, a young man in delicate health, who, when detained with him in quarantine at the mouth of the Thames, sickened and died. The description of the quarantine burying-ground, in which his remains were deposited, is suited to remind the reader of some of the descriptions of similar places given by Dickens.

"We went," says Kitto, in his journal, "in a boat of the vessel, to a kind of low island devoted to the burial of persons dying in quarantine. The coffin was plain, without a plate, and with pieces of ropes for handles, but had the honor of being covered with the ensign of the doctor's ship as a pall. The grave-place, overgrown with long, reedy grass, was not more than a few paces from the water's edge; and its uses were indicated only by what the captain calls '*wooden tombstones*,' of which there are only two, both dated 1832, and all of wood, painted of a stone color, the first I have seen in England. S—— was carried to his last home by the sailors of our vessel.

On arriving at the grave, we found it of dark clay, with water at the bottom; a wet ditch being near, above its level. It was also too small, and we had to wait till it was enlarged; and then, the coffin being brought to the side, ready to be let down, the doctor's head servant took out a prayer-book, and, all uncovering, read a part of the burial service. We waited till the grave was filled up and banked over; and then, with a sigh, not the last, returned to the boat. On our return, the flags, which had hitherto been floating half-mast high, were raised to their usual position."

Kitto's fellow-traveller, whose dust he saw thus consigned to the dark, obscure burial-yard at the mouth of the Thames, had been engaged to a young lady, on whom, after his release from quarantine, the deaf man waited, to communicate to her the fate of her lover. The two widowed hearts drew kindly together; and in course of time the lady became Mrs. Kitto, — a match from which her husband, now entering on a literary life of intense labor, derived great comfort and support.

Never did literary man toil harder or more incessantly. His career as an author commenced in 1833, and terminated at the close of 1853; and during that period he produced twenty-one separate works, some of them of profound research and great size. Among these we may enumerate the "Pictorial Bible," the "Pictorial History of Palestine," the "History of Palestine from the Patriarchal Age to the Present Time," the "Cyclopædia of Biblical Literature," the "Lost Senses," "Scripture Lands," and the "Daily Bible Illustrations." And in order to produce this amazing amount of elaborate writings, Dr. Kitto used to rise, year after year, at four o'clock in the morning, and toil on till night. But the overwrought brain at length gave way, and in his fiftieth year he broke down and died. Could he have but retained the copyright of his several works, he would have been a wealthy man; he would at least have left a competency to his family. But commencing without capital, and compelled, by the inevitable expense of a growing family, to labor for the booksellers, he was ever engaged in "providing," according to Johnson, "for the day that was passing over him," and died, in consequence, a poor man. And his widow and family have, we understand, a direct interest in the sale of the well-written and singularly interesting biographic work to which we are indebted for the materials of our article, and which we can recommend with a good conscience to the notice of our readers. We know not a finer example than that which it furnishes of the "pursuit of knowledge under difficulties," nor of a devout and honest man engrossingly engaged in an important work, in which he was at length to affect the thinking of his age, and to instruct and influence its leading minds. It may be interesting to remark how such a man received the first decided direction in his course of study; and so the following extract, with which we conclude, of a letter on the subject from a gentleman much before the public at the present time, from his, we believe, honest and fearless report on the mismanagement of our leading officers in the Crimea during the campaign now brought happily to a close, may be regarded by our readers as worthy of perusal: —

"My first meeting with Kitto," says Sir John M'Neill, "was at Tabreez, in 1829. He was going with Mr. and Mrs. Groves and their two sons to Bagdad, where Mr. Groves intended to establish himself as a missionary. Kitto was then acting as tutor to the two boys, who were lively and intelligent; and I was struck with the singularity of his position, as the deaf and almost dumb teacher of boys who were very far from being either deaf or dumb. This circumstance, and the loneliness of mind which was a necessary consequence of his inability to communicate with the persons whom he was thrown amongst at Tabreez, led me to put some questions to him in writing, with the view of drawing him into conversation; but I found great difficulty in comprehending his answers, in consequence of the peculiarity of his voice and enunciation. With the assistance of his pupils, however, who spoke with great rapidity on their fingers, and appeared to have no difficulty in understanding what he said, I succeeded in engaging him in such conversation as could be so carried on. I found his intelligence and his information vastly greater than I had anticipated. He had evidently the greatest avidity for information; but was restrained from pressing his inquiries, apparently by his modesty, and the fear of being considered obtrusive or troublesome. Finding him well read and deeply interested in the Scriptures, I directed his attention to the many incidental allusions in the Bible to circumstances connected with Oriental habits and modes of life, which had become intelligible to me only after I had been for some time in the East. I remember he was particularly interested in something I had said in illustration of the importance attaching to the fact that 'Jacob digged a well.' I had explained to him, that, in arid countries, where cultivation could only be carried on by means of irrigation, the land was of no value unless when water could be brought to irrigate it; and that in Persia the theory of the law still is, that he who digs a well in the desert is entitled to the land which it will irrigate. He came to me more than once for fuller information upon this subject, and was greatly delighted with some illustrations of Scripture which I pointed out to him in 'Morier's Second Journey to Persia.' I refer to these circumstances because I believe that they relate to the first steps of that inquiry which he prosecuted so assiduously and successfully during the remainder of his life, and to which he constantly recurred almost every time I met him afterwards, either in Asia or in England."

IX.

THE IDEALISTIC SCHOOL.

It is not often in these latter days that a metaphysical question is forced on the notice of the public. The muse of abstract thought — the genius that asserts as her special province the region of "being and knowing" — has been dozing for at least an age in a state of partial hybernation, sucking her paws in closets and class-rooms, and getting so marvellously thin and spiritual under the process, that her attenuated form has long since failed to make any very distinct impression on the retina of the community. The case was widely different once. During the latter half of the last century no other class of questions possessed half so great an interest in Scotland as metaphysical ones. Metaphysical had succeeded to theological disquisition, and was pursued with equal earnestness; partly, no doubt, because the metaphysics of the age had set the theology of the age that had gone before virtually on its trial, but in great part also because the largest minds of the time had given themselves to the work; and, further, because the limited character of that cycle in which the mental philosophy is doomed to expatiate was not yet known. Early in the present century the interest had in some degree begun to flag, and the keen eye of Jeffrey was one of the first to detect the slacking of the tide. And in his ingenious critique on "Stewart's Life of Reid," he attempted to render a reason for it. The age had already started forward in that course of natural, physical, and mechanical experiment in which such distinguished trophies have since been won, and which have given its peculiar character to the time; and it had become impatient, said the critic, of barren, non-productive observation. And it was a grand distinction, he held, between the physical and the metaphysical walks, that, while *experiment* reigned paramount in the one, and formed the all-potent key by which man could lay open at will the arcana of nature, and arm himself with her powers, *observation* only could be employed in the other, — a mere passive faculty, that had an ability of seeing, but none whatever of controlling. Hence, he argued, the unproductive character of metaphysical science, and the natural preference which the public had begun to manifest, on ascertaining such to be its character, for pursuits through which solid benefits were to be secured. "In the proper experimental philosophy," he said, "every acquisition of knowledge is an increase of power, because the knowledge is necessarily derived from some intentional disposition of materials, which we may always command in the same manner. In the philosophy of observation, it is merely a gratification of our curiosity. The phenomena of the human mind are almost all of the latter description. We feel, and perceive, and remember, without any purpose or contrivance of ours, and have evidently no power over the mechanism by which those functions are performed. We cannot decompose our perceptions in a crucible, nor divide our sensations by a prism; nor can we by act and contrivance produce any combination of thoughts or emotions besides those with which all men have been provided by nature. No metaphysician expects by analysis to discover a new power, or to excite a new sensation, in the mind, as a chemist discovers a new earth or a new metal; nor can he hope by any process of synthesis to exhibit a mental combination different from any that nature has produced in the minds of other persons."

Certainly metaphysical found in physical science at the beginning of the present century a formidable rival, that could reward her followers much more largely than she could; and even ere the retirement of Dugald Stewart, her decline in interest and influence, which the keen eye of Jeffrey had remarked at an earlier period, might be seen by all. The genius of Thomas Brown created a diversion in her favor; but he sank and died in middle life, and his science in Scotland might be said to die with him. His successor in the moral philosophy chair of our university was at least his equal in genius; but the bent of Wilson was literary, not scientific; and the enthusiasm which he excited among his pupils was an enthusiasm for the sensuous, not the abstract. But while all must agree in the *fact* remarked by Jeffrey, many may fail to acquiesce in the cause which he assigns for it. Pursuits not more profitable than metaphysical ones have been eminently popular in the age just gone by, and are so still. We know not that we should instance theology, seeing that on theological truth man's most important interests may be regarded as suspended; but we surely may instance that department of philosophic criticism in which Jeffrey himself won his laurels. We may instance, besides, at least two of the natural sciences, astronomy and geology, neither of them more rich of dowry than metaphysical science itself, and which cannot be advantageously prosecuted save at a much greater expense. And yet both have been zealously cultivated, especially the latter, in the age during which metaphysics have been neglected. We must look for some other cause; nor do we think it ought to be difficult to find. Metaphysical pursuit fell into abeyance in this country, not because it rested on a mere basis of observation, not experiment, or because it led to no such tangible results as the pursuit of the physical sciences; but simply in consequence of a thorough divorce which took place, through the labors of some of the most acute and ingenious metaphysicians the world ever saw, between the deductions of the science and the conclusions of common sense. Reid, who raised one of the most vigorous protests ever made on the other side, has well remarked that "it is genius, and not the want of it, that adulterates philosophy, and fills it with error and false theory." And certainly none but very superior men could have run their science so high and dry upon the beach, that, with all the interest which attaches to its objects, men have preferred leaving it there to taking the trouble of getting it afloat again. We have before us Brown's "Philosophy of the Human Mind," open at one of the most ingenious portions of the work, that on the phenomena of simple suggestion, and would cite one of his views by way of example.

Hume had previously shown that there is no other visible connection between cause and effect than that of invariable contiguity. Cause and effect were Siamese twins persistently seen together, but with the connecting ligament, if any such really existed, invariably concealed. And Brown, following close in the wake of the elder dialectician, deliberately erased the very words from his metaphysic vocabulary, and substituted *antecedent* and *consequent* instead. The very terms *cause* and *effect* vanished from his speculations, and with the terms the doctrine they involved; and he taught, instead, that power is nothing more than the relation of one object or event as antecedent to another object or event, its immediate and invariable consequent. Hume, whose vigorous common-sense was ever raising protests against his ingenuities, and in whose ever-recurring *asides*, if we may so speak, the germ of the Scotch philosophy may be found, had stopped short when he showed that no known argument existed by which it could be proved that effects were the *necessary* results of causes, and that it could only be shown instead, and thus simply as a matter of experience, not reason, that they were always associated with causes, — always tagged to them in the exhibiting areas of space and time, as the cart is tagged to the horse, or as a train of railway carriages is tagged to the engine. And in summing up these links of the associative faculty, which keeps up the ever-moving train of thinking in the human mind, and constitutes one thought master of the ceremonies in introducing another, he enumerated, as distinct and separate, first, the link of contiguity in time and place; and, second, the link of cause and effect. And well he might. Let a misemployed ingenuity compound them as they may, they are wide as the poles asunder. They are separated by the entire breadth of the human intellect; nay, by the entire breadth of the brute and human intellect united. The prevailing link of association in the mind of the highest philosopher is the link of causation. It was the link that connected the sublime thoughts of Newton, when, sitting in his arbor, he saw the apple fall from the tree, and traced in profound meditation the effect of the great law to which it owed its fall, from the earth to the moon, and from thence to the sun and all the planets. And, on the other hand, the link of mere contiguity is the prevailing link in those minds in which intellect is feeblest; it was the link on which the ideas of Dame Quickly were suspended. Her recollections hung upon the parcel-gilt goblet, the sea-coal fire, and the chance visit of goodwife Keech, the butcher's wife. Nay, even the inferior animals are not too low to be under its influence. The horse quickens his pace when some contiguous object reminds him of the neighborhood of his stable, with its corn and hay; and the cat learns to associate the dinner-bell with the dinner which it precedes. And yet we find one of the most ingenious of the idealistic metaphysicians fusing these two widely-distant links of association into one, — the prevailing Newtonian link into the prevailing link of the cat and horse; or, as he himself expresses it, suppressing the link of causation as superfluous, and leaving instead, in conformity with his adoption of the doctrine of Hume (though Hume himself avoided the absurdity), only the link of contiguity in time and space. The "olde polde-headed manne," who held that Tenterden steeple was the cause of the Goodwin Sands, because the steeple had been built in their neighborhood, he said, just immediately before they began to form, has been a standing joke in English literature for the last four hundred years. He made the mistake of substituting contiguity in time and place for causality, and has become a jest in consequence. But what shall be said of a scheme of metaphysics that does deliberately and knowingly, in order to preserve the consistency of a foregone conclusion, what the "polde-headed manne" did in his simplicity and ignorance? The shrewd natural philosopher who saw in the slow deposition of a few particles of earth or mud in still water, formed by the opposing action of two currents, a future sandbank, and, reasoning from cause to effect, was reminded, through the associative link thus furnished, of the brown wastes of the Goodwin Sands strewed with wrecks, and with the white surf beating over them, and the garrulous old woman to whom a print of Tenterden steeple suggested the contiguous sand-pit along whose margin she had been accustomed to pick up bits of broken planks for her fire, would be, on the showing of Dr. Brown, under the influence of identical suggestions; for contiguous cause and contiguous steeple he has virtually placed in the same category. Is there any wonder that a busy age should leave philosophers who argued after such a fashion, however nice their genius or however excessive their ingenuity, to milk their rams unheeded (we borrow the old illustration), and that only a few ill-employed students should be found idle enough to hold the pail? And yet, such is no extreme illustration of the idealistic philosophy.

It is, in truth, the grand objection to this philosophy, that it sets itself in direct opposition to mind engaged in all the practical walks. Let us adduce another instance. It is one of the fundamental principles of an ingenious metaphysician of the present time, a principle in which he is virtually at one with Berkeley, that being is to be regarded as tantamount to knowing; and that whatever is not an object of consciousness cannot be regarded as existent. Berkeley held that the absolute existence of unthinking beings, without any relation to their being perceived, was wholly unintelligible; and we at once grant that a bar of metal kept in the fire until it glows a bright red has no consciousness of redness, that the caloric with which it is charged has no sense of heat, and, further, that the bar itself has no feeling whatever of expansion or solidity. Redness, heat, expansion, and the idea of solidity are all impressions of sentient existence, — accidents or qualities to be seen, felt, or conceived of. But it does not follow, that, because a heated bar of iron is not conscious of heat, solidity, or redness, it is not therefore a heated bar of iron; or that because the senses can testify to its existence only as the senses of the living can testify of the existence of what is non-vital and non-sentient, it has therefore no existence as a non-vital, non-sentient substance. The leap in the logic seems most extraordinary, from the fact of the non-sentient character of the heated bar to the non-existence of the heated bar. And yet such virtually was the conclusion of Berkeley. "Some truths are so near and obvious to the mind," he said, "that a man need only open his eyes to see them. And such," he added, "I take this important one to be, namely, that all the choir of heaven and furniture of earth — in a word, all those bodies which compose the mighty framework of the world — have not any substance without a mind; that their being is to be perceived or known: to be convinced of which, the reader need only reflect and try to separate in his own thoughts the being of a sensible thing from its being perceived." In this last sentence the sophism seems to lie. It confounds conceiving with existing, light with eye and the optic nerve, and caloric and solidity with feeling and the tactile sense. It would date the beginning of the sun, not from that early period during which the sun influenced the yearly motions of our planet, but from the long posterior period during which eyes began to exist. And such essentially is the philosophy of that other ingenious metaphysician of our own time to which we refer. "He" also "goes so far as to affirm," says Mr. Cairns, in his admirable pamphlet, "that thought and existence are identical. Knowledge of existence, he says, — the apprehension of one's self and other things, — is alone true existence." Yes, true *rational* existence; but, judged by the common-sense of mankind, it would be an eminently irrational existence that would deny the reality of existence of any other kind, — that would recognize the *bona fide* being of an Edinburgh professor, but deny, in an argument four hundred pages long, that the university in which he lectured had any being whatever. And if, while such a teacher of moral philosophy, seated in its logic chair, mayhap, was lecturing in one room on the general nonentity of things, there was a professor of natural science demonstrating in another, on evidence which no ingenuous mind could resist, that, during immensely protracted periods, this old earth of ours had moved round the sun in a state so nearly approximating to the incandescent, that its diurnal motion propelled outward its matter at the meridian, so that its equatorial diameter still exceeds its polar one, in consequence, by about twenty-six miles, — that for periods more than equally protracted, when it became a home of sentient existence, its highest creatures were in succession but trilobites, fishes, reptiles, birds, and mammals, and that not until comparatively of yesterday did its rational existence come into being, — we could not regard such neighborhood as other than formidable to the logician to whom this brief latter day would be the only one recognized as a reality. It would be such a neighborhood as that of a disciple of Newton busied in weighing and measuring the planets or calculating the return of a comet on the parallax of a fixed star, to an old sophist engaged in showing his lads, on what he deemed excellent grounds, that if a tortoise which crept a hundred yards in an hour had got the start by a few furrows' breadth of Achilles, who ran a mile in five minutes, the fleet warrior might be engaged for ever and ever in vain attempts to come up with it.

38

One of two things would of necessity occur in a state of matters so little desirable, — either the pupils of the logician would become such mere triflers in argument as the Jack Lizard of Steele's essay, who, when his mother scalded her fingers, angered the honest woman by assuring her there was no such thing as heat in boiling water; or they would learn to despise both their professor and his science. It gives us sincere pleasure to find that the Edinburgh University is in no such danger. So long as the logic chair remained vacant, we purposely abstained from making any allusion to the subject, in the fear that any expression of opinion, even in a matter so impersonal as the respective merits of two schools of philosophy, might and would be misinterpreted. But we are in no such danger now; and we must be permitted to express our sincere pleasure that the election of Tuesday has resulted in the selection of an asserter of the Scotch school of philosophy to teach in the leading Scotch university. Nor are we influenced by any idle preference for the mere name Scotch. We know not that so large an amount of ingenuity has yet been expended on that common-sense school of which Reid was the founder, and Beattie, Hamilton, and Dugald Stewart the exponents, as on the antagonistic school, which at least equally distinguished Scotchmen, such as Hume and Thomas Brown, have illustrated and adorned. George Primrose, in the "Vicar of Wakefield," found that the best things remained to be said on the wrong side; and so, in the determination of astonishing the world, he set himself to dress up his three paradoxes. And, unquestionably, the paradoxes of the idealistic philosophy have been admirably dressed. But the Scotch philosophy has at least this grand advantage over the opponent school, that all its principles and deductions can be brought into harmony with those of all the other departments of science. It is not a jarring discord in the great field of mental exertion, — a false bar, to be slurred over or dropped in the general concert, — but a well-toned and accordant part, consistent with the harmony of the whole. It was acknowledged by Hume, that it was only in solitude and retirement that he could yield any assent to his own philosophy. Nor was he always true to it even in solitude; for in solitude he wrote his admirable political essays, and his "History of England." And the Scotch school is simply an appeal, on philosophic grounds, from Hume the metaphysical dreamer, wrapped up in the moonshine of sceptical speculation, to Hume the practical politician and shrewd historian. And we know no man better fitted to be an exponent of this true and solid school, or whose mind partakes more of the character of that of its founder Reid, than the gentleman on whom the choice of the council has fallen. We trust he has a long career of usefulness before him; and have every reason to hope that his expositions will be found not unworthy of the chair of Hamilton, nor of a philosophy destined ultimately, we cannot doubt, to give law in the regions of mental philosophy, at a time when the ingenuities of its opponents shall have shared the fate of the paradoxes of George Primrose.

X.

THE POESY OF INTELLECT AND FANCY.

It has been well said of singing a song, — in reference, of course, to the extreme commonness of musical accomplishment in a low degree, and its extreme rarity in a high one, — that it is what every one can do, and not one in a thousand can do well. A musical ear is, like seeing and hearing, one of the ordinary gifts of nature, just because music was designed to be one of the ordinary delights of the species; but while the class capable of being delighted is a very large one, the class capable of delighting is one of the smallest. A not large apartment could contain all the first-class singers in the world; and, mayhap, judged by men of the highest degree of taste, a closet roomy enough to contain Jenny Lind might be found sufficient to accommodate for a time its *preëminent* musical talent. And it is so as certainly with poetry as with music. There are a few men in every community wholly destitute of both the musical and the poetic sense, just as in every community there are a few men born blind and a few more born deaf; but, with these exceptions, all men have poetry and music in them, — music enough, if their education has not been wholly neglected, to derive pleasure from music, and poetry enough to derive pleasure from poetry. And in due accordance with this fact, we find that in what man's Creator appointed from the beginning to be the commonest of all things, religion, he has made large use of both. Every church has its music, and a large portion of the divine revelation has been made in poetry. But if the great musicians who can exquisitely delight be few, the great poets are still fewer. There is but one Jenny

Lind in the world; but then the world has not had a Shakspeare for the last two hundred and forty years; and, though greatly more than a century has elapsed since Dryden took tale, in his famous epigram, of all the great epic poets, and found them but three, no one has since been able to add a fourth to the list. Of all rare and admirable gifts, the poetic faculty in the high and perfect degree is at once the most admirable and the most rare. It may, however, be very genuine and exquisite, though not full-orbed, as in a Homer or a Milton. Nature, when she makes a poet of the first class, adds a powerful imaginative faculty, and a fancy of great brilliancy, to an understanding the profoundest; she takes all that makes the great philosopher and all that is peculiar to the true poet, and, adding them together, produces, once in a thousand years or so, one of her fully-rounded and perfect intellects. But a man may have much though he may not have all; nay, a very few faculties, if of a rare order and wisely employed, may well excite admiration and wonder. Tannahill could achieve only a song; but as the songs which he did achieve were very genuine ones, with the true faculty in them, Scotland seems to be in no danger of forgetting them. Beranger, the greatest of living song-writers, is a man of similar faculty with Tannahill. He is known as a song-writer, and as that only; but never had France such songs before, and France knows how to value them. The one thing which Beranger can do, no other man can do equally well; and not a few of the fairest names among the poets of antiquity are those of poets equally limited, apparently, in what they were fitted to produce, but also equally exquisite in the quality of their productions. Anacreon has left only little odes, and Pindar only great ones; but scholars tell us that it is almost worth while acquiring Greek in order to be able to read them. Ancient Rome has immortalized her Lucretius for his single faculty of transmuting not very good philosophy into very noble verse, and modern Italy her Petrarch for his rare skill in turning a sonnet. In short, almost all the poets of the second order have been poets, not full-orbed in their brightness, like the sun or the great outer planets of the system, but, like the inner planets, and like the moon ere her full term has come, mere segments and crescents of glory.

There can be no very adequate division made of these partially orbed poets; and yet they naturally enough divide into two classes, — a class in whom intellect is comparatively strong and genius weak, and, *vice versa*, a class in whom intellect is comparatively weak and genius strong. Pure intellect dissociated from the poetic faculty can of course accomplish but little in the fields of poesy. And yet, such is the power of determination, diligence, and high culture, that a little it has accomplished. If it has not produced brilliant poems, it has at least produced pointed stanzas and pleasing stories, narrated in easy and elegant verse. We greatly question whether Hayley was born a poet; but his "Triumphs of Temper," though they triumphed over the temper of Byron, certainly did not triumph over ours. On the contrary, we found the piece, in its character as a metrical tale, at least as readable as if it had been written in good prose; and there are even some of its stanzas which we still remember. The few lines in which the father of the heroine is described may not be poetry, but they are nearly as good as if they were. There are not many characters better hit off in a few lines, in the whole round of English verse, than that of

"The good Sir Gilbert, to his country true,
A faithful Whig, who, zealous for the state,
In freedom's service led the loud debate;
Yet every day, by transmutation rare,
Turned to a Tory in his elbow-chair,
And made his daughter pay, howe'er absurd,
Passive obedience to his sovereign word."

But of all the achievements of the prose men in the province of verse, that of Swift is the greatest. Dryden was quite in the right when he said that the young clergyman was no poet; and yet the "no poet" has so fixed his name in the poesy of the country that in no general biography of the English poets do we find his life omitted, and in no general collection of English poetry do we fail to find his verses. The works of a class of writers not certainly so devoid of poetry as Swift and Hayley, but who were rather men of fine taste and vigorous intellect than of high poetic genius, represent in large measure the common staple of English poesy during the earlier and middle part of the last century. Not only the Broomes, Fentons, and Lytteltons, but even the Armstrongs and Akensides, belonged to this class. The men who assisted Pope in translating the "Odyssey;" the man who wrote that work on the Conversion of St. Paul which still maintains its place in what may be termed the higher literature of the "Evidences;" in especial, the men who produced the "Pleasures of the Imagination" and the "Art of Preserving Health,"— had all very vigorous minds. Akenside would have made a first-class metaphysical professor, particularly in the æsthetic department; and Armstrong could have effectually grappled with very severe and rugged subjects; but the poetic faculty that was in them was very subordinate to their intellect. It was true so far as it extended, but embroidered only thinly and in a threadbare way the strong tissue of their thinking. And yet both the "Art of Preserving Health" and the "Pleasures of the Imagination" are noble poems. The latter is the better known of the two: Thomas Brown used to repeat almost the whole of it every season in his class, as at once good poetry and good metaphysics. But the former deserves to be known as well. The man who could transmute such a subject into passable poetry, and render his composition readable as a whole, — and much of the poetry is more than passable, and the piece as a whole eminently readable, — must be regarded as having accomplished no ordinary achievement. It is, however, from the strong intellect displayed in the staple texture of the piece, rather than from its poetic embroidery, that it derives its merit.

The second class — the class composed of men whose poetic genius overrode their intellect — is not so largely represented in English poetry as the other. It may be safely said, however, that in the writings of men of the last century, such as Collins, Chatterton in his Rowley poems, and perhaps Meikle, we find more of poetry than of pure intellect; and in writings of men of the present century, such as those of Keats, Wilson, and perhaps Leigh Hunt, we find *much* more. In the writings of Wilson there is often scarce tissue enough to support the load of gorgeous embroidery that mantles over it. In especial in his "Isle of Palms" do we find the balance of the poetry preponderately cast against the intellect. It is, as a poem, in every respect the antipodes of the "Art of Preserving Health." In Keats the preponderance is also very marked. What a gorgeous gallery of poetic pictures that "Eve of St. Agnes" forms, and yet how slim the tissue that lies below! How thin the canvas on which the whole is painted! For vigorous sense, one deep-thoughted couplet of Dryden would make the whole kick the beam. And yet what can be more exquisite in their way than those pictures of the young poet! Even the old worn-out gods of Grecian mythology become life-like when he draws them. They revive in his hands, and become vital once more. In "Rimini" we detect a similar faculty. A man of profound, nay, of but rather strong intellect, would scarce have chosen such a repulsive story for poetic adornment; but, once chosen, only a true poet could have adorned it so well.

Such are specimens of the class of poets which we would set off against that to which the Lytteltons, Akensides, and Armstrongs belonged, and at whose head Pope and Dryden took their stand. And it is a class that, comparatively at least, — the sum total of the poetic stock taken into account, — is largely represented at the present time.

We shall not repeat the nickname which has been employed to designate them; for, believing, whatever may be their occasional aberrations, that they possess "the vision and the faculty divine," we shall not permit ourselves to speak other than respectfully of them. We could fain wish that they oftener rejected first thoughts, and waited for those second ones which, according to Bacon, are wiser: we could fain wish that what was said of Dryden —

> "Who either knew not, or forgot,
> That greatest art, — the art to blot,"—

could not be said so decidedly of them. But we must not forget that their compositions, though not without fault in their character as wholes, and often primed in, as a painter might say, on too thin a groundwork, contain some of the most brilliant passages in the wide range of modern poetry. To this school Gerald Massey, a name already familiar to most of our readers, has been held to belong. He has less of its peculiar faults, however, than any of its other members, with certainly not less of its peculiar beauties. With all the marked individuality of original genius, he reminds us more of Keats than of any other English poet; but with the same rare perception of external beauty, and occasionally the same too extreme devotion to it, he adds a lyrical power and a depth of feeling which Keats did not possess. And from these circumstances we augur well of his future. It is ever the tendency of genuine feeling to pass from the surface of nature to its depths; and though, as we see exemplified in the songs of Burns, the true lyrist may find in description adequate employment for his peculiar powers, it is always in preparation for some burst of sentiment, or by way of garnishing to some striking thought. Mr. Massey's new poem "Craigcrook Castle" furnishes admirable illustrations of the various phases of his genius. The plan of the work is one of which our literature has furnished many examples, from the times of the "Canterbury Tales" down to those of the "Queen's

Wake," and which is taken up year after year in the Christmas stories of the writers connected with the "Household Words." There is a meeting of friends at the hospitable board, over which Jeffrey once presided, and at which a man of similar literary tastes and feelings presides now; and each guest, in passing the evening, brings forward his contribution of song or story. The introduction, with none of the cadences of Keats, reminds us in every line of that poet's delight in sensuous imagery and influences, and of the crust of rich thought, if we may so express ourselves, that mantled over the surface of his poetry. The advent of the morning at Craigcrook we find thus described: —

"The meek and melting amethyst of dawn
Blushed o'er the blue hills in the ring o' the world;
Up purple twilights come the shining sea
Of sunlight breaking in a silent surge,
Whence morning, like the birth of beauty, rose;
While at a rosy touch, the clouds, that lay
In sullen purples round the hills of Fife,
Adown her pathway spread their paths of gold.

* * * *

"Sweet lilies of the valley, tremulous fair,
Peep through their curtains, clasped with diamond dew
By fairy jeweller's working while they slept;
The arch laburnum droops her budding gold
From emerald fingers with such taking grace;
The fuchsia fans her fairy chandelry,
And flowering current crimsons the green gloom;
The pansies, pretty little puritans,
Come peering up with merry, elvish eyes;
At summer's call the lily is alight;
Wallflowers in fragrance burn themselves away
With the sweet season on her precious pyre;
Pure passionate aromas of the rose,
And purple perfume of the hyacinth,
Come like a color through the golden day."

There is much of Keats in this passage, and yet Keats was not in the mind of the writer: the similarity of result is an effect, evidently, not of imitation, but of a similarity of genius. The following passage, much in the same vein, has been greatly criticized, and yet none but a true poet could have produced it. It is a remarkable picture of a remarkable man, with points about it which might easily be laid hold of in a mocking spirit, but which impart not a little of its character and individuality to the portrait. We quote from the second edition : —

> " We gathered all within the house, and there
> Shook off the purple silence of the night.
> Cried one, Come, let us a symposium hold,
> And each one to the banquet bring their best
> In song or story : all shall play a part.
> So, for a leader simple and grand, we chose
> Our miracle-worker in midwifery, — he
> Who wrestled with the fiend of corporal pain,
> And stands above the writhing agony
> Like Michael with the dragon 'neath his heel;
> Who is in soul Love riding on a lion;
> In body, a Bacchus crowned with the head of Jove :
> The keen life looks out in his lighted face
> So fulgent, that the gazer brightens too;
> He bravely towers above our fume and fret,
> Like the old hills, whose feet are in the surge,
> And on their lifted brows the eternal calm;
> For he is one of those prophetic spirits
> That, ere the world's night, dreams of things to come."

There may be faults here, as the reviewers suggest, — nay, it may be all fault; but it certainly does remind us of those aberrations of genius specially described by the poet as " glorious faults, that critics dare not mend." In illustration of the lyrical spirit and deep tenderness of Mr. Massey, we give the following extracts from a series of simple triplets on the death of a beloved child : —

> " Within a mile of Edinburgh town
> We laid our little darling down, —
> Our first seed in " God's acre " sown.
>
> " The city looketh solemn and sweet;
> It bares a gentle brow to greet
> The mourners mourning at its feet.

"The sea of human life breaks round
This shore o' the dead with softened sound;
Wild flowers climb each mossy mound
To place in resting hands their palm,
And breathe their beauty, bloom, and balm,
Folding the dead in fragrant calm.

* * * *

"Lone mother, at the dark grave-door
She kneeleth, pleading o'er and o'er;
But it is shut for evermore.

"She toileth on — the mournfulest thing —
At the vain task of emptying
The cistern where the salt tears spring.

* * * *

"The spirit of life may leap above,
But in that grave her prisoned dove
Lies cold to th' warm embrace of love;

"And dark though all the world is bright,
And lonely with a city in sight,
And desolate in the rainy night.

"Ah, God! when in the glad life-cup
The face of Death swims darkly up,
The crowning flower is sure to droop!

"And so we laid our darling down,
When summer's cheek grew ripely brown;
And still, though grief has milder grown,
Unto the stranger's land we cleave,
Like some poor birds that grieve and grieve
Round the robbed nest, so loth to leave."

There are one or two obscurities of figure here that crave a second thought to unlock them; but nothing can be more sadly tender than the whole, and there is poetry in every stanza. Gerald Massey is still a young man, and much of his time in the past must have been spent in shaking off the stiff soil that clogs round for a time the thoughts and expressions of untutored genius. A man still under thirty, who never attended any school save a penny one for a brief period, and who at eight years of age was sent to toil in a silk manufactory from five o'clock in the morning till half-past six at night, may well be regarded as still but partially developed; and we are convinced the world has not yet seen his best. He has but to give his intellect as full scope as his fancy and imagination, and to bestow upon his pieces that elaboration and care which high excellence demands from even the happiest genius, in order to become one of the enduring lights of British song.

XI.

THE UNTAUGHT POETS.

In more than one respect the untaught poets of England have fared better than those of our own country. In the first place, Southey, perhaps the raciest English writer of his day, wrote their history, and made not a few of them known who had succeeded but indifferently in making known themselves; and in the second, we find from his narratives that, with few exceptions, their poetry served them as a sort of stepping-stone, by which they escaped upwards from the condition of hard labor and obscurity, to which they seemed born, into a sphere of comparative affluence and comfort. For one of the first of their number, John Taylor, the "Water Poet," — a man who was certainly not a water-poet in the teetotal sense, — nothing could have been done. He was a bold, rough, roystering fellow, quite as famous for his feats and wagers as for his rhymes. On one occasion he navigated his cockle-shell of a wherry all the way from London Bridge to York; on another, he rowed it across the German Sea from London to Hamburg; on yet another, in 1618, he undertook to travel from London to Edinburgh, and thence into the Highlands, "not carrying money to or fro, neither begging, borrowing, nor asking meat, drink, or lodging;" and what he undertook to do he did, and bequeathed to us, in his history of his "Pennyless Pilgrimage," the best account extant of hunting in the Highlands by the "Tinckhell," and of the "wolves and wild horses" which, at even that comparatively recent period, abounded in the ruder districts of Scotland. It would have been scarce possible to elevate such a man, even had a very generous patronage been the order of the age; but Taylor had all his days enough to eat and drink, and died the keeper of a thriving public house, much frequented, during the times of the Commonwealth, by the cavaliers. And no sooner did men of this class arise, to whom a judicious patronage could be extended, than they were admitted to its benefits. Stephen Dick, the "Thresher," was rather a small poet, but he was an amiable and conscientious man; and, mainly through the exertions of the Rev. Mr. Spence, Professor of Poetry in the University of Oxford, he obtained orders in the English Church, and was preferred to a not uncomfortable living. Dodsley, still known by his "King and Miller of Mansfield," was elevated, through the exercise of a genial patronage, from his original place as a table-boy, to be one of the most respectable London booksellers of his day, — a man whose name still imparts a recognizable bibliographical value to the works to which it is attached. The shoemaker Woodhouse, and the tobacco-pipe-maker Bryant, were also fortunate in their patrons; Gifford was eminently so; all seems to have been done for Ann Yearsley, the poetical milkwoman, that her own unhappy temper allowed; and in our own times, John Clare was kindly and liberally dealt with; though not more in his case than in that of his predecessor Duck could the degree of favor with which he was treated ward off the cruel mental malady that darkened his latter years. With, in short, the exception of one of the best, and in every respect most meritorious and deserving of the class, — poor Robert Bloomfield, who was suffered to die in great poverty, — we know not a single untaught English poet who gave evidence of the possession of the true faculty, however narrow its scope, and had at the same time character enough to be capable of being benefited by a liberal patronage, that failed to receive the encouragement which he deserved. And we find Southey laying down very admirably, in combating a remark of the elder Sheridan, — whom he terms an ill-natured, perverse man, — the generous principle on which this had been done. "Wonder," says the author of the first "Pronouncing Dictionary," — a man whom the greater lexicographer, Johnson, described as not only naturally dull, but as also rendered, through dint of immense effort on his own part, vastly duller than nature had made him, — "wonder, usually accompanied by a bad taste, looks only for what is uncommon; and if a work comes out under the name of a thresher, a bricklayer, a milkwoman, or a lord, it is sure to be eagerly sought after by the million." "Persons of quality," remarks the poet-laureate, "require no defence when they appear as authors in these days; and, indeed, as mean a spirit may be shown in traducing a book because it is written by a lord, as in extolling it beyond its deserts for the same reason. But when we are told that the thresher, the milkwoman, and the tobacco-pipe-maker did not deserve the patronage they found, when it is laid down as a maxim of philosophical criticism that poetry ought never to be encouraged unless it is excellent in its kind; that it is an art in which inferior execution is not to be tolerated, a luxury, and must therefore be rejected unless it is of the very best, — such reasoning may be addressed with success to cockered and sickly intellects, but it will never impose upon a healthy understanding, a generous spirit, or a good man. . . . . If the poet be a good and amiable man," continues Southey, "he will be both the better and the happier for writing verses. 'Poetry,' says Landon, 'opens many sources of tenderness that lie forever in the rock without it. . . . . The benevolent persons who patronized Stephen Duck did it, not with the hope of rearing a great poet, but for the sake of placing a worthy man in a station more suited to his intellectual endowments than that in which he was born. Bryant was befriended in a manner not dissimilar, for the same reason. In the case of Woodhouse and Ann Yearsley the intention was to better their condition in their own way of life. And the Woodstock shoemaker was chiefly indebted for the patronage which he received to Thomas Warton's good nature; for my predecessor Warton was the best-natured man that ever wore such a wig." There is the true English generosity of sentiment here, — a generosity which, in such well-known cases as that of Henry Kirke White and John Jones, was actually exemplified by Southey himself; and his remark regarding the humanizing influence of poesy on even its humbler cultivators will scarce fail to remind some of our readers of the still happier one which our countryman Mackenzie puts into the mouth of "old Ben Silton." "There is at least," said the stranger, "one advantage in the poetical inclination, that is an incentive to philanthropy. There is a certain poetic ground on which a man cannot tread without feelings that enlarge the heart. The causes of human depravity vanish before the romantic enthusiasm he professes; and many who are not able to reach the Parnassian heights may yet approach so near as to be bettered by the air of the climate."

The untaught poets of Scotland have fared much more hardly than those of the sister country. Some of them forced their way through life simply as energetic, vigorous men. Allan Ramsay throve as a tradesman, and built for himself a house in Edinburgh, which continues to attract the eye of the stranger by its picturesqueness, and which few literary men of the present day could afford to purchase. And Falconer, though he died a sailor's death in the full vigor of his prime, had first risen from the forecastle to the quarter-deck as a bold and skilful seaman.

Allan Cunningham, too, made his way good as a hard-working business man. But, if unable to help themselves after the manner of Falconer, Cunningham, and Ramsay, the untaught poets of Scotland received but little help from the patronage of their countrymen. The aristocracy of Scotland made Burns a gauger, and employed one of the noblest intellects which his country ever produced in "searching," as he himself in bitter mirth expressed it, "auld wives' barrels." And neither Alexander Wilson nor poor Tannahill ever received even the miserable measure of patronage that gave Burns seventy pounds a year, and demanded, in return, that he should waste three fourths of his time in a half-reputable and uncongenial employment. Poor Tannahill, the harmless, the gentle, the affectionate, was left to perish unhappily when he was but little turned of thirty; and Wilson, a stronger, though not a finer spirit, quitted his country in disgust, and made himself an enduring fame in the United States as a naturalist, by the great work which Prince Charles Lucien Bonaparte did not disdain to complete. We cannot point to a single untaught poet in the literary history of our country that ever enjoyed a pension. Pensions were reserved for the friends and relatives of the statesmen to whom Toryism in Edinburgh and elsewhere built senseless columns. But though the untaught poets of Scotland fared thus differently from those of England, it was certainly not because they deserved less. On the contrary, if we except Shakspeare, — one of those extraordinary minds that, according to Johnson, "bid help and hinderance alike vanish before them," — our untaught Scotchmen have been men of larger calibre, and greater masters of the lyre, than the corresponding class in England. Passing over the John Taylors and Ned Wards as deserving of no special remark, we would stake Ramsay with his "Gentle Shepherd" against his brother poet and brother bookseller Dodsley with his "Miller of Mansfield" and his "Toy Shop," taking odds of ten to one any day; Bloomfield, though a worthy personage, and possessed of the true faculty, was a small man compared with Robert Burns; and the Ducks, Woodhouses, Bryants, and Bennets were slim and stunted of stature compared with the Falconers, Tannahills, Wilsons, Allan Cunninghams, and Hoggs. In this, as in other walks, though English genius of the highest class takes the first place in the literature of the world, its genius of the second class fails to equal second-class genius in Scotland. There have been poets among our countrymen whose lives no one thinks of writing, and whose verses have failed to attract any very large share of notice who possessed powers greatly superior to most of the authors enumerated by Southey in his Essay on the Uneducated Poets, and who, had they written in England, would have been extensively known. To one of these, still among us, we find pleasing reference made in the correspondence of Jeffrey. "The greater part of your poems," we find him saying, in a note to the self-taught poet Alexander Maclagan, "I have perused with singular gratification. I can remember when the appearance of such a work would have produced a great sensation, and secured to its author both distinction and more solid advantages." And in another note, written in reference chiefly to a second and enlarged edition of Mr. Maclagan's poems, and which occurs in the volume of "Correspondence," edited by Lord Cockburn, we find the distinguished critic specifying the pieces which pleased him most. "I have already," says his lordship, "read all [the poems] on the slips, and think them, on the whole, fully equal to those in the former volume. I am most pleased, I believe, with that which you have entitled 'Sisters' Love,' which is at once very touching, very graphic, and very elegant. Your 'Summer Sketches' have beautiful passages in all of them, and a pervading joyousness and kindliness of feeling, as well as a vein of grateful devotion, which must recommend them to all good minds. The 'Scorched Flowers,' I thought the most picturesque."

We have read over Mr. Maclagan's works, — both the volume of poems which so gratified the taste of Jeffrey,[1] and an equally pleasing volume, of subsequent appearance, dedicated to the Rev. Dr. Guthrie, and devoted to the cause of ragged schools.[2] The general strain of both is equally pleasing; though we know not whether we do not prefer the simplicity and pathos of some of the "Ragged School Rhymes" to even those compositions of the earlier volume on which Jeffrey has stamped his *imprimatur*. Let us, however, ere quoting from the latter work, submit to the reader a few stanzas of the piece which most pleased the critic. It is a younger sister that thus addresses — in strains that, for their quaint beauty, remind us of some of the happier pieces of Marvell — a sister older than herself, but still young, that had been to her, in her state of orphanage, as a mother.

"Lo! whilst I fondly look upon
Thy lovely face, drinking the tone
Of thy sweet voice, my early known, —
My long, long loved, — my dearest grown, —
I feel thou art
A joy, — a part
Of all I prize in soul and heart.

"Sweet guardian of my infancy,
Hast thou not been the blooming tree
Whose soft green branches sheltered me
From withering want's inclemency?
No cloud of care
Nor bleak despair
Could blight me 'neath thy branches fair.

"And thou hast been, since that sad day
We gave our mother's clay to clay,
The morning star, the evening ray,
That cheered me on life's weary way, —
A vision bright,
Filling my night
Of sorrow with thy looks of light.

1 Sketches from Nature, and other Poems. By Alexander Maclagan.
2 Ragged School Rhymes. By Alexander Maclagan.

"Yet there were hours I'll ne'er forget
Ere sorrow and thy soul had met, —
Ere thy young cheeks with tears were wet,
Or grief's pale seal was on them set, —
        Ere hope declined,
        And cares unkind
Threw sadness o'er thy sunny mind.

"In glorious visions still I see
The village green, the old oak tree,
The sun-bathed banks where oft with thee
I've hunted for the blaeberrie,
        Where oft we crept,
        And sighed and wept,
Where our dead linnet soundly slept.

"Again I see the rustic chair
In which you swung me through sweet air,
Or twined fair lilies with my hair,
Or dressed my little doll with care;
        In fancy's sight
        Still rise its bright
Blue beads, red shoes, and boddice white.

"And at the sunsets in the west,
And at my joy when gently prest
To the soft pillow of thy breast,
Lulled by thy mellow voice to rest,
        Sung into dreams
        Of woods and streams,
Of lovely buds, and birds, and beams.

"When wintry tempests swept the vale,
When thunder and the heavy hail
And lightning turned each young cheek pale,
Thine ever was the Bible tale
        Or psalmist's song
        The wild night long,
Fresh from the heart where faith is strong.

"Now summer clouds, like golden towers,
Fall shattered into diamond showers:
Come, let us seek our wildwood bowers,
And lay our heads among the flowers;
        Come, sister dear,
        That we may hear
Our mother's spirit whispering near."

These stanzas are, as the great Scotch critic well remarked, at once "touching," "graphic," and "elegant," and certainly exhibit no trace of what Johnson well terms the "narrow conversation" to which untaught men in humble circumstances "are inevitably condemned." But regarding the difficulties with which Mr. Maclagan has had to contend, we must quote from himself: "That a working-man," we find him saying, "should write and publish a volume of verse, is no phenomenon: many of the brightest lights of literature in all countries have toiled for years at the press, the plough, the loom, and the hammer. That wealth and education in themselves have never made a true minstrel, is proverbial; nevertheless, they are powerful allies in his favor. Take, for instance, a youth from school, ten years of age, and bind him at thirteen or fourteen to a laborious trade. See him working ten hours a day for years without intermission, struggling to unravel, meanwhile, the mysteries of literature, science, and art, without assistance or encouragement, and you will find that he has many hard battles to fight before he can hope to attain even standing-room in the literary arena. Such, literally, has been the position of the author of the present volume." Let us remark, however, that untaught men possessed of the true poetic faculty are usually, in one important respect, happier in their genius than untaught men whose intellect is of the reflective cast, and their bent scientific. The poets are developed much earlier, and lose less in life. Ramsay began to publish his poems, in detached broad-sheets, in his five-and-twentieth year; Burns in his twenty-sixth year had written the greater part of his Kilmarnock volume, including his "Twa Dogs," "Halloween," and the "Cottar's Saturday Night;" Alexander Wilson produced his "Watty and Meg" at the same age; and the writings of both Tannahill and Allan Cunningham saw the light ere either writer was turned of thirty. But self-taught men of science have usually to undergo a much longer probationary period ere they can elevate themselves into notice. James Ferguson was nearly forty before he began to give public lectures on his favorite subjects, astronomy and mechanics. Franklin was in his forty-third year ere he had demonstrated the identity of lightning with the electric spark; and not until he had attained the same age did Sir William Herschel render himself known as a great astronomer and the discoverer of a new planet. Both in national and individual history, poetry is of early and science of late growth. The self-taught poet is not unfrequently developed at as early an age as men of a similar cast of genius who have enjoyed all the advantages of complete culture; judging from the experience of the past, he need not lose a single year of life; whereas the self-taught man of science may deem himself more than usually fortunate if he does not lose at least ten.

We have said that in some respects we prefer Mr. Maclagan's second publication, the "Ragged School Rhymes," to his first. It is, in the main, a more earnest, and, in the poetic sense, more truthful work. When the poet, in his earlier volume, sings, as he does at times, though rarely, of drinking "cronies" and usages, we know that he is catching but the dying echoes of a bypast time, when there was not a little staggering on the top of Parnassus, and Helicon used to run at times, like a town cistern on an election day, whiskey punch by the hour. But there is none of this in the other volume. The distress which it exhibits, the sympathy which it expresses, the views of nature which it embodies, are all realities of the present day. The earlier volume, however, contains more thinking; and the possession of both are necessary to the man desirous of rightly appreciating the untaught poet Maclagan. We find some little difficulty in selecting from the "Ragged School Rhymes" an appropriate specimen, not from the poverty, but from the wealth, of the volume. We fix, however, on the following, as suited to remind the reader of that passage in one of the larger poems of

Langhorne which, according to Sir Walter Scott, powerfully elicited the sympathy of Burns, though we are pretty certain Mr. Maclagan had not the passage in his eye when he wrote. Indeed, the latter part of his poem could have been written in only the present age: —

THE OUTCAST.

" And did you pity me, kind sir?
Say, did you pity me?
Then, oh how kind, and oh how warm,
Your generous heart must be!
For I have fasted all the day,
Ay, nearly fasted three,
And slept upon the cold, hard earth,
And none to pity me;
And none to pity me, kind sir,
And none to pity me.

" My mother told me I was born
On a battlefield in Spain,
Where mighty men like lions fought,
Where blood ran down like rain!
And how she wept, with bursting heart,
My father's corse to see,
When I lay cradled 'mong the dead,
And none to pity me;
And none to pity me, kind sir,
And none to pity me.

" At length there came a dreadful day, —
My mother too lay dead, —
And I was sent to England's shore
To beg my daily bread, —
To beg my bread; but cruel men
Said, Boy, this may not be,
So they locked me in a cold, cold cell,
And none to pity me;
And none to pity me, kind sir,
And none to pity me.

"They whipped me,—sent me hungry forth;
  I saw a lovely field
Of fragrant beans; I plucked, I ate:
  To hunger all must yield.
The farmer came,—a cold, a stern,
  A cruel man was he;
He sent me as a thief to jail,
  And none to pity me;
And none to pity me, kind sir,
  And none to pity me.

"It was a blessed place for me,
  For I had better fare;
It was a blessed place for me,—
  Sweet was the evening prayer.
At length they drew my prison bolts,
  And I again was free,—
Poor, weak, and naked in the street,
  And none to pity me;
And none to pity me, kind sir,
  And none to pity me.

"I saw sweet children in the fields,
  And fair ones in the street,
And some were eating tempting fruit,
  And some got kisses sweet;
And some were in their father's arms,
  Some on their mother's knee;
I thought my orphan heart would break,
  For none did pity me;
For none did pity me, kind sir,
  For none did pity me.

"Then do you pity me, kind sir?
  Then do you pity me?
Then, oh how kind, and oh how warm,
  Your generous heart must be!
For I have fasted all the day,
  Ay, nearly fasted three,
And slept upon the cold, hard ground,
  And none to pity me;
And none to pity me, kind sir,
  And none to pity me."

XII.

OUR NOVEL LITERATURE.

WHAT are the most influential writings of the present time, — the writings that tell with most effect on public opinion? Not, certainly, the graver or more eleaborate productions of the press. Some of these in former times exerted a prodigious influence. There were four great works, in especial, that appeared at wide intervals during the seventeenth and eighteenth centuries, — the last of their number about eighty years ago, — that revolutionized, on their respective subjects, the thinking of all Europe; and these were, the "Laws of Peace and War," the "Essay on the Human Understanding," the "Spirit of Laws," and the "Wealth of Nations," — all works of profound elaboration, that contain the thinking of volumes condensed into single pages. At an earlier period there were theological works that stirred men's minds to their utmost depths, and changed the political relations of states and kingdoms over all Christendom. Such was the influence exerted by the treatises of Luther, whose written "words were half-battles;" and by those "Institutes of Calvin" that gave form and body to the thinking of half the religious world. But whether it be that we live in an age too superficial to produce, or too busy too read, such works, or at once superficial and busy both, without either the works to read or the time to read them in, it is certain that almost all power has passed away from the grave and the elaborate to the light and the clever, and that what would have been pronounced about a century ago the least influential kinds of writing must now be recognized as by far the most influential. Had one said to a literary man in the early days of Johnson, "Pray, what do you regard as the least important departments of your literature, both in themselves and their effects, and that tell least on the public mind?" the reply would probably have been, "Why, the writing in our newspapers and our novels." And now the same reply would serve at least equally well to indicate the kinds of writing that are most telling and influential. None others exert so great a power over the general mind of the community as novels and newspaper articles. And the mode of piecemeal publication recently resorted to by our more popular novelists gives to the effect proper to their compositions as pictures of great genius and power the further effect of pamphlets or magazines: they are at once novels and newspaper articles too.

Considerably more than a century has passed, however, since a judicious critic might have seen how very influential a class of compositions well-written novels were to become. "The Life and Adventures of Robinson Crusoe" appeared as far back as the year 1719, and at once rose to the popularity which it has ever since maintained. But it failed to attract the notice of the critics. The men who sat in judgment on the small elegances of the wits of the reign of George I., and marked how sentences were balanced and couplets rounded, could not stoop to notice a composition so humble as a novel, more especially a novel written by a self-taught man. But his singularly vivacious production forced a way for itself, leaving the fine sentences and smart couplets to be forgotten. In a short time it was known all over Europe; several translations appeared simultaneously in France, much about the period when Le Sage was engaged in writing, in one of the smaller houses of one of the most neglected suburbs of Paris, his "Gil Blas" and his "Devil on Two Sticks;" and such was the rage of imitation which it excited in Germany, that no fewer than forty-one German novels were produced that had Robinson Crusoes for their heroes, and fifteen others, that, though equally palpable imitations, had heroes that bore a different name. Eight years after the publication of Defoe's great work, there appeared an English novel of a more extraordinary form, and of higher literary pretensions, in the "Travels of Gulliver;" and it too at once attained to a popularity which has never since flagged or diminished. Thirteen years more elapsed, and Richardson had produced his "Pamela," and, shortly after, Fielding his "Joseph Andrews." Smollett came upon the scene with his "Roderick Random" in eight years more. There followed in succession, after the lapse of about ten other years, the "Rasselas" of Johnson and the "Candide" of Voltaire, — both works which spread over the world; and in yet seven other years Goldsmith's "Vicar of Wakefield" appeared, and attained to even a more extensive popularity than either. And yet still, after the teaching of nearly half a century, — nay, after nearly two centuries had elapsed since a novel was recognized as the most popular and influential of all the works ever produced by Spain, — grave and serious people continued to speak of novels as mere frivolities, that were to be in every case eschewed by the young, but were scarce of importance enough to be heeded by the old at all. Nor even yet, — after the novels of Scott have, if we may so express ourselves, taken possession of the world, — after the most potent work of Germany, the "Wilhelm Meister" of Goethe, has appeared, like that of Spain, in the form of a novel, — after the modern novels of France have been measuring lances with even its priesthood, and approving themselves, in at least the larger towns, the mightier power of the two, — and after, in our own country, it has been accepted altogether as a marvel that history, in the case of Macaulay's, should have its thirty thousand subscribers, but as quite an expected and ordinary thing that fiction, in Dickens's current work, should have at least an equal number, — the old estimate in the minds of many has been suffered to remain uncorrected, and the novel is thought of rather as a light, though not always very laudable toy, than as a tremendously potent instrument for the origination or the revolutionizing of opinion. Some of our great lawyers could make sharp speeches, about two years ago, against what they termed the misrepresentations of "Bleak House," evidently regarding it, as they well might, as the most formidable series of pamphlets against the abuses of chancery, and the less justifiable practices of the legal profession, that ever appeared. We are by no means sure, however, that the church is as thoroughly awake to the tendencies of his present work as members of the legal faculty, wise in their generation, were to the design of his last.

Most of the novelists have been hostile to virtue of a high or severe kind in general; and there were few of eminence produced in our own country that did not leave on record their dislike of evangelism in particular. We are afraid Byron was in the right in holding that Cervantes laughed away the chivalry of Spain: Spain produced no heroes after the age of Don Quixote. As for Le Sage, Vinet is at least as just in his criticism as Byron in his, when he says that "his novels do not contain a single honest character, — nothing but knaves and weaklings, and even the very weaklings are far from being honest." "In a word," we find the critic again remarking, "'Gil Blas' is but a paraphrase of the celebrated maxim of Rochefoucauld, — 'Virtue is only a word; it is nowhere found on the earth; and we must be resigned." Most of the modern novelists of France stand on a still lower level than that of their great master, Le Sage. He did not inculcate virtue, and they teach positive vice. Nor is Goethe a safer guide. The "Sorrows of Werter" and "Wilhelm Meister's Apprenticeship" are both very mischievous books. The novelists of our own country have been more mixed in their character. Defoe we must regard as, with all his faults, a well-meaning man, who had been an object of persecution himself, and had learned to sympathize with the persecuted. The Scotch were very angry with him for the part he took in the Union; but that did not prevent his doing justice, in his history, to their long struggles for ecclesiastical independence; and religion never comes across him in his novels, — some of them quite loose enough, — but he has always a good word to say in its behalf. He was no very profound theologian: Friday, in the dialogue parts of "Crusoe," is nearly as subtle a divine as his master; and when poor Olivia Primrose instances, as a proof of her large acquirements in controversy and her consequent ability of converting 'Squire Thornhill, that she had read all the "Religious Courtship," — another of Defoe's works, — we at once agree that the worthy doctor, her father, did quite right in sending her off to "help her mother in making the gooseberry pie." Swift, clergyman as he was, manifested, however, a very different spirit from that of Defoe: in proportion as he knew more he reverenced less; and there is perhaps nothing in our literature more essentially profane than his essay on the "Mechanical Operation of the Spirit," and his "Tale of a Tub." Richardson, no doubt, deemed himself a friend to virtue and religion. He patronized both after a sort, and many good ladies and clergymen were moved, in consequence, to patronize him; and yet, as Vinet pointedly says of the general literature of France in that age, his "very morality was in fact immoral." We know not whether we would not give "Tom Jones" as readily into the hands of a young person as the virtuously written "Pamela." There is more of a wholesome, generous, unselfish spirit about the scapegrace, than in the demure, designing girl, who, after behaving herself well for a time, sets her cap to catch her master, and is at length rewarded with a fine house, a fine coach, and Mr. Booby. And yet Fielding, like his hero, *was* a sad scapegrace. He had a respect for what he deemed religion. We see it in his novels even. Of the few thoroughly honest men he ever drew, — and, unlike Le Sage, he did occasionally draw honest men, — two are clergymen, —

Dr. Harrison in "Amelia," and the world-renowned Parson Adams in "Joseph Andrews;" and both are represented, though in the case of the latter with many a ludicrous accompaniment, as at least as good and sincere Christians as Fielding could make them. Nay, curiously enough, one of the novelist's last works, a work which he did not live to finish, was a defense of religion against Bolingbroke, and a very ingenious one. But alas for a Christianity such as that of Whitefield when it came across him! If the devoted missionary could have been annoyed by anything, it would have been by the ruthless humor with which his brother and his brother's wife are introduced by name into "Tom Jones," as the landlord and landlady of the Bell public house in Gloucester; and the terms in which the lady is spoken of as "a very sensible person," who, though at first the preacher's "documents" made so much impression on her "that she put herself to the expense of a long hood in order to attend the extraordinary movements of the Spirit," got tired of emotions, "which proved to be not worth a farthing," and at once "laid by the hood, and abandoned the sect."

Smollett was of a similar spirit. We know nothing better on the subject in our language than the essay in which he argues against Shaftesbury that ridicule is not the test of truth; but no little ridicule does he himself heap on Methodism in his "Humphrey Clinker." There is no bitterness in his exhibition; his untaught Methodist preacher is not a disagreeable fool, like the Rev. Mr. Chadband, or a greedy rogue, like the Methodist preacher in "Pickwick," whom old Weller treats to a ducking; but, on the contrary, a thoroughly honest fellow, and, in his own proper sphere, a sensible and useful one. He is, in short, no other than the faithful Clinker himself. But he never associates religion of any earnestness save with characters of humble parts and acquirements, and always accompanied with points of extreme ludicrousness. Goldsmith was of a more genial temperament than Smollett.

His Vicar is one of the most thoroughly honest men that ever lived, and has all the religion that poor Goldie could give him. It was not until a later time, however, and in Scotland too, — for we need not reckon on the now forgotten novel of Mrs. Hannah More, — that religious characters were most largely introduced into our novel literature. Scott, Lockhart, Wilson, Galt, Ferrier, have all brought religion in review before the public in their novels, — some of them with great power, some with considerable truth, some with truth and with power too; and at least one novelist of considerable ability, the excellent authoress of "Father Clement," made it her leading subject. They all at least knew more of religion than the earlier novelists; and, save when carried away, as in the case of Scott, by Jacobite predilections, or in that of Lockhart, by moderate ones, did it more justice. Even in some of Scott's pictures there is wonderful truth. The few words in which poor Nanty Ewart is made, in his remorse, to describe his father, are those of a great master of character. "There was my father (God bless the old man!), a true chip of the old Presbyterian block, walked his parish like a captain on the quarter-deck, and was always ready to do good to rich and poor. Off went the laird's hat to the minister as fast as the poor man's bonnet. When the eye saw him, — Pshaw! what have I to do with that now? Yes, he was, as Virgil hath it, '*Vir sapientia et pietate gravis*.'" Still more distinctive is he, however, when he speaks of him in connection with two charitable ladies of the Roman Catholic Church. "These Misses Arthciret," says Nanty, "feed the hungry and clothe the naked, and such like acts, *which, my poor father used to say, were filthy rags; but he dressed himself out with as many of them as most folk*." There is not such a stroke as this in all Dickens. The writer who could draw such a feature with a single dash of the pencil well knew what he was about.

But it would be easy to multiply remarks such as these

on the novelists. The fact of their mighty influence on opinion cannot, we think, be challenged; and so it is of great importance that the influence should be a good one, or at least so far negatively good as not to be hurtful. We are aware that there are very excellent people who would altogether *taboo* this class of works; they would fain render them the subject of a sort of Maine law, make the open perusal of them unlawful, and severely punish all smuggling. But their attempts hitherto have been attended with but miserable success. We have often had occasion to know, that, even among their own children, they succeeded with only the very stupid ones, who have no turn for reading; and that model-grown men or women of their training, ignorant of our novel literature, are usually scarce less ignorant of literature of any other kind, and yet not a whit better than their neighbors. Besides, even were the case otherwise, — even were they to be really successful in their own little spheres, — the great fact of the influence and popularity of the genuine novel would still remain untouched. Dickens would have his thirty thousand subscribers for every new work, and at least his half million of readers; and the proprietor of the Scott novels would continue to sell sixty thousand volumes yearly. Further, the novel *per se*, the novel regarded simply as a literary form, is *morally* as unexceptionable as any other literary form whatever, — as unexceptionable as the epic poem, for instance, or the allegory, or the parable. The "Vicar of Wakefield," as a form, is as little blamable as the "Deserted Village," or "Waverley" as "Marmion" or the "Lay of the Last Minstrel." And so we must hold, that, on every occasion in which the *form* is made the vehicle of truth, — truth of external nature, truth of character, historic truth in at least its essence, and ethical truth in its bearings on the great problem of society, — it should be received with merited favor, — not frowned upon or rejected. We have been much pleased, on this principle, with the novels of a writer to whom we ought to have referred approvingly long ago, — the authoress of "Mrs. Margaret Maitland,"— one of the most thoroughly truthful writers of her class, and one of the most pleasing also. We have now before us what may be regarded as a continuation of her first work, — in "Lilliesleaf,"— a concluding series of passages in the life of "Mrs. Margaret Maitland." It is, of course, a formidable matter to introduce a second time to the public any character that had on its first appearance engaged and interested it. Shakspeare could do it with impunity. Falstaff, on even his third appearance, — an appearance, however, which, had the great dramatist been left to himself, he would never have made, — is Falstaff still. But even Scott has been but partially successful in an attempt of the kind. The Cœur de Lion of the "Talisman" is not at all so interesting a personage as the Cœur de Lion of "Ivanhoe." And so we took up these new volumes with some little solicitude regarding Mrs. Margaret. The old lady has, however, acquitted herself admirably, — in some passages more admirably, we will venture to say, in the face of an opposite opinion which we have seen elsewhere expressed, than on her first appearance. In the early part of the first volume we were, indeed, sensible of an air of languor, and the narrative moved on too slowly, — Mrs. Maitland seemed to have grown greatly older than when we had last seen her; though even in this part of the work we found some very admirable things, — among the rest, a true life-picture of the ancient dowager lady of Lilliesleaf, with her broken health and failed understanding, ever carping and fault-finding; and, while beyond the reach of all advice herself, always obtruding her worse than useless advices on other people, who did not want them, and could not take them, and had no need of them. As the work goes on, however, the interest increases; there are new characters introduced, truthful glimpses of the Scotch people given, the incidents thicken, and the narrative, though always quiet, as becomes the grave and gentle narrator, gathers headway, and grows more rapid. We know few things more masterly than the character of Rhoda, a wild, clever, ill-taught girl, brought up by a reckless, extravagant father, who, after utterly neglecting her himself, introduces her into the house of her half-sister, an excellent but somewhat proud and cold woman, who evinces but little sympathy for her provoking and haughty but very unhappy relation. Mrs. Margaret, however, after encountering many a rebuff, at length wins her; and there are few things finer in our novel literature than the scene in which she does so: —

" As I was going to my bed, I tarried in the long gallery, where Miss Rhoda's door opened into, to look at the bonnie harvest moon mounting in the sky, the which was so bright upon the fields and the garden below the window, that I could not pass it by without turning aside to glance upon the grand skies, and the warm earth with all routh and plenty yet upon her breast, that were both the handiwork of the Lord. I had put my candle upon a table at the door of my own room; and as I was standing here, I heard a sound of crying and wailing out of Miss Rhoda's room. It was not loud, but for all that it was very bitter, as if the poor bairn was breaking her heart. Now, truly, when I heard that, I never took two thoughts about it, nor tarried to ponder whether I would be welcome to her or no; but hearing that it was her voice, and that she was in distress, I straightway turned and rapped at the door.

" The voice stoppit in a moment; so quick I scarce could think it was real; and then I heard a rustling and motion in the room. I thought she might be feared, seeing it was late; so I said, 'It is me, my dear; will you let me speak to you?' It was all quiet for a moment more, and then the door was opened in an impatient way, and I entered in. Rhoda was there, turning her back upon me; and there was no light but the moonlight, which made the big room, eerie though it was, so clear that you could have read a book. The curtains of the bed were drawn close, as Cecy had drawn them when she sorted the room for the young lady, and Rhoda's things were lying about on the chairs; and through the open door of the small room that was within there was another eerie glint of the white moonlight; and pale shadows of it, that, truly, I liked not to look upon, were in the big mirror that stood near. It was far from pleasant to me, — and I was like to be less moved by fancy than a young thing like Rhoda, — the look this room had.

"'My dear bairn,' said I, being more earnest than I ever was with her before, 'will you let me hear what ails you? I ken what trouble is myself; and many a young thing has told her trouble to me. And you are lone and solitary and motherless, my poor bairn; and I am an aged woman, and would fain bring you comfort if it was in my power. Sit down here, and keep no ill thought in your heart of me; for I ken what it is to be solitary and without friends mysel.'"

"She stood awhile, and would not mind what I said, nor the hand I put upon her arm. And then she suddenly fell down upon her knees in a violent way, and laid her face upon the sofa, and cried. Truly, I kent not of such tears. I have shed heavy ones, and have seen them shed; but I kent not aught like the passion and anger and fierceness of this.

"'I can't tell you what grieves me,' she said, starting up, and speaking in her quick way, that was so strange to me, — 'a hundred thousand things — everything! I should like to go and kill myself — I should like to be tortured — oh! anything — anything rather than this!'

"'My dear, is it yourself you are battling with?' said I; 'for that is a good warfare, and the Lord will help you if you try it aright. But if it is not yourself, what is it, my bairn?'

"She flung away out of my hand, and ran about the room like a wild thing. Then she came, quite steady and quiet, back again. 'Yes,' she said, 'I suppose it is myself I am fighting with. I am a wild beast, or something like it; and I am biting at my cage. I wish you would beat me, or hurt me, — will you? I should like to be ill, or have a fever, or something to put me in great pain. For you are a good old lady, I know, though I have been very rude to you. No, I am sure I cannot tell *you* what grieves me; for I cannot fight with you. It is all papa's fault, — that is what it is! He persuaded me that people would pay attention to me here. But I am nobody here, — nobody even takes the trouble to be angry with *me!* And I cannot hate you all, either, though I wish I could. Oh! — old lady, go away!'

"'Na, Miss Rhoda,' said I; 'I am not going away.'

"'That ridiculous Scotch, too!' cried out the poor bairn, with a sound that was meant for laughter. 'But I can't laugh at it; and sometimes I want to be friends with you. How do you know that I never had a mother? for it is quite true I never had one, — never from the first day I was in the world. And I love papa with my whole heart, though he is not good to me; and I hate every one that hates him; and I will not consent to live as you live here, however good you may pretend to be.'

"'But, Miss Rhoda,' said I, 'what ails you at the way we live here?'

"'It is not living at all,' said the poor bairn. 'I never can do anything very well when I try; but I always want to be something great. I cannot exist and vegetate as you quiet people do. What is the good of your lives to you? I am sure I cannot tell; but it will kill me.'

"'You have never tried it, my dear,' said I; 'so whether it will kill you or no, you can very ill ken. But till me how you would like to be great.'

"'Why should I speak of such things? You would not understand me,' said Rhoda. 'I would like to be a great writer, or a great painter, or a great musician, — though I never would be a servant to the common people, and perform upon a stage. I know I could do something, — indeed, indeed, I know it! And you would have me take prim walks, and do needlework, and talk about schools and stuff, and visit old women. Such things are not for me.'

"'Such things have been fit work for many a saint in heaven, my dear,' said I; 'but truly I ken no call that has been made upon you, either for one thing or another. Great folk, so far as I have heard, are mostly very well pleased with the common turns of this life to rest themselves withal; and truly it is my thought, that the greater a person is, the less he will disdain a quiet life, and kindness, and charity. But it has never been forbidden you, Miss Rhoda, to take your pleasure; and I wot well it never will be.'"

This surely is powerful writing, — so entirely worthy of Mrs. Margaret Maitland, that we know not whether we could quite equal it by any extract of the same length from her former work. There is much quiet power, too, in the sketches given of external nature in the present volumes, and much originality of observation. We know not that we ever before met in books with what we may term the echo of that peculiar sound characteristic of a furzy moor under a hot sun which is so well described as in the following passage. All our readers must remember the incessant "crack, crack, crack," which they have so often heard when the sun was hot and high, mingling, amid the long broom or prickly whins, with the chirp of the grasshopper and the hum of the bee: —

"Naw, we had scarce ended our converse, when, looking out at the end window, I saw Rhoda coming her lane along the road; and, seeing she might be solitary in her own spirit among such a meeting of near friends, I went out to the door to bring her in myself. It was a very bonny day, as I have said, and, the bairns being round upon the lawn at the other side, there was but a far-off sound of their voices, and everything else as quiet as it could be under the broad, warm, basking sun, — *so quiet, that you heard the crack of the seed husks on a great bush of gorse near at hand, — a sound that ever puts me in mind of moorland places, and of the very heart and heat of sunny days.* Rhoda, poor bairn, was in very deep black, as it behoved her to be, and was coming, in a kind of wandering, thoughtful way, her lane down the bright sandy road, and below the broad branches of the chestnut trees, that scarce had a rustle in them, so little air was abroad; and the bit crush of her foot upon the sand was like to a louder echo of the whins, and made a very strange kind of harmony in the quietness."

This wholesome and very interesting novel is calculated to exert a salutary influence, and to yield, besides, much pleasure in the perusal. Like all the other works of its authoress, it is thoroughly truthful: there is no exaggeration of character or incident; events such as it narrates occur in real life; and the men and women which it portrays may be met in ordinary society, though the better ones are unluckily not very common. And yet a wild romance, full of all sorts of marvels and monstrosities, could scarce amuse so much even a youthful reader, far less readers of sober years. In nothing, however, has the work more merit than in its representations of the religious character. Here, also, there is no exaggeration. The natural temperament is exhibited as exerting its inevitable influence. Rhoda's half-sister, Grace, for instance, though one of the excellent, is not at all so lovable a person as Mrs. Margaret, just because *in her* religion was set on what was originally a more wilful and less loving nature; and we find this thoroughly truthful distinction maintained throughout. In short, this latest production of Mrs. Margaret Maitland is a book which may be safely placed in any hands; and, seeing that novels must and will exist, and must and will exercise prodigious influence, whether the religious world gives its consent or no, we think the good people should by all means try whether they cannot conscientiously patronize the good ones.

XIII.

EUGENE SUE.

It is not from the formal histories of a country, as history has hitherto been written, that the manner and morals of its people may best be learned. Its works of fiction, if they have been produced by the hand of a master, and have dealt with the aspects of contemporary society, are vastly more true to the lineaments of its internal life than its works of sober fact. Smollett's "History of the Reign of George II." is a dull record, that bears on its weary series of numbered paragraphs no distinguishable impress of the character of the age; whereas Smollett's "Humphrey Clinker" is one of the most admirable pictures of English society during that reign which anywhere exists. The severe history, with all its accuracy of names and dates, wants truth; the amusing novel, that seems but to play with ideal characters, is, in all its multitudinous lights and shadows, a true portraiture of the time. And the rule seems general. Does the student wish to acquaint himself with the aspect of English society in the days of our great grandfathers? — he will gain wonderfully little by poring over heavy sections in the "Annual Registers" of Dodsley, but a very great deal in the study of the graphic sketches of Richardson and Fielding. The "Waverley" of Scott is truer beyond comparison to the real merits of the Rebellion of 1741 than the authentic history of Home, though Home was himself an actor in many of the scenes which he describes.

It is partly at least from a consideration of this kind that we have placed at the head of our article the name of one of the most popular French novelists of the present day, — a writer whose fictions have been introduced nearly as extensively to the people of London, through the medium of cheap translations, as to those of Paris in the original French, and which are widely circulated over the Continent generally. His novels, with all their extravagances, give a striking picture of the state of society among at least the city-reared masses of France, and are singularly efficient vehicles in spreading over Europe the contagion of their principles. We find in them more of the philosophy of the late movement in Switzerland against the Jesuits, though they contain not a single allusion to that event, than in any of the narratives of the outbreak which we have yet seen. They serve to show how opinion among the anti-Jesuit party came first to be formed, the nature, too, of that opinion, and how it happens that they are not merely an anti-Jesuit, but also an anti-evangelistic and anti-tolerant party. Their views and principles are exactly those of Eugene Sue; and their numbers bid fair to increase over Europe, wherever the influence of his writing shall be found to prevail. But a brief sketch of some of the leading characters in one of his latest and most characteristic works — the "Wandering Jew," of which we perceive a cheap English translation has just appeared — may better serve to show what his fictions teach than a general reference to their tendency or effects. Rome, in the course of its history, has been signally damaged by two great revolutions in religious opinion, — the Reformation of Luther, and the great revolt of Voltaire. The revived Christianity of the New Testament was the formidable antagonist with which it had to deal in the one case, and a singularly enthusiastic and fanatical infidelity the enemy with which it had to contend in the other; and, for a time, the injury which it received seemed in both cases equally severe. But they were in reality very different in their nature. The wound dealt by infidelity was a flesh wound, and soon healed; whereas the blow dealt by the revived Christianity amputated the members on which it took effect, and separated them forever from the maimed and truncated carcass. Infidelity dips its idle bucket into the sea of superstition, and labors to create a chasm, where, in the nature of things, no chasm can exist; there is a momentary hollow formed, but the currents come rushing in from every side, and fill it up. But evangelism not only scoops out the hollow, but also occupies it, leaving no vacuum for aught else to flow in. France, in less than an age after the canonization of her atheists, had again become popish; the tides flowed in, and the vacuum was annihilated: whereas evangelistic Scotland is as little popish now as she was two centuries ago; for in her that perilous space which must be occupied either by religion or superstition was thoroughly filled by the doctrines of the New Testament. The remark bears very directly on the nature of the warfare waged on Rome and the Jesuits by Eugene Sue. His labors, like those of Voltaire, serve but to create a vacuum, abhorrent to the nature of man.

The chief group in his recent novel, round which all its other groups are made to revolve, and on whose designs their destiny is made to hang, is the Society of the Jesuits. We see them pursuing their schemes of ambition and aggrandizement undeterred by any sense of justice, and without any feeling of pity or remorse. And the picture, we are afraid, is scarce exaggerated. As exhibited in this work of fiction, there is no part of it so black as to be without its counterpart in real history. There are two grand circumstances which have conspired to render the Jesuits what they are, — the specific nature of their principles, and their generic character as a society. An able man, possessed of much power, who held by the principles of the Jesuits, and cared not what means he employed in effecting his ends, would be eminently dangerous. Their principles are, in fact, the principles of the great bad man, who subordinates to his designs whatever is venerable in morals or sacred in religion, and regards the end as justifying the means. The Machiaevel-taught despot, whether he be a Charles I. or a Louis XIV., is, to the extent of his principles, a Jesuit on his own behalf. But then the individual bad man has what the bad society has not, — he has human feelings; and these often create a diversion against his principles in favor of his suffering fellows. Even a Nero could weep. But societies have no tears: they are abstract embodiments of their principles; and if their principles be bad, it is in vain to look for protection against them to their feelings. They don't feel. Even when their principles are not ostensibly bad, — when the cord by which they are united is a mere love of gain, — it is too much their tendency, as well described by Cowper, to become cruel and unjust: —

> "Man in society is like a flower
> Blown in its native bed: 'tis there alone
> His faculties, expanded in full bloom,
> Shine out, — there only reach their proper use.
> But man associated and leagued with man
> By regal warrant, or self-joined by bond
> For interest's sake, or swarming into clans
> Beneath one head, for purposes of war,
> Like flowers selected from the rest, and bound
> And bundled close, to fill some crowded vase,
> Fades rapidly, and, by compression marred,
> Contracts defilement not to be endured."

But when their end is not vulgar gain, but power, however

attained, and the aggrandizement of a false and bloody church, — when their principles, untrue to the first laws of morals, strike at the very foundations of all justice, and are, in short, what Pascal has so well described, — and when to all this the inevitable lack of human feeling is added, — the result is, not a corporation of ordinary and every-day iniquity, but a society without parallel in the annals of the world, — the Society of the Order of Jesus. And so Eugene Sue has not done them less than justice in his fiction. Moliere, in one of his dramas, introduces a character who, after he had been guilty of almost every crime, — after he had abandoned his wife, cheated his friends, deceived and insulted his father, and made an open profession of his atheism, — completes the climax of his infamy by becoming hypocrite. Eugene Sue, in holding up the Jesuits to abhorrence, improves on the design. Such is the character which he gives to but the second worst Jesuit in the piece. In early life the Jesuit had been a traitor to his country, and had fought against it; he had been the ungenerous enemy of a brave and honest man who abhorred his treachery, and had pursued with bitter hatred his unprotected wife and defenceless children. His prevailing passion was a vulgar love of power; and in order to obtain it, there was no intrigue too mean for him to stoop to, no crime too atrocious for him to perpetrate; but, with all his baseness and villany, he is drawn as not wholly devoid of human feeling: his mother on her death-bed enjoins that he should visit her; and it is with reluctance, and hesitatingly, that he sets aside the dying injunction, and sets out in an opposite direction on some business of the Society; and this one touch of inoperative human feeling is rendered a sufficiently grave fault in the hands of the novelist to reduce him from a first to a second place in the community of Loyola. The first place is assigned to a wretch whom we recognize as actually a man and not a demon, when we find that he has a frame which can be acted upon by poison and the cholera, but not before. In the development of the plot, we see the machinations of the Society involving in ruin all that is good and lovable among the *dramatis personæ* of the piece: the just, the generous, the honorable, — the unsuspecting maiden, the kind master, the attached father, the devoted friend, — all become, in turn, the victims of the meanest and basest villany; and Jesuitism, devoid of all tinge of pity and remorse, exults over them as they perish. We do not wonder how the admirers of such a work should learn to hate the Jesuits. It seems suited to accomplish, amid the superficiality of the present age, in the innumerable class of French novel-readers, the effects which were produced in a higher order of minds, rather more than a century and a half ago, by the "Provincial Letters of Pascal." The English reader who has read the "Wandering Jew" will be better able to estimate from the perusal than before the intense hatred of the Jesuits which animated, in their late outbreak, the insurgent Switzers of Vaud and Argovia.

But we can see no elements of permanency in the principles marshalled against them, either as embodied in the characters of Eugene Sue, or as illustrated from time to time by the minute portions of passing history. The controversy does not lie between truth and error, but between antagonist errors. The determined assailants of priestly superstition and villany are themselves the asserters of principles which, if reduced to practice, would subvert all public morals; and for the false belief which they would so fain extinguish, they would substitute an unnatural vacuum, into which other false beliefs would assuredly crowd. Nay, in the fictions of Eugene Sue we already see the phantoms of a false faith crowding into the gap. All the honest devotees which he draws are exhibited as weak in proportion to the strength of their religious feelings. Their religion is represented as forming a mere handle by which they are converted into the tools of designing hypocrites; and yet, in the supernatural machinery of the piece, we see, as in the athestic poetry of Shelley, the elements of a new religion coming into view, and embodying, in an incipient state, not a few of the worst errors of Rome. One of the leading characters in the novel — a young lady of high birth and talent, whose destruction the Jesuits at length effect, and are rendered detestable by effecting — is represented as adorned by qualities the most generous and lovable. We must select one trait of many, not merely as a specimen of the character, but of the art also with which the novelist addresses himself to the independent feelings of the French people, which have been so prominently developed since the Revolution. The heroine of the following passage is, as we have said, a lady of birth and fortune; and it is a poor journeyman mechanic, of spirit and talent, however, who is the second actor in the scene: —

"When Adrienne entered the saloon, Agricola was examining a magnificent silver vase, which bore the words, 'Jean Marie, working-chaser, 1823.' Adrienne trod so lightly, that she had approached the blacksmith without his being aware of it.

"'Is not that a handsome vase, sir?' she said, in a silver-toned voice.

"Agricola started, and replied, in confusion, 'Very handsome, mademoiselle.'

"'You see that I am an admirer of what is just and right,' said Adrienne, pointing to the words engraved on the vase. 'A painter puts his name to a picture, a writer to his book; and I hold that a workman who distinguishes himself in his trade should put his name to his workmanship. When I bought this vase it bore the name of a wealthy goldsmith, who was astonished at my fantasies, for I caused him to erase it, and to insert that of the maker of this wonderful piece of art; so that if the workman lack riches, his name at least will not be forgotten. Is this just, sir?'

"'As a workman, mademoiselle, I feel sensible of this act of justice.'

"'A skilful artisan merits esteem and respect. But take a seat, sir.'"

This is a fine trait, and the character of Adrienne is

mainly composed of such; but the author takes particular care to inform us that she is not a Christian; and when we come to learn her views on marriage, we find that they are exactly those of Mary Wolstonecraft. The sentiments which she is made to express in the following scene are not unworthy of being examined. They are not simply those of a writer of fiction, struck out at a sitting, and then given to the world merely to amuse it, and keep up the interest of his work: they are, on the contrary, widely disseminated over the cities of Europe, and very extensively acted upon. Socialism in our own country ostensibly adopts them as its own; and there are many not Socialists, who, though the usages of society prevent their acting upon them, have not hesitated to adopt them. We need scarce remind the reader that the subject is one upon which the Saviour has authoritatively spoken, and that if he be Truth, the modern theory is a lie: —

"'Something is wanting to consecrate our union; and in the eyes of the world there is only one way, — by marriage, which is binding for life.'

"Djalma looked at the young girl with surprise.

"'Yes, for life; and yet who can answer for the sentiments of a whole life? A Deity able to look into futurity could alone bind irrevocably certain beings together for their happiness. But, alas! the future is impenetrable to us; therefore we can only answer for our present sentiments. To bind ourselves indissolubly is a foolish, selfish, and impious action, — is it not?'

"'That is sad to think of,' said Djalma, after a moment's reflection, 'but it is true.' He then regarded her with an expression of increasing surprise.

"Adrienne hastily resumed, in a tender tone, — 'Do not mistake my meaning, my friend. The love of two beings who, like ourselves, after a patient investigation of heart and mind, have found in each other all the assurances of happiness, — a love, in short, like ours, is so noble, so divine, that it must be consecrated from above. I am not of the religion of my venerable aunt; but I worship God, from whom we derive our ardent love. For this he must be piously adored. It is therefore by invoking his name with deep gratitude that we ought to promise *not* to love each other forever, — *not* to remain always together.'

"'What!' cried Djalma.

"'No,' resumed Adrienne, 'for no one can take such an oath without falsehood or folly; but we can, in the sincerity of our hearts, swear to do faithfully everything in our power to preserve our love. Indissoluble ties we ought not to accept; for if we should always love each other, of what use are they? and if not, our chains are then only an instrument of odious tyranny. Is it not so, my friend?'

"Djalma did not reply; but with a respectful gesture he signed to the young girl to continue.

"'And, in fine,' resumed she, with a mixture of tenderness and pride, 'from respect to your dignity as well as my own, I would never promise to observe a law made by man against women with brutal selfishness, — a law which seems to deny to woman mind, soul, and heart, — a law which she cannot obey without being a slave or a perjurer, — a law which deprives her of her maiden name, and declares her, as a wife, in a state of incurable imbecility, by subjecting her to a degrading state of tutelage; as a mother, refuses her all right and power over her children; and as a human being, subjects her son even to the will and pleasure of another human being, who is only her equal in the sight of God. You know how I honor your noble and valiant heart; I am not, therefore, afraid of seeing you employ those tyrannical privileges against me; but I have never been guilty of falsehood in my life, and our love is too holy, too pure, to be subjected to a consecration which must be purchased by a double perjury.'"

Such are the principles of this Parisian heroine, and such are some of the plausibilities with which she defends them. There are two other female characters in the work, twin sisters, of great beauty, whom the Jesuits also succeed in destroying; and they, too, are devoid of religion. Unlike Adrienne, however, they are not intellectually infidel, — they have simply never heard of Christianity; and when they pray, it is to their deceased mother. Yet another of the female characters, a poor seamstress, possessed, however, of a cultivated mind and a noble heart, finds no time to attend to the duties of religion; and when, through the machinations of the Jesuits, she becomes destitute and wretched, she proposes to go out of the world by her own act, as convinced that she is in the right in doing so, as if, wearied and overcome by sleep, she had prepared to go to bed. She is joined in the purpose of death by her sister; and the scene throws light on the acts of social suicide so common in France, and of which we have had a few instances of late years in our own country.

"The sisters embraced each other for some minutes amid a profound and solemn silence.

"'O heavens,' cried Cephysi, 'how cruel, to love each other thus, and be compelled to part forever!'

"'To part!' exclaimed the Mayeux, while her pale face was suddenly lighted up with a ray of divine hope; —'to part! Oh no, sister, no: what makes me so calm is, that I feel certain we are going to another world, where a happier life awaits us. Come, hasten; come where God reigns alone, and where man, who on this earth brings about the misery and despair of his fellow-creatures, is nothing. Come, let us depart quickly, for it is late.'

"The sisters, having laid the charcoal ready for lighting, proceeded with incredible self-possession to stop up the chinks in the door and windows; and during this sinister operation, the calmness and mournful resignation of these two unfortunate beings did not once forsake them."

We had intended referring to several other points in this mischievous work of fiction, which at once serves to exhibit the opinions entertained by no inconsiderable proportion of the anti-Jesuit party on the Continent, and to spread these opinions more widely. Wherever we find the devotional feeling introduced, some disaster is sure always to follow. One of the best characters in the novel is a highly intellectual and generous manufacturer, more bent on ministering to the happiness of his workmen than on the accumulation of gain. He provides them with comfortable dwellings, extends their leisure hours, gives them a share in the profits of his trade, — conducts his manufactory, in short, on the model of the philanthropic economist; and all this when he is an avowed Freethinker; but, falling into bad health, and meeting with a crushing disappointment, he becomes a devotee, loses all his interest in the welfare of his workmen, becomes enfeebled in body and mind, and the Jesuits ruin him. The wife of a brave and faithful soldier, a thoroughly excellent man, but devoid of all sense of religion, has also the misfortune, though a very honest and good sort of person, to be devout; and the weakness, like the dead fly in the apothecary's ointment, imparts a dangerous taint to the whole character. And thus the lesson of the tale runs on. We see in it the secret of the hostility entertained to evangelism by the insurgents of Vaud and Argovia, and which rendered them not less tolerant of a vital Protestantism than even the Jesuits whom they so determinedly opposed. We see in it, too, the grand error of Voltaire repeated, — miserable attempts to create a blank where, in the nature of things, no blank can exist; and an utter ignorance of the great fact, that the religion of the New Testament is the only efficient antidote against superstition, and a widely-circulated Bible the sole permanent protection against the encroachments of an ambitious priesthood. It would be bold to conjecture what the rising crop of opinion, so thickly sown over Europe, is ultimately to produce. There exists a widely-extended belief that Popery, when its final day has come, is to have infidelity for its executioner. Do we see in works such as those of Eugene Sue the executioner in training? or is the old cycle again to revolve, and the blank formed by infidelity to be filled up by superstition? We would fain see a safer *exposé* of the Jesuits than the fiction of the insidious novelist, — an *exposé* at once so just to the order that they could raise no effectual protest against it, and so true to the interests of religion and the nature of man that it could contain no elements of reaction favorable to the body it assailed. When are we to have a translation of the "Provincial Letters" at once worthy of Pascal and of the existing emergency?

XIV.

THE ABBOTSFORD BARONETCY.

The intimation in our last of the death of Lieutenant-Colonel Sir Walter Scott, and the extinction of the Abbotsford baronetcy, must have set not a few of our readers athinking. The lesson of withered hopes and blighted prospects which it reads is, sure enough, a common one, — a lesson for every-day perusal in the school of experience, and which the history of every day varies with new instances. But in this special case it reads with more than the usual emphasis. The literary celebrity of the great poet and novelist of Scotland, — the intimate knowledge of his personal history which that celebrity has induced, and which exists coextensive with the study of letters, — the consequent acquaintance with the prominent foible that stood out in such high relief in his character from the general groundwork of shrewd good sense and right feeling, — have all conspired to set the lesson, as it were, in a sort of illuminated framework. Sir Walter says of Gawin Douglas, — in his picture of the "noble lord of Douglas blood," whose allegorical poem may still be perused with pleasure, notwithstanding the veil of obsolete language which mars its sentiment and obscures its imagery, — that it "pleased him more"

—— "that in a barbarous age
He gave rude Scotland Virgil's page,
Than that beneath his rule he held
The bishopric of fair Dunkeld."

Not such, however, was the principle on which Sir Walter estimated his own achievements or prospects. It pleased

42

him more to contemplate himself in the character of the founder, as seemed likely, of a third-rate border family, — of importance enough, however, to occupy its annual line in the almanac, — than that his name should be known as widely as even Virgil's own. And the ambition was one to which he sacrificed health, and leisure, and peace of mind, with probably a few years of life itself, and undoubtedly the very wealth which for this cause alone he so anxiously strove to realize. Never was there one who valued money less for its own sake; but it flowed in upon him, and, save for his haste to be rich that he might be a landholder on his family's behalf, Sir Walter would have died a man of large fortune, quite able to purchase three such properties as that of Abbotsford. And in last week's obituary we see the close of all he had toiled and suffered for, in the extinction of the family in which he had so fondly hoped to *live* for hundreds of years. One is reminded by the incident of some of the more melancholy strokes in his own magnificent fictions. He describes, for instance, in the introduction to the "Monastery," a weather-wasted stone fixed high in the wall of an ancient ecclesiastical edifice, and bearing a coat-of-arms which no one for ages before had been able to decipher. Weathered as it was, however, it was all that remained to testify of the stout Sir Halbert Glendinning, who had so bravely fought his way to a knighthood and the possession of broad lands, but whose wealth and honors, won solely by himself, he had failed to transmit to other generations, and whose extinct race and name had been lost in the tomb for centuries. Henceforth the honors of the Abbotsford baronetcy will be exhibited on but a hatchment whitened with the painted tears of the herald. A sepulchral tablet in Dryburgh Abbey will form, if not their only record, as in the imaginary case of the knight of Glendinning, at least their most striking memorial.

It is a curious enough fact, that Shakspeare, like Sir Walter Scott, cherished the ambition of being the founder of a family. "All his real estate," says one of his later biographers, Mr. C. Knight, "was devised to his daughter, Susanna Hall, for and during the term of her natural life. It was then entailed upon her first son and his heirs-male; and, in default of such issue, on her second son and his heirs-male; and so on, in default of such issue, to his granddaughter, Elizabeth Hall; and, in default of such issue, to his daughter Judith and her heirs-male. By this strict entailment," remarks the biographer, "it was manifestly the object of Shakspeare to found a family; but, like many other such purposes of short-sighted humanity," it is added, "the object was not accomplished. His elder daughter had no issue but Elizabeth, and she died childless. The heirs-male of Judith died before her. And so the estates were scattered after the second generation; and the descendants of his sister were the only transmitters to posterity of his blood and lineage." We see little of the great poet's own character in his more celebrated writings; he was too purely dramatic for that; and, like the "mirror held up to nature" of his own happy metaphor, reflected rather the features of others than his own. It is, however, a curious fact, that in the portion of his writings which *do* most exhibit him — his sonnets — there is no pleasure on which he dwells half so much as the pleasure of living in one's posterity. And, in urging the young friend to whom these exquisite compositions are addressed to marry, he rings the changes on this motive alone throughout twenty sonnets together. We rather wonder how the circumstance should have escaped the thousand and one critics and commentators who have written on Shakspeare, but certain it is that an intense appreciation of the sort of prospective, shadowy immortality that posterity confers on the founder of a family forms one of the most prominent features of the poetry in which he most indulged his own feelings, and that with this marked appreciation the provisions of his will thoroughly harmonize. He tells his friend that the sear leafless autumn of old age, and the "hideous winter" of death, draw near, when beauty "shall be o'ersnowed," and "bareness left everywhere;" and that unless the odors of the summer flowers continue to survive, distilled by the art of the chemist, they shall be as if they had never been, — things without mark or memorial.

"Then, were no summer's distillation left
A liquid prisoner, pent in walls of glass,
Beauty's effect with beauty were bereft,
Nor it, nor no remembrance what it was:
But flowers distilled, though they the winter meet,
Leese but their show; their substance still lives sweet."

And then the poet, with the happy art in which he excelled all men, applies the figure by urging his young and handsome friend to live in his posterity, as the vanished flowers live in their distilled odors; and expatiates on the solace of enduring throughout the future in one's offspring: —

"Be it ten for one,
Ten times thyself were happier than thou art,
If ten of thine ten times re-figured thee;
Then, what could Death do, if thou shouldst depart,
Leaving thee living in posterity?
Be not self-willed, for thou art much too fair
To be Death's conquest, and make worms thine heir."

What strange vagaries human nature does play in even the greatest minds! Shakspeare was thoroughly aware that his verse was destined to immortality. We have his own testimony on the point to nullify the idle conjectures of writers who have set themselves to criticize his works, without having first taken, as would seem, the necessary precaution of reading them. He tells us in his sonnets, that "not marble, nor the gilded monuments of princes," would outlive "his powerful rhime." And, again, addressing his friend, he says: —

"I'll live in this poor rhime
While Death insults o'er dull and speechless tribes;
And thou in this shalt find thy monument,
When tyrants' crests and tombs of brass are spent."

And yet again, with still greater beauty, if not greater energy, he says: —

> " Your life from hence immortal life shall have,
> Though I, once gone, to all the world must die.
> The earth can yield me but a common grave,
> While you entombèd in men's eyes shall lie.
> Your monument shall be my gentle verse,
> Which eyes not yet created shall o'er-read,
> And tongues to be your being shall rehearse,
> When all the breathers of this world are dead;
> You still shall live (such virtue hath my pen)
> Where breath most breathes, — e'en in the mouths of men."

And yet this great poet, so conscious of the enduring vitality that dwelt in his verse, could find more pleasure in the idea of living in future ages in his descendants, — a sort of pleasure in which almost every Irish laborer may indulge, — than in being one of the never-dying poets of his country and the world. What may be termed the human instinct of immortality, — the natural sentiment which, when rightly directed, rests on that continuity of life in the individual in which the dark chasm of the grave makes no break or pause, — may be found, though wofully misdirected, both in the sentiment that rejoices in the prospect of posthumous celebrity, always so shadowy and unreal, and the sentiment that gloats over the fancied, delusive life which one lives in one's descendants. Shakspeare felt himself sure of posthumous celebrity; and finding it, like every sublunary good, when once fairly secured, valueless and unsatisfactory, he fixed his desires with much solicitude on the other earthly immortality, and sought to live in his offspring. It would have been well had the instinct been better directed, both in Sir Walter and his great prototype the dramatist of Avon. It would be also well, with such significant lessons before us, to be reading them aright. They tell us that the longings after immortality, in which it is the nature of man to indulge, are not to be satisfied by the world-wide, ever-enduring fame of the poet, and that the humbler and not less unsubstantial shadow of future life which one lives in one's children and their descendants is at least not more satisfying in its nature, and that it lies greatly more open than the other to the blight of accident and the influence of decay.

Judging from the history of the past, there is no class of men less entitled to indulge in the peculiar hope of Shakspeare and Sir Walter Scott than the greater poets, — men whose blow of faculty, ratiocinative and imaginative, has attained to the fullest development at which, in the human species, it ever arrives. Has the reader ever bethought him how exceedingly few of the poets of the two last centuries have bequeathed their names to posterity through their descendants? No doubt by much the greater part of them — ill-hafted in society, and little careful how they guided their course — were solitary men, who, without even more than their characteristic imprudence, could not have grappled with the inevitable expense of a family. Thus it was that Cowley, Butler, and Otway died childless, with Prior and Congreve, Gay, Phillips, and Savage, Thomson, Collins, and Shenstone, Akenside, Goldsmith, and Gray. Pope, Swift, Watts, and Cowper were also unmated, solitary men; and Johnson had no child. Even the poets in more favorable circumstances, who could not say, in the desponding vein of poor Kirke White, —

> "I sigh when all my happier friends caress, —
> They laugh in health, and future evils brave;
> Them shall a wife and smiling children bless,
> While I am mould'ring in the silent grave," —

even of this more fortunate class, how very few were happy in their offspring! The descendants of Dryden, Addison, and Parnell did not pass into the second generation; those of Shakspeare and Milton became extinct in the second and the third. It would seem as if we had an illustration, in this portion of the literary history of our country, of Doubleday's curious theory of population.

The human mind attained in these remarkable men to its full intellectual development, as the rose or the carnation, under a long course of culture, at length suddenly *stocks*, and doubles, and widens its gorgeous blow of a thousand petals; and then, when in its greatest perfection, transmission ceases, and there is no further reproduction of the variety thus amplified and expanded to the full. Nature does her utmost, and then, stopping short, does no more.

Abbotsford, a supremely melancholy place heretofore, will be henceforth more melancholy still. Those associations of ruined hopes and blighted prospects which cling to its picturesque beauty will now be more numerous and more striking than ever. The writings of Scott are the true monuments of his genius; while Abbotsford, on which he rested so much, will form for the future a memorial equally significant of his foibles and his misfortunes, — of bright prospects suddenly overcast, and sanguine hopes quenched in the grave forever. Is the reader acquainted with the poem in which the good Isaac Watts laments the untimely death of his friend Gunston, — a man who died childless, in the vigor of early manhood, just as he had finished a very noble family seat? The verse flows more stiffly than that of Shakspeare or Sir Walter Scott, for Watts was not always happiest when he attempted most; and there is considerable more poetry in his hymns for children than in his "Pindaric Odes" or his "Elegies." Still, however, his funeral poem on his friend brings out not unhappily the sentiment which must breathe for the future from the deserted halls of Abbotsford: —

> "How did he lay the deep foundations strong,
> Marking the bounds, and reared the walls along,
> Solid and lasting, where a numerous train
> Of happy Gunstons might in pleasure reign,
> While nations perished and long ages ran, —
> Nations unborn and ages unbegan;
> Nor time itself should waste the blest estate,
> Nor the tenth race rebuild the ancient seat.
> How fond our fancies are! * * *

And must this building, then, — this costly frame, —
Stand here for strangers? Must some unknown name
Possess these rooms, the labors of my friend?
Why were these walls raised for this hapless end,
Why these apartments all adorned so gay,
Why his rich fancy lavished thus away?
The unhappy house looks desolate and mourns,
And every door groans doleful as it turns."

We find we cannot better conclude our desultory remarks than in the words of the London "Morning Herald," whom we find thus referring to the death of the Lieutenant-Colonel, Sir Walter: —

"The deceased Baronet was the last of a family which it cost one precious life to create, and for whose perpetuation its founder would have accounted no purchase too dear, and reckoned no sacrifice too costly. It was not sufficient for the head of that house, whose last member has so recently quitted the earth, that he stood foremost in the ranks of celebrated men during life, — that he secured immortality upon his departure. Beyond the prodigal gifts of Heaven he esteemed the factitious privileges of earth, and treated lightly an imperishable wealth, for the sake of dross as poor as it was passing. The memoirs of the first Sir Walter — albeit penned by no unloving hand — leave painful impressions upon the minds of all who have made for themselves the character of the great magician, as far as it was possible, from his undying works. If the history teaches anything at all, it is one of the saddest lessons that can be brought home to humanity, — that of gigantic powers ill used, of insatiable though petty ambition derided and destroyed. The vocation of Sir Walter Scott was to enlighten and instruct mankind: *he* believed it was to found a family, and to become a great landed proprietor. To achieve the ignoble mission, the poet and the novelist embarked the genius of a Shakspeare, and the result is now before us. The family is extinct; the landed proprietor was a bankrupt in his prime. Who that has read the life of Sir Walter but has wept at his misfortunes, and marvelled at the sacrifices heaped upon sacrifices, freely made, in furtherance of a low and earthly seeking? Heaven pointed one way, human frailty another. 'Be mighty amidst the great,' said the former; 'be high amongst the small,' whispered the latter. He obeyed the latter, and lo the consequence! The small know him not: amidst the great he still continues mighty. The history of Scott is the history of mankind. We cannot violate the will, expressed or understood, of Heaven, and be happy. We cannot sinfully indulge a single passion, and not be disappointed. The spiritual and moral laws which regulate our life are as constant and invariable as any to be found in matter. Had Scott not enlisted every hope, thought, and energy in his miserable aim at power and position, he would in all probability have been alive to-day. He was a hale and hearty man when the failure of the booksellers compelled him to those admirable and superhuman exertions which crushed and killed him. That failure would have been nothing to the poet, if he had not involved himself in trade in order the more rapidly to secure the purpose which he had at heart, — for which he wrote and lived. '*The spirits of the wise sit in the clouds and mock us.*' All that Scott bargained for at the outset of life he possessed for an instant before he quitted it. He cared not to be renowned, — he wished to be rich. To be spoken of as the master of prose and verse was nothing, if the term could not be coupled with that of master of Abbotsford. The dream was realized. Money came in abundance, and with it lands and increasing possessions. The mansion of the laird rose by degrees, and child after child promised to secure lands and house, as the founder would have them, in the immediate possession of a Scott. Then came, as if to complete the fabric and to insure the victory, honors and titles fresh from the hand of Majesty itself. Nothing was wanting; all was gained, and yet nothing was acquired. The gift melted in the grasp; the joy passed away in the possession. With his foot on the topmost step of the ladder, Scott fell. His ambition was satisfied, but Providence was avenged. All that could be asked was given, but only to show how vain are human aspirations, — how less than childish are misdirected aims. Scott lived to see his property, his house and lands, in the hands of the stranger; we have lived to see his children one by one removed. Is there no lesson here?"

The End.

Gould and Lincoln's Publications.

(*LITERARY.*)

THE PURITANS; or, the Court, Church, and Parliament of England. By SAMUEL HOPKINS. 3 vols., 8vo, cloth.

HISTORICAL EVIDENCES OF THE TRUTH OF THE SCRIPTURE RECORDS, STATED ANEW, with Special Reference to the Doubts and Discoveries of Modern Times. Bampton Lecture for 1859. By GEORGE RAWLINSON. 12mo, cloth.

CHRIST IN HISTORY. By ROBERT TURNBULL, D. D. A New and Enlarged Edition. 12mo, cloth.

THE CHRISTIAN LIFE; Social and Individual. By PETER BAYNE. 12mo, cloth.

ESSAYS IN BIOGRAPHY AND CRITICISM. By PETER BAYNE. Arranged in two series, or parts. 2 vols., 12mo, cloth.

THE GREYSON LETTERS. By HENRY ROGERS, Author of the "Eclipse of Faith." 12mo, cloth.

CHAMBERS' WORKS. *CYCLOPÆDIA OF ENGLISH LITERATURE.* Selections from English Authors, from the earliest to the present time. 2 vols., 8vo, cloth.

——— *MISCELLANY OF USEFUL AND ENTERTAINING KNOWLEDGE.* 10 vols., 16mo, cloth.

——— *HOME BOOK;* or, Pocket Miscellany. 6 vols., 16mo, cloth.

THE PREACHER AND THE KING; or, Bourdaloue in the Court of Louis XIV. By L. F. BUNGENER. 12mo, cloth.

THE PRIEST AND THE HUGUENOT; or, Persecution in the age of Louis XV. By L. F. BUNGENER. 2 vols, 12mo, cloth.

MISCELLANIES. By WILLIAM R. WILLIAMS, D. D. 12mo, cloth.

ANCIENT LITERATURE AND ART. Essays and Letters from Eminent Philologists. By Profs. SEARS, EDWARDS, and FELTON. 12mo, cloth.

MODERN FRENCH LITERATURE. By L. RAYMOND DE VERICOUR. Revised, with Notes by W. S. CHASE. 12mo, cloth.

BRITISH NOVELISTS AND THEIR STYLES. By DAVID MASSON, M. A., Author of Life of Milton. 16mo, cloth.

REPUBLICAN CHRISTIANITY; or, True Liberty, exhibited in the Life, Precepts, and early Disciples of the Redeemer. By E. L. MAGOON. 12mo, cloth.

THE HALLIG; or, the Sheepfold in the Waters. Translated from the German of BIERNATSKI, by Mrs. GEORGE P. MARSH. 12mo, cloth.

MANSEL'S MISCELLANIES; including "Prolegomina Logica," "Metaphysics," "Limits of Demonstrative Evidence," "Philosophy of Kant," etc. 12mo. *In press.*